Microbiology
of
Chlamydia

Editor

Almen L. Barron, Ph.D.

Professor and Chairman
Department of Microbiology and Immunology
University of Arkansas for Medical Sciences
Little Rock, Arkansas

CRC Press
Taylor & Francis Group
Boca Raton London New York

CRC Press is an imprint of the
Taylor & Francis Group, an **informa** business

CRC Press
Taylor & Francis Group
6000 Broken Sound Parkway NW, Suite 300
Boca Raton, FL 33487-2742

Reissued 2019 by CRC Press

© 1988 by Taylor & Francis Group, LLC
CRC Press is an imprint of Taylor & Francis Group, an Informa business

No claim to original U.S. Government works

A Library of Congress record exists under LC control number:

Publisher's Note
The publisher has gone to great lengths to ensure the quality of this reprint but points out that some imperfections in the original copies may be apparent.

Disclaimer
The publisher has made every effort to trace copyright holders and welcomes correspondence from those they have been unable to contact.

ISBN 13: 978-0-367-22674-9 (hbk)
ISBN 13: 978-0-367-22707-4 (pbk)
ISBN 13: 978-0-429-27652-1 (ebk)

Visit the Taylor & Francis Web site at http://www.taylorandfrancis.com and the CRC Press Web site at http://www.crcpress.com

PREFACE

Organisms classified in the genus *Chlamydia* were once considered to be unusual viruses and are now known to be obligate intracellular procaryons with a unique reproductive cycle. Older medical knowledge associated these organisms with trachoma, inclusion conjunctivitis, lymphogranuloma venereum, and psittacosis. Current medicine is concerned with their role in sexually transmitted disease, infertility in females, pneumonitis of the newborn, and possibly pneumonia in the absence of an avian reservoir. In this book we attempt to state the microbiology of *Chlamydia* as we understand it with regard to their nature as microorganisms and as pathogens. The editor is grateful to the contributors for their cooperation, patience, and commitment to excellence.

Almen L. Barron

THE EDITOR

Almen L. Barron Ph.D. is Professor and Chairman, Department of Microbiology and Immunology, University of Arkansas for Medical Sciences, Little Rock. A Canadian by birth, he received his Ph.D. at Queen's University, Kingston, Ontario in 1953. He joined the Department of Bacteriology, University of Buffalo (later State University of New York at Buffalo) in 1954 to work on the Salk vaccine field trial under the direction of Dr. David T. Karzon. He remained at Buffalo until 1974 when he assumed his present position. Research at Buffalo was mainly focused on Echoviruses and viral immunology. In 1964 he received a Fulbright Research Scholar award and worked in the laboratory of the late Professor Hans Bernkopf, Hebrew University — Hadassah Medical School, Jerusalem, Israel. It was there that his interest in *Chlamydia* was kindled. Early research was on chlamydial hemagglutinin and biological properties. Later studies involved the role of *Chlamydia* in genital tract infections using animal models, which have continued to the present. Other activities have included co-editing *Microbiology: Basic Principles and Clinical Applications* (Macmillan) with Dr. Noel R. Rose, membership on Bacteriology and Mycology Study Section, NIH, and service on the editorial boards for *Infection and Immunity* and *Proceedings of the Society for Experimental Biology and Medicine*.

CONTRIBUTORS

Gerald I. Byrne, Ph.D.
Associate Professor
Department of Medical Microbiology
University of Wisconsin Medical School
Madison, Wisconsin

Thomas P. Hatch, Ph.D.
Associate Professor
Department of Microbiology and
  Immunology
University of Tennessee, Memphis
Memphis, Tennessee

Cho-chou Kuo, M.D., Ph.D.
Professor
Department of Pathobiology
University of Washington
Seattle, Washington

Akira Matsumoto, D.Sc.
Associate Professor
Department of Microbiology
Kawasaki Medical School
Kurashiki, Okayama, Japan

James W. Moulder, Ph.D.
Professor Emeritus
Department of Molecular Genetics and
  Cell Biology
University of Chicago
Chicago, Illinois

Wilbert J. Newhall V, Ph.D.
Assistant Professor
Department of Medicine
Indiana University School of Medicine
Indianapolis, Indiana

Roger G. Rank, Ph.D.
Associate Professor
Department of Microbiology and
  Immunology
University of Arkansas for Medical
  Sciences
Little Rock, Arkansas

Julius Schachter, Ph.D.
Professor
Departments of Laboratory Medicine, and
  Epidemiology and International Health
University of California at San Francisco
San Francisco, California

Richard S. Stephens, Ph.D., M.S.P.H.
Assistant Adjunct Professor
Department of Laboratory Medicine
University of California at San Francisco
San Francisco, California

Johannes Storz, D.V.M., Ph.D.
Professor and Department Head
Department of Veterinary Microbiology
  and Parasitology
School of Veterinary Medicine
Louisiana State University
Baton Rouge, Louisiana

Michael E. Ward, Ph.D.
Senior Lecturer
Department of Medical Microbiology
Southampton University Medical School
Southampton, England

Dwight M. Williams, M.D.
Associate Professor
Department of Medicine
University of Texas Health Science
  Center and Audie L. Murphy V.A.
  Hospital
San Antonio, Texas

This book is dedicated to my grandson, Jared Joseph Barron, born June 11, 1986, from Zaydie with love.

TABLE OF CONTENTS

CHLAMYDIA AS ORGANISMS

CHLAMYDIA AS PATHOGENS

Chlamydia as Organisms

Chapter 1

CHARACTERISTICS OF CHLAMYDIAE

James W. Moulder

TABLE OF CONTENTS

I. INTRODUCTION

The chlamydiae have been assigned to the order *Chlamydiales*, which is comprised of one family, *Chlamydiaceae*, with a single genus, *Chlamydia*, and two species, *C. trachomatis* and *C. psittaci*.[1-3] The description of the genus *Chlamydia* in *Bergey's Manual of Systematic Bacteriology* follows this classification,[4] and it provides a generally satisfactory characterization of the chlamydiae infecting humans, other mammals, and birds. However, the present state of chlamydial taxonomy is unsatisfactory in at least two respects. First, it takes no account of the many reports of chlamydia-like organisms living intracellularly in invertebrate hosts, and second, being based solely on phenotypic similarities and differences, its relation to the evolutionary history of chlamydiae is uncertain.

This chapter will describe the definitive properties of the genus *Chlamydia*, how its species and biovars are distinguished one from the other, and how recent advances in chlamydial biology and bacterial phylogeny allow construction of a plausible evolutionary history of chlamydiae. Finally, it will consider ways in which the familiar *C. trachomatis* and *C. psittaci* and the as yet poorly characterized chlamydia-like inhabitants of invertebrates may all be accommodated within the order *Chlamydiales*, perhaps by the creation of new families and genera.

II. PROPERTIES OF THE GENUS *CHLAMYDIA*

A. Definition of the Genus *Chlamydia*

The genus *Chlamydia* is defined by the properties listed in Table 1. Presently, only strains of *C. psittaci* and *C. trachomatis* are known to fit this definition. With one exception, the chlamydia-like organisms of invertebrates have not been propagated in the laboratory. So, apart from morphology, their characteristics are largely unknown.

1. Obligate Intracellular Habitat

Although no chlamydia has so far been observed to grow extracellularly, either in nature or in the laboratory, serious and sustained efforts to achieve host-free multiplication have not been made. With our rapidly increasing understanding of chlamydial biology, such efforts may soon become worthwhile.

2. Developmental Cycle

Like many other intracellular parasites,[5] chlamydiae have evolved morphologically distinct infectious and reproductive forms.[6-8] Chlamydial elementary bodies never divide. Their role is to carry the infection from one cell (or one host) to another, where they reorganize into reticulate bodies which multiply by binary fission in membrane-bound intracytoplasmic vacuoles or inclusions. Reticulate bodies do not infect new host cells. Instead, they reorganize into new generations of elementary bodies to complete the developmental cycle. The most critical structural difference between elementary bodies and reticulate bodies appears to be the extent to which their outer membrane proteins are complexed by disulfide cross-linking. These proteins are extensively cross-linked in elementary bodies but not in reticulate bodies.[9-15] There is a strong temptation to ascribe the many biological differences (Table 2) between the two chlamydial cell types to this structural difference, but direct evidence for a cause and effect relationship is largely lacking.

3. Gram Negative Envelope Without Peptidoglycan

Envelopes of both elementary bodies and reticulate bodies resemble those of host-independent Gram-negative bacteria in that they are made up of an inner cytoplasmic membrane and an outer membrane,[7,8,17,18] are disrupted by polymyxin B and ethylenediaminetetra-

Table 1
CHARACTERS THAT DEFINE THE GENUS *CHLAMYDIA*

Obligate intracellular habitat.
Developmental cycle with morphologically distinct infectious and reproductive forms.
Gram negative envelope without peptidoglycan.
Genus-specific lipopolysaccharide.
Patches of hexagonally arrayed cylindrical projections.
Utilization of host ATP for synthesis of chlamydial protein.
Small genome.

Table 2
BIOLOGICAL DIFFERENCES BETWEEN
ELEMENTARY BODIES AND RETICULATE BODIES

Property	Elementary body	Reticulate body
Infectivity	Yes	No
Multiplication	No	Yes
Inhibition of phagosome-lysosome fusion	Yes	No
Toxic for mice	Yes	No
Toxic for macrophages	Yes	No
ATP transport	No	Yes
Protein synthesis	No	Yes

Modified from Moulder, J. W., *ASM News,* 50, 353, 1984. With permission.

acetate,[19,20] and contain an outer membrane protein that accounts for half of the total protein of that membrane.[10,18,21,22] Chlamydial cell envelopes differ from those of typical Gram-negative bacteria in that they have no peptidoglycan. Electron micrographs show that there is no peptidoglycan layer between the inner and outer membranes,[18] and chemical analysis reveals no muramic acid or any other amino sugar that might have replaced it in the peptidoglycan subunit.[23,24]

Absence of peptidoglycan could mean, as has been suggested for the peptidoglycan-less budding bacteria,[25] that the chlamydiae branched off the main eubacterial tree before peptidoglycan was invented. However, it is more likely that chlamydiae have evolved from ancestors with peptidoglycan because they appear to have retained vestiges of a former peptidoglycan-containing state. Chlamydiae have penicillin-binding proteins similar in location, size, and affinity for the antibiotic to those of host-independent Gram-negative bacteria.[23] In low concentration, penicillin inhibits the growth and division of reticulate bodies and prevents their reorganization into elementary bodies.[26] Growth and multiplication of both *C. trachomatis* and *C. psittaci* are also blocked by D-cycloserine,[27,28] another inhibitor of peptidoglycan synthesis,[29] although *C. trachomatis* strains are usually much more susceptible. Penicillin inhibits the transpeptidation reaction responsible for the closing of the peptide cross-links in peptidoglycan,[29,30] whereas D-cycloserine inhibits both the formation of D-alanine from L-alanine and the synthesis of D-alanyl-D-alanine.[29] Sensitivity of chlamydial multiplication to these two antibiotics implies the presence of a D-alanyl-D-alanine sequence somewhere in the chlamydial cell. For this reason, it has been suggested that chlamydial envelopes contain D-alanyl-D-alanine peptides that are cross-linked to structures other than peptidoglycan.[24] The chlamydial susceptibility to inhibitors of peptidoglycan synthesis in the absence of peptidoglycan is without known parallel. The peptidoglycan-less budding bacteria are, for example, relatively resistant to both penicillin and D-cycloserine.[25]

4. Genus-Specific Lipopolysaccharide

All isolates of *C. trachomatis* and *C. psittaci* so far examined contain a lipid-soluble complement-fixing antigen that is present at all times in the developmental cycle.[31] This genus-specific antigen (formerly called the group antigen) strongly resembles the lipopolysaccharides (LPSs) of host-independent Gram negative bacteria in its location in the outer membrane of the chlamydial cell envelope,[32] in its chemical structure,[33-35] and in its biological activity.[36,37] There is also strong immunologic cross-reaction between the chlamydial LPS and the innermost core of the LPS from *Salmonella* mutants in which that structure is exposed.[38-40] Monoclonal antibodies against chlamydial LPS reveal at least three antigenic domains, two of which are shared with the LPSs of some free-living Gram-negative organisms and one of which is unique to the LPS of chlamydiae.

5. Patches of Hexagonally Arrayed Cylindrical Projections

On the outer membranes of elementary bodies from both *C. psittaci* and *C. trachomatis* there are patches of hexagonally arrayed cylindrical projections that are without obvious counterpart in other bacteria.[41-43] These projections are 10 to 25 nm high and about 25 nm in diameter. On a single chlamydial cell there is never more than one patch, with about 20 cylindrical projections roughly 50 nm apart, center to center. The cylinders extend all the way through the outer membranes, and it has been suggested that they are transmembrane pores connecting the interior of the chlamydial cell with the external environment.[41,44] Other unusual architectural features of the chlamydial surface have also been described, but it has not yet been demonstrated that they occur throughout the genus.

6. Utilization of Host ATP for Synthesis of Chlamydial Protein

In adapting to intracellular life, chlamydiae appear to have evolved mechanisms for exploiting the energy-rich compounds of their hosts and to have subsequently lost whatever energy-producing systems they might once have had.[45] Host-free chlamydiae have no respiratory enzymes other than the pyridinoproteins,[46] they catabolize glutamate, glucose, and pyruvate to a limited extent but without producing useful energy,[47,48] and infected host cells do not develop novel energy-generating mechanisms.[49,50] However, host-free reticulate bodies of both *C. psittaci* and *C. trachomatis* move ATP in and ADP out of their intracellular space by means of an ATP-ADP exchange system,[51] and use the host-derived ATP for synthesis of chlamydial protein.[52] Such a mechanism for transport of intact ATP into a cell has otherwise been described only in rickettsiae.[53]

7. Small Genome

Both chlamydial species have genomes consisting of double-stranded DNA molecules with average lengths of 346 μm.[54,55] These lengths correspond to a molecular weight of 660×10^6. Another estimate of chlamydial genome size based on the rate of reassociation of disassociated DNA gives a comparable value.[56] Although the chlamydial genome is much larger than that of the largest viruses, it is among the smallest of all procaryotic genomes (Table 3). Only the *Mycoplasma* genome is smaller. It may be that, once the chlamydiae learned to use host ATP, they no longer needed the genes and gene products associated with energy generation, that these genes disappeared without unfavorable consequences, and that the chlamydial genome shrank accordingly. Perhaps other groups of dispensable genes met similar fates. Lwoff suggested a long time ago that,[62] in the presence of a required metabolite, auxotrophic mutants should have the growth advantage over their prototrophic parents because they have fewer biosynthetic functions to perform. Subsequent discovery of mechanisms for preventing unneeded synthetic activities by means of feedback inhibition and gene repression have cast doubts on Lwoff's suggestion. However, there is still no better explanation for the consistently lower size of genome among procaryotes that live in and on eucaryotic cells (Table 3).

Table 3
SOME COMPARATIVE GENOME SIZES

Bacterial genus or virus	Genome size (M daltons)	Relative size	Ref.
Chlamydia	660	1	54,55
Vaccinia virus	160	0.24	57
Mycoplasma	500	0.76	58
Coxiella	1040	1.6	59
Rickettsia	1100	1.7	60
Neisseria	1300	2.0	56
Escherichia	2840	4.2	61

Modified from Moulder, J. W., *ASM News*, 50, 353, 1984. With permission.

Table 4
DIFFERENTIATION OF *C. TRACHOMATIS* AND *C. PSITTACI*

Character	*C. trachomatis*	*C. psittaci*
Inclusion morphology	Oval, vacuolar	Variable, dense
Accumulation of glycogen in inclusion[a]	Yes	No
Synthesis of folates[b]	Yes	No
Natural hosts[c]	Mice, humans	Birds, nonhuman mammals

[a] Revealed by staining with iodine.
[b] Revealed by growth inhibition with sulfadiazine. There are some exceptions.
[c] Conventional view of host range. There may be exceptions.

Modified from Moulder, J. W., Hatch, T. P., Kuo, C.-c., Schachter, J. and Storz, J., *Bergey's Manual of Systemic Bacteriology*, Vol. 1, Krieg, N. R., Ed., Williams & Wilkins, Baltimore, Md., 1984, 729. With permission.

B. Differentiation of the Two Species of *Chlamydiae*

Page separated *Chlamydia* into two species, *C. trachomatis* and *C. psittaci*, on the basis of these characters: morphology of the inclusion, accumulation of glycogen in the inclusion, and susceptibility to growth inhibition by sulfadiazine (Table 4).[2] Another character not used by Page, natural host range, is also useful.

1. Inclusion Morphology

Individual elementary and reticulate bodies of one species are not readily distinguished from those of the other by either light or electron microscopy. Nevertheless, the intracytoplasmic inclusions in which chlamydiae multiply are obviously different in the two species.[63] Inclusions of *C. trachomatis* are round or oval, relatively rigid, and often not completely filled with chlamydial cells. There tends to be only one inclusion per host cell, and the host cell nucleus is frequently displaced to the periphery. In contrast, *C. psittaci* inclusions are irregular or diffuse, not noticeably rigid, and usually packed with chlamydial cells. There may be several inclusions in one host cell, and the nucleus is not displaced.

2. Accumulation of Glycogen in Inclusions

A second difference between inclusions of the two chlamydial species is that glycogen accumulates in inclusions of *C. trachomatis* but not in those of *C. psittaci*.[63] The presence of glycogen is customarily demonstrated by staining with iodine, but the red-staining substance in the inclusions of *C. trachomatis* has been unequivocally identified as glycogen.[64] It is generally agreed that glycogen accumulates extracellularly, that is, inside the inclusion but outside the chlamydial cells. The largest accumulations of glycogen are found late in the developmental cycle at about 48 to 72 hr after infection.[65-69] There is little doubt that glycogen is synthesized by chlamydial enzymes and is not the product of a host response to infection. Accumulation of glycogen is inhibited by penicillin[65] and chloramphenicol.[69] *C. trachomatis*-infected cells incorporate adenosine diphosphate glucose, the bacterial glycogen precursor, into glycogen in preference to uridine diphosphate glucose, the mammalian substrate.[70] The role of glycogen synthesis in *C. trachomatis* metabolism and why it is deposited in the extracellular phase (inside the inclusion, outside the chlamydiae) of the inclusion is unknown.[64] The apparent invariant coupling of glycogen accumulation with inclusion morphology suggests both a role for glycogen and a way to explain how chlamydial cells of like morphology produce unlike inclusions. Perhaps glycogen modifies some physical property of the fluid phase of the inclusion in such a way as to produce inclusions of the *C. trachomatis* type.

3. Susceptibility to Growth Inhibition by Sulfadiazine

Susceptibility to sulfadiazine, a manifestation of the ability to synthesize folate,[29] is also significantly associated with inclusion morphology, although not as closely as glycogen accumulation. Most chlamydial strains, classified as *C. trachomatis* on the basis of inclusion morphology, are sensitive to sulfadiazine, whereas most strains similarly designated *C. psittaci* are not.[28] However, there are exceptions. For example, the 6BC strain of *C. psittaci*, widely used in studies on chlamydial biology, is sulfonamide-susceptible,[28] and some isolates of the lymphogranuloma venereum biovar of *C. trachomatis* are partially[71] or completely resistant.[72]

4. Natural Host Range

The orthodox view of the natural host range of *Chlamydia* is that *C. trachomatis*, with the exception of the mouse biovar, is a uniquely human pathogen, whereas *C. psittaci* is a parasite of birds and nonhuman mammals. Conventional wisdom holds that strains of *C. psittaci* indigenous to other mammals are negligible sources of human disease, whereas avian strains may be transmitted to people and cause psittacosis, a chlamydial pneumonia in which person-to-person transfer rarely occurs.

Although this view holds in the great majority of cases, there are exceptions. Strains of *C. psittaci* that are the agents of disease in nonhuman mammals may on occasion cause serious disease in people. The recent well-documented case of acute placentitis and spontaneous abortion due to ovine *C. psittaci* in a farm woman who had helped with lambing is a good example.[73,74] There is also a strain(s) of *C. psittaci* (TWAR) that violates not one but two of the tenets of the conventional wisdom. It produces both conjunctivitis and pneumonia in humans in the absence of any demonstrated nonhuman reservoir.[75,76] Isolates from Taiwan and India appear identical. Serological surveys indicate that antibodies to this agent(s) are prevalent in populations all over the world.[75,76]

C. Differentiation of the Three Biovars of *C. trachomatis*

C. trachomatis has been further divided into biovars, but *C. psittaci* has not.[4] This is not because *C. psittaci* is the more homogeneous of the two species, but rather because it is so heterogeneous that rational subdivision is presently impossible. Each of the three biovars of

Table 5

DIFFERENTIATION OF THE BIOVARS OF *C. TRACHOMATIS*

	Biovar		
Characteristic	Trachoma	Lymphogranuloma venereum	Mouse
Behavior in natural hosts			
Host range	Humans	Humans	Mice
Preferred site of infection	Squamocolumnar epithelial cells	Lymph nodes	Lungs
Behavior in laboratory animals			
Intracerebral lethality for mice	No	Yes	No
Follicular conjunctivitis in primates	Yes	No	No
Behavior in cell culture			
Plaques in mouse fibroblasts	No	Yes	Yes
Entry into host cells markedly enhanced by			
Centrifugation onto cell sheet	Yes	No	No
Treatment of host cells with DEAE[a]	Yes	No	No

[a] Diethylaminoethyl dextran

Modified from Moulder, J. W., Hatch, T. P., Kuo, C.-c., Schachter, J., and Storz, J., *Bergey's Manual of Systemic Bacteriology*, Vol. 1, Krieg, N. R., Ed., Williams & Wilkins, Baltimore, Md., 1984, 729. With permission.

C. trachomatis, mouse, lymphogranuloma venereum (LGV), and trachoma, exhibit the species-defining properties listed in Table 3. The three biovars may be distinguished one from the others according to the criteria of Table 5.

1. Differentiation of Biovar Mouse from Biovars LGV and Trachoma

In addition to a lesser degree of DNA homology,[78] mouse may be readily distinguished from LGV and trachoma in at least two other important ways. First, its natural hosts are mice, not people, and, second, its antigens (the genus-specific antigen excepted) cross-react only minimally with antigens of the other two biovars. There is no cross-reaction at all in the serological reactions usually used to identify chlamydiae,[79,80] but when the major outer membrane protein of the mouse biovar is denatured with sodium dodecyl sulfate or oxidized with periodate, it reacts with monoclonal antibodies that recognize epitopes on the major outer membrane proteins of the LGV and trachoma biovars.[81]

2. Distinction Between Biovars LGV and Biovars Trachoma

These two biovars are much closer to each other than either of them is to mouse. Not only do their DNAs exhibit nearly complete homology,[78] their antigens also extensively cross-react at the species-specific, subspecies-specific, and serovar-specific levels when tested with monoclonal antibodies by micro-immunofluorescence.[80] These antigens are located mainly, if not exclusively, on the major outer membrane protein.[82,83] There are also structural differences between the outer membrane proteins of biovars LGV and trachoma.[84,85] Each biovar is uniquely defined by its behavior in natural human hosts, in laboratory animals, and in cell culture (Table 5).

In humans, the LGV biovar infects mainly cells of the lymphatic system, and, although its clinical manifestations are protean, they are all lumped together as a single disease entity, lymphogranuloma venereum.[4,86,87] In contrast, the trachoma biovar infects chiefly squamocolumnar epithelial cells in various tissues and organs of its human hosts to give rise to a whole spectrum of pathology that is described in terms of a number of distinct disease entities such as follicular conjunctivitis (trachoma), urethritis, cervicitis, salpingitis, and infant pneumonia[4,86,87] (Refer also to Chapter 8 "Overview of Human Diseases").

Table 6
CHLAMYDIA-LIKE MICROORGANISMS IN INVERTEBRATES

Common name	Host Latin binomial	Phylum	Ref.
Hydra	*Hydra viridis*	Coelenterata	91
Clam	*Mercenaria mercenaria*	Mollusca	92
Scallop	*Argopecten irradians*	Mollusca	93
Tellina	*Tellina tenuis*	Mollusca	94
Oyster	*Crassostrea angulata*	Mollusca	95
Spider	*Coelotes luctuosus*	Arthropoda	96
Scorpion	*Buthus occitanus*	Arthropoda	97
Isopod	*Porcellio scaber*	Arthropoda	98
Crab	*Cancer magister*	Arthropoda	99

In laboratory animals, the two biovars may be unequivocally separated.[86] Only the LGV biovar kills mice by the intracerebral route, and only the trachoma biovar causes a follicular conjunctivitis when instilled into the eyes of nonhuman primates.

In cell culture, the LGV biovar forms plaques on monolayers of susceptible cells such as the mouse L cell,[88] but the trachoma biovar has not been observed to produce plaques on any known host cell. Failure to produce plaques is an expression of the low efficiency with which this biovar establishes secondary infections in the cells adjacent to the primarily infected cell. Even with susceptible cell lines such as HeLa[221] or McCoy, infection is so inefficient that entry-promoting procedures are used both in initial infection of cell cultures with clinical specimens and in subsequent serial propagation. These procedures include centrifugation of the inoculum onto the host-cell monolayer[66] and pretreatment of the monolayer with polyanions.[89] Entry of the LGV biovar may be modestly enhanced (less than two-fold), but entry of the trachoma biovar is increased 10- to 100-fold.[90]

III. CHLAMYDIAL PHYLOGENY

Each step toward a better understanding of evolutionary relationships among the chlamydiae broadly defined satisfies a deep-seated human longing to know where things come from. For those of us concerned with pathogenesis of chlamydial disease and host resistance to infection, there are also practical benefits. Good phylogenetic information provides a rational basis for choosing the best animal models of human chlamydial disease and for predicting the behavior of one chlamydial agent of disease from that of another.

A. Some Matters of Phylogenetic Importance
1. Invertebrate Hosts

A casual search of the literature unearthed nine reports of chlamydia-like organisms living intracellularly in invertebrate hosts (Table 6). Only one of these microorganisms has been cultured outside its natural hosts,[94] and with the exception of the clam agent,[100] the evidence for kinship with *Chlamydia* is entirely morphological. The presence of a developmental cycle is inferred from electron microscopic observation of cell types resembling elementary bodies and reticulate bodies. The clam agent contains the genus-specific antigen[100,101] and its inclusions stain positively with iodine.[100] The agents from isopods[98] and hydras[91] do not react with genus-specific antibody. Rare observation of a chlamydia-like organism in an invertebrate host might be ascribed to chance association with a chlamydia of vertebrate provenance, but the repeated observation of such organisms in a wide range of invertebrates cannot be brushed aside. These invertebrate-dependent agents must be considered chlamydiae *sensu lato*, and a place must be found for them in chlamydial phylogeny.

2. Absence of Peptidoglycan

Bacterial cell constituents of limited taxonomic distribution are useful markers in the identification and classification of bacteria.[102] Negative markers, the absence of widely occurring cell constituents in a limited number of organisms, are less reliable because the negative markers could have been lost independently in different organisms, as peptidoglycan was undoubtedly lost in *Mycoplasma*[58] and *Chlamydia*. However, bacteria without peptidoglycan are rare,[103] and peptidoglycan-less bacteria with penicillin-binding proteins are rare indeed. Penicillin susceptibility in the absence of peptidoglycan has so far been reported only in *Chlamydia*. Properly interpreted, lack of peptidoglycan should be a useful chemotaxonomic marker.

3. DNA Hybridization

DNA hybridization is most useful as an indicator of phylogenetic relationships when the organisms being compared are relatively closely related.[102] Among the biovars of *C. trachomatis*, strains of biovar LGV and biovar trachoma have near 100% homologous DNAs, but a single strain of the mouse biovar is only 30 to 60% homologous with the other two biovars.[78] Within *C. psittaci*, a limited number of isolates of both avian and mammalian origin show almost complete DNA homology.[104,105] Perhaps more extensive comparisons will reveal heterogeneity of hybridization in *C. psittaci* as well as in *C. trachomatis*. The DNAs of *C. trachomatis* and *C. psittaci* hybridize only to the extent of about 10%,[105] an unexpectedly low figure considering the many unusual and perhaps unique phenotypic characters they share in common (Table 1). This situation is uncommon but not unique. For example, the genus *Legionella* contains a large number of species with great phenotypic similarity and only minimal DNA homology.[106]

4. 16S Ribosomal RNA

Characterization and comparison of the nucleotide sequences in 16S ribosomal RNA permits direct estimation of genealogical relationships among even distantly related organisms because the sequences in these molecules change so slowly over evolutionary time.[107] The recent sequencing of the 16S ribosomal RNA of *C. psittaci* (strain 6BC) by Weisburg et al.[108] and comparison of the chlamydial sequence with those of other eubacteria reveal that the ancestral chlamydial branch probably split off the main eubacterial trunk at an unexpectedly early date. The 16S ribosomal RNA of *C. psittaci*, although definitely eubacterial in nature, is not closely related to that of any other eubacterium for which a sequence is available. Chlamydiae thus stand in contrast to the rickettsiae, which by ribosomal RNA sequence comparisons, are members of one of the major eubacterial subdivisions and closely related to the plant-associated genera *Rhizobium* and *Agrobacter*.[109]

On the basis of ribosomal RNA comparison, the nearest known relative of *C. psittaci* is a free-living aquatic bacterium *Planctomyces staleyi*, which also has no close relatives.[110] It has motile buds, numerous fimbriae, and a holdfast.[111] It was first isolated from a wood piling in Portage Bay of Lake Union, immediately adjacent to the University of Washington Health Sciences Center. The close proximity of the isolation site to one of the great centers of chlamydial research must be taken as entirely fortuitous. The definite but not too close relation between this budding bacterium and chlamydiae might be dismissed as trivial were it not for one additional resemblance — *Planctomyces*, like *Chlamydia*, has no peptidoglycan.[25] Additional information will be needed to solidify this relationship, but even the mere possibility is fascinating.

B. A New Family Tree for Chlamydiae

I constructed my first chlamydial family tree 20 years ago.[112] With the accumulation of new knowledge of chlamydiae in particular and bacterial phylogeny in general, it is time

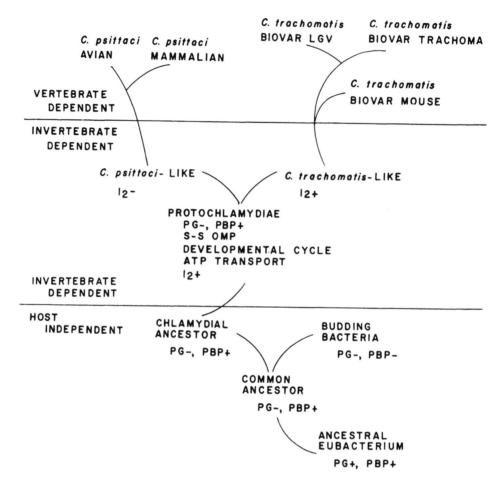

FIGURE 1. A family tree for chlamydiae, (PG) peptidoglycan, (PBP) penicillin-binding proteins, (S-S-OMP) disulfide bond-linked outer membrane proteins, ($I_2 +$) iodine-positive inclusions (glycogen in inclusions).

we had a new one. The new family tree (Figure 1) was constructed by going from the ancestral eubacterium to extant chlamydiae with the smallest number of branchings that accommodate all pertinent knowledge. Like the old family tree, the new one is largely conjectural, but it is consistent with known facts, and it makes some predictions that will be confirmed or refuted as new information becomes available.

1. A Single Origin for Chlamydiae

A eubacterial ancestor is assumed from the ribosomal RNA structure. A single origin for all present-day members of the genus *Chlamydia* is chosen because the chances of organisms with the same set of rare and perhaps unique phenotypic characters arising more than once are small and because their 165 rRNA gene sequences differ by less than 5%.[108]

2. Lack of Peptidoglycan

There are two possibilities: divergence from the eubacterial mainstream before the appearance of peptidoglycan, or loss of peptidoglycan at the time of or after divergence. I prefer the latter because chlamydiae have penicillin-binding proteins and are penicillin-susceptible. Their putative distant relatives, the budding bacteria, are not penicillin-sensitive[25] and are presumed to have lost their penicillin-binding proteins. The new family tree predicts that freshwater and marine habitats contain bacteria, living free or as ectoparasites of invertebrates, that are more closely related to *Chlamydia* than *Planctomyces*. Perhaps it was just such chlamydial forebears that first became infected with the viruses now found in both invertebrate-[92,94] and vertebrate-dwelling[113] chlamydiae. It would have been much easier for these viruses to have first associated themselves with free-living hosts.

3. Invasion of the Intracellular Habitat

Because invertebrates were here first and in greater number and variety, they were probably the first hosts for chlamydiae *sensu lato*. There is no way of knowing which of the characters associated with extant chlamydiae appeared as preadaptations before invasion of the intracellular habitat, and which appeared during subsequent adaptation to life inside cells. However, it is likely that loss of peptidoglycan and reversible, disulfide bond-based, cross-linking of the outer membrane proteins[11,14,15] were indispensible preadaptations for intracellular life.

4. Branching of the Genus into Species

It is assumed that present-day chlamydiae, including the invertebrate-dependent ones, have descended from an iodine-positive ancestor and that *C. psittaci* and *C. psittaci*-like organisms have lost the ability to accumulate glycogen in their inclusions. It may be that the iodine stain is for chlamydiae what the Gram stain is for eubacteria in general, another example of an in itself trivial marker serendipitously separating the two major divisions of a taxon. Because the results of DNA hybridization suggest that *C. trachomatis* and *C. psittaci* diverged a long time ago, I show them separating before the colonization of vertebrate hosts. The clam chlamydia is iodine-positive, so we know that there are glycogen accumulators in invertebrates. The family tree predict that iodine-negative, *C. psittaci*-like chlamydiae will also be found in invertebrate hosts.

5. Branching within the Chlamydial Species

Branching among extant vertebrate-inhabitating chlamydiae is based partly on DNA hybridization and partly on the comparison of phenotypes made earlier in this chapter. *C. psittaci* is separated into avian and mammalian branches on the basis of lack of serological cross-reaction (other than the group antigen) between the two groups.[114,115]

IV. CHLAMYDIAL NOMENCLATURE AND TAXONOMY

A. Importance of a Rational Nomenclature and Taxonomy

As pointed out elsewhere,[116] bacterial classifications are devised by and for microbiologists, not by and for bacteria. It is entirely permissible for there to be as many different schemes of chlamydial phylogeny as there are chlamydiologists, but we must all agree on one and only one name for each taxon according to internationally agreed-upon rules of nomenclature.[117] However, the closer a system of nomenclature follows natural taxonomic relations, the more useful it is likely to be. The importance of having a simple and generally accepted nomenclature cannot be overemphasized. The ground-clearing suggestions of Page[1,2] that all the then-known chlamydiae be gathered together in a single genus with two species could not have come at a more opportune time, just when our knowledge of the biology and clinical importance of chlamydiae was undergoing the explosive increase that has continued until today.

Chlamydial nomenclature is fortunate in that it is not overburdened with taxa. One genus and two species will probably not suffice for all the different kinds of chlamydiae *sensu lato*. However, new taxa should not be proposed until the right kind of information is at hand, information that will permit construction of a system of nomenclature consistent with evolutionary relations.

B. Future of Chlamydial Nomenclature

1. Order and Family

It now appears likely that extant chlamydiae have a long evolutionary history and that modern methods of establishing genealogical lineages will sooner or later show that they are related to both free-living and parasitic bacteria without strong phenotypic resemblance to chlamydiae. What should be included in the order *Chlamydiales* and what should not? Where should the line be drawn? The first edition of *Bergey's Manual of Systematic Bacteriology*[118] states that members of the order are " . . . coccoid microorganisms whose obligately intracellular mode of multiplication within cytoplasmic vacuoles is characterized by change of small, rigid-walled infectious forms (elementary bodies) into larger, flexible-walled noninfectious forms (elementary bodies) that divide by binary fission. The developmental cycle is complete when daughter cells reorganize into elementary bodies which survive intracellularly to infect new host cells." To this statement might now be added "Members of the order are devoid of peptidoglycan. The mechanical strength of the cell envelope derives from the major outer membrane protein which is extensively cross-linked with disulfide bonds in the elementary body and much less so in the reticulate body." In my opinion, this constitutes a minimum phenotypic definition for the order *Chlamydiales* that will accommodate any organism at all closely related to *C. trachomatis* and *C. psittaci*.

At present only one family (*Chlamydiaceae*) is recognized.[118] If future work reveals the existence of major subsets in *Chlamydiales*, then, and only then, will there be a need for more families.

2. Genus and Species

Let us assume for the moment that the invertebrate-dwelling chlamydia-like organisms meet the suggested criteria for inclusion in the order *Chlamydiales*. Are they then to be classified in the existing genus *Chlamydia* or will they have to be accommodated in new genera? The clam chlamydia contains the *Chlamydia*-specific antigen, accumulates glycogen in its inclusions, and has been provisionally designated as a strain of *C. trachomatis*.[100] On the other hand, the agents from isopods[98] and hydras[91] do not react with *Chlamydia*-specific antibody. Should the definition of the genus *Chlamydia* be broadened to include these agents or should one or more new genera be created? Perhaps DNA hybridization and 16S ribosomal RNA oligonucleotide sequence comparison will provide the answers. Since most of the invertebrate-dependent chlamydiae have not been cultivated outside their natural hosts, measurements of DNA homology will require developments of *in situ* hybridization techniques, perhaps similar to the one used for analysis of persistent virus infections.[119]

Within *C. trachomatis*, there is some justification for splitting off the mouse biovar as a separate species. DNA hybridization sets it apart from the other two biovars,[78] and there would be some advantage in having the residual *C. trachomatis* taxon represent only strains of human origin. Within *C. psittaci*, there has been a long-standing feeling that the species is excessively heterogeneous and should be subdivided. However, despite repeated demonstration of phenotypic heterogeneity,[120] evidence for corresponding genetic heterogeneity is still lacking.

V. SUMMARY

The characteristics of a few strains of *Chlamydia* that cause disease in humans, farm animals, and birds have been reasonably well established. There remain, however, many inadequately characterized members of the genus — the "human psittacosis" strain(s), for example. It now appears that chlamydiae have an ancient evolutionary lineage with many previously unrecognized close and distant relatives. They are actors on a far broader stage than any of us had imagined. Delineation of the interrelationships among the chlamydiae most broadly defined will be an interesting new chapter in chlamydial biology.

REFERENCES

1. **Page, L. A.,** Revision of the family *Chlamydiaceae* Rake *(Rickettsiales)*: unification of the psittacosis-lymphogranuloma venereum-trachoma group in the genus *Chlamydia* Jones, Rake and Stearns 1945, *Int. J. Syst. Bacteriol.*, 16, 223, 1966.
2. **Page, L. A.,** Proposal for the recognition of two species in the genus *Chlamydia* Jones, Rake, and Stearns 1945, *Int. J. Syst. Bacteriol.*, 18, 51, 1968.
3. **Storz, J. and Page, L. A.,** Taxonomy of the chlamydiae: reasons for classifying organisms of the genus *Chlamydia*, family *Chlamydiaceae* in a separate order, *Chlamydiales* ord. nov., *Int. J. Syst. Bacteriol.*, 21, 332, 1971.
4. **Moulder, J. W., Hatch, T. P., Kuo, C.-C., Schachter, J., and Storz, J.,** Genus *Chlamydia*, in *Bergey's Manual of Systematic Bacteriology*, Vol. 1, Krieg, N. R., Ed., Williams & Wilkins, Baltimore, Md., 1984, 729.
5. **Moulder, J. W.,** Comparative biology of intracellular parasitism, *Microbiol. Rev.*, 49, 298, 1985.
6. **Litwin, J.,** The growth cycle of the psittacosis group of organisms, *J. Infect. Dis.*, 105, 129, 1959.
7. **Higashi, N.,** Electron microscopic studies on the mode of reproduction of trachoma virus and psittacosis virus in cell cultures, *Exp. Mol. Pathol.*, 4, 24, 1965.
8. **Friis, R. R.,** Interaction of L cells and *Chlamydia psittaci*: entry of the parasite and host responses to its development, *J. Bacteriol.*, 110, 706, 1972.
9. **Tamura, A. and Manire, G. P.,** Preparation and chemical composition of the cell membranes of developmental reticulate forms of meningopneumonitis organisms, *J. Bacteriol.*, 94, 1184, 1967.
10. **Newhall, W. J. V. and Jones, R. B.,** Disulfide-linked oligomers of the major outer membrane protein of chlamydiae, *J. Bacteriol.*, 154, 998, 1983.
11. **Hatch, T. P., Allan, I., and Pearce, J. H.,** Structural and polypeptide differences between envelopes of infective and reproductive life cycle forms of *Chlamydia*, *J. Bacteriol.*, 157, 13, 1984.
12. **Bavoil, P., Ohlin, A., and Schachter, J.,** Role of disulfide bonding in outer membrane structure and permeability in *Chlamydia trachomatis*, *Infect. Immun.*, 44, 479, 1984.
13. **Allan, I., Hatch, T. P., and Pearce, J. H.,** Influence of cysteine deprivation on chlamydial differentiation from reproductive to infective life-cycle forms, *J. Gen. Microbiol.*, 131, 3171, 1985.
14. **Hackstadt, T., Todd, W. J., and Caldwell, H. D.,** Disulfide-mediated interactions of the chlamydial major outer membrane protein: role in the differentiation of chlamydia? *J. Bacteriol.*, 161, 25, 1985.
15. **Hatch, T. P., Miceli, M., and Sublett, J. E.,** Synthesis of disulfide bonded outer membrane proteins during the development cycle of *Chlamydia psittaci* and *Chlamydia trachomatis*, *J. Bacteriol.*, 165, 379, 1986.
16. **Moulder, J. W.,** Looking at chlamydiae without looking at their hosts, *ASM News*, 50, 353, 1984.
17. **Tamura, A., Matsumoto, A., Manire, G. P., and Higashi, N.,** Electron microscopic observations on the structure of envelopes of mature elementary bodies and developmental reticulate forms of *Chlamydia psittaci*, *J. Bacteriol.*, 105, 355, 1971.
18. **Caldwell, H. D., Kromhaut, J., and Schachter, J.,** Purification and partial characterization of the major outer membrane protein of *Chlamydia trachomatis*, *Infect. Immun.*, 31, 1161, 1981.
19. **Matsumoto, A., Higashi, N., and Tamura, A.,** Electron microscopic observations on the effects of polymyxin B on cell walls of *Chlamydia psittaci*, *J. Bacteriol.*, 104, 357, 1973.
20. **Narita, T., Wyrick, P. B., and Manire, G. P.,** Effect of alkali on the structure of cell envelopes of *Chlamydia psittaci* elementary bodies, *J. Bacteriol.*, 125, 300, 1976.
21. **Salari, S. H. and Ward, M. E.,** Polypeptide composition of *Chlamydia trachomatis*, *J. Gen. Microbiol.*, 123, 197, 1981.
22. **Hatch, T. P., Vance, D. W., Jr., and Al-Hossainy, E.,** Identification of a major envelope protein in *Chlamydia* spp., *J. Bacteriol.*, 146, 426, 1981.
23. **Barbour, A. G., Amato, K.-I., Hackstadt, T., Perry, L., and Caldwell, H. D.,** *Chlamydia trachomatis* has penicillin-binding proteins but not detectable muramic acid, *J. Bacteriol.*, 151, 420, 1982.
24. **Garrett, A. J., Harrison, M. J., and Manire, G. P.,** A search for the bacterial mucopeptide component muramic acid in *Chlamydia*, *J. Gen. Microbiol.*, 80, 315, 1974.
25. **König, E., Schlesner, H., and Hirsch, P.,** Cell wall studies on budding bacteria of the *Planctomyces/Pasteuria* group and on a *Prosthecomicrobium* species, *Arch. Microbiol.*, 138, 200, 1984.
26. **Matsumoto, A. and Manire, G. P.,** Electron microscopic observations on the effects of penicillin on the morphology of *Chlamydia psittaci*, *J. Bacteriol.*, 101, 278, 1970.
27. **Moulder, J. W., Novasel, D. L., and Officer, J. E.,** Inhibition of the growth of agents of the psittacosis group by D-cycloserine and its specific reversal by D-alanine, *J. Bacteriol.*, 85, 707, 1963.
28. **Lin, H.-S. and Moulder, J. W.,** Patterns of response to sulfadiazine, D-cycloserine, and D-alanine in members of the psittacosis group, *J. Infect. Dis.*, 116, 372, 1966.
29. **Gale, E. F., Cundliffe, E., Reynolds, P. E., Richmond, M. H., and Waring, M. J.,** *The Molecular Basis of Antibiotic Action*, 2nd ed., John Wiley & Sons, New York, 1981, Chaps. 1, 3.

30. **Frere, J.-M. and Joris, B.,** Penicillin-sensitive enzymes in peptidoglycan biosynthesis, *CRC Crit. Rev. Microbiol.,* 11, 299, 1985.

31. **Reeve, P. and Taverne, J.,** Some properties of the complement-fixing antigens of the agents of trachoma and inclusion blennorrhea and the relation of the antigen to the developmental cycle, *J. Gen. Microbiol.,* 27, 501, 1962.

32. **Dhir, S. P. and Boatman, E. S.,** Location of polysaccharide on *Chlamydia psittaci* by silver-methenamine staining and electron microscopy, *J. Bacteriol.,* 111, 267, 1972.

33. **Dhir, S. P., Hakomani, S., Kenny, G. E., and Grayston, J. T.,** Immunochemical studies on chlamydial group antigen (presence of a 2-keto-3-deoxy-carbohydrate as immuno-dominant groups), *J. Immunol.,* 109, 116, 1972.

34. **Nurminen, M., Wahlström, E., Kleemola, M., Leinonen, M., Saikku, P., and Mäkelä, P. H.,** Immunologically related ketodeoxyoctanate-containing structures in *Chlamydia trachomatis,* Re mutants of *Salmonella* species, and *Acinetobacter calioaceticus* var. *anitratus, Infect. Immun.,* 44, 609, 1984.

35. **Nuriminen, M., Rietschel, E. T., and Brade, H.,** Chemical characterization of *Chlamydia trachomatis* lipopolysaccharide, *Infect. Immun.,* 48, 573, 1985.

36. **Lewis, V. J., Thacker, W. L., and Mitchell, J. H.,** Demonstration of chlamydial endotoxin-like activity, *J. Gen. Microbiol.,* 114, 214, 1979.

37. **Nuriminen, M., Leinonen, M., Saikku, P., and Mäkelä, P. H.,** The genus-specific antigen of *Chlamydia:* resemblance to the lipopolysaccharide of enteric bacteria, *Science,* 220, 1279, 1983.

38. **Caldwell, H. D. and Hitchcock, P. J.,** Monoclonal antibody against a genus-specific antigen of *Chlamydia* species: location of the epitope on the chlamydial lipopolysaccharide, *Infect. Immun.,* 44, 306, 1984.

39. **Thornley, M. J., Zamze, S. E., Byrne, M. D., Lusher, M., and Evans, R. T.,** Properties of monoclonal antibodies to the genus-specific antigen of *Chlamydia* and their use for antigen detection by reverse passive haemagglutination, *J. Gen. Microbiol.,* 131, 7, 1985.

40. **Brade, L., Nurminen, M., Mäkelä, P. H., and Brade, H.,** Antigenic properties of *Chlamydia trachomatis* lipopolysaccharide, *Infect. Immun.,* 48, 569, 1985.

41. **Matsumoto, A.,** Electron microscopic observations of surface projections on *Chlamydia psittaci* elementary bodies, *J. Bacteriol.,* 150, 358, 1982.

42. **Gregory, W. W., Gardner, M., Byrne, G. I., and Moulder, J. W.,** Arrays of hemispheric projections on *Chlamydia psittaci* and *Chlamydia trachomatis* observed by scanning electron microscopy, *J. Bacteriol.,* 138, 241, 1979.

43. **Nichols, B. A., Setzer, P. Y., Pang, F., and Dawson, C. R.,** New view of the surface projections of *Chlamydia trachomatis, J. Bacteriol.,* 164, 344, 1985.

44. **Louis, C., Nicolas, G., Eb, F., Lefebvre, J.-F., and Orfila, J.,** Modifications of the envelope of *Chlamydia psittaci* during its developmental cycle: freeze-fracture studies of complementary replicas, *J. Bacteriol.,* 141, 868, 1980.

45. **Moulder, J. W.,** *The Biochemistry of Intracellular Parasitism.* The University of Chicago Press, Chicago, Ill., 1962, Chap. 4.

46. **Weiss, E. and Kiesow, L. A.,** Incomplete citric acid cycle in agents of the psittacosis-trachoma group *(Chlamydia), Bacteriol. Proc.,* 85, 1965.

47. **Weiss, E.,** Adenosine triphosphate and other requirements for utilization of glucose by agents of the psittacosis-trachoma group, *J. Bacteriol.,* 90, 243, 1965.

48. **Weiss, E. and Wilson, N. N.,** Role of exogenous adenosine triphosphate in catabolic and synthetic activities of *Chlamydia psittaci, J. Bacteriol.,* 97, 719, 1970.

49. **Gill, S. D. and Stewart, R. B.,** Respiration of L cells infected with *Chlamydia psittaci, Can. J. Microbiol.,* 16, 1079, 1970.

50. **Moulder, J. W.,** Glucose metabolism of L cells before and after infection with *Chlamydia psittaci, J. Bacteriol.,* 104, 1189, 1970.

51. **Hatch, T. P., Al-Hossainy, E., and Silverman, J. A.,** Adenine nucleotide and lysine transport in *Chlamydia psittaci, J. Bacteriol.,* 150, 662, 1982.

52. **Hatch, T. P., Miceli, M., and Silverman, J. A.,** Synthesis of protein in host-free reticulate bodies of *Chlamydia psittaci* and *Chlamydia trachomatis, J. Bacteriol.,* 162, 938, 1985.

53. **Winkler, H. H.,** Rickettsial permeability: an ADP-ATP transport system, *J. Biol. Chem.,* 251, 389, 1976.

54. **Sarov, T. and Becker, Y.,** Trachoma agent DNA, *J. Mol. Biol.,* 42, 581, 1969.

55. **Matsumoto, A. and Higashi, N.,** Electron microscopic observations of DNA molecules of the mature elementary bodies of *Chlamydia psittaci, Annu. Rep. Inst. Virus Res. Kyoto Univ.,* 18, 51, 1973.

56. **Kingsbury, D. T.,** Estimate of the genome size of various microorganisms, *J. Bacteriol.,* 98, 1400, 1969.

57. **Joklik, W. K.,** Some properties of vaccinia virus nucleic acid, *J. Mol. Biol.,* 5, 265, 1962.

58. **Razin, S.,** The mycoplasmas, *Microbiol. Rev.,* 42, 414, 1978.

59. **Myers, W. F., Baca, O. G., and Wisseman, C. L., Jr.,** Genome size of the rickettsia *Coxiella burnetti, J. Bacteriol.,* 144, 460, 1980.

60. **Myers, W. F. and Wisseman, C. L., Jr.,** Genetic relatedness among the typhus group of rickettsiae, *Int. J. Syst. Bacteriol.,* 30, 143, 1980.
61. **Kornberg, A.,** *DNA Replication,* W. H. Freeman & Co., San Francisco, Calif., 1980, 20.
62. **Lwoff, A.,** L'evolution physiologique. Etude des pertes de fonctions chez les microorganismes, *Actualities Scientific et Industrielles* No. 970, Hermann et Cie, Paris, 1944, 238.
63. **Gordon, F. B. and Quan, A. L.,** Occurrence of glycogen in inclusions of the psittacosis-lymphogranuloma venereum-trachoma agents, *J. Infect. Dis.,* 115, 186, 1965.
64. **Garrett, A. J.,** Some properties of the polysaccharide from cell cultures infected with TRIC agent *(Chlamydia trachomatis), J. Gen. Microbiol.,* 90, 133, 1975.
65. **Bernkopf, H., Mashiah, P., and Becker, Y.,** Correlation between morphological and biochemical changes and the appearance of infectivity in FL cell cultures infected with the trachoma agent, *Ann. N.Y. Acad. Sci.,* 98, 62, 162.
66. **Gordon, F. B. and Quan, A. L.,** Isolation of the trachoma agent in cell culture, *Proc. Soc. Exp. Biol. Med.,* 118, 354, 1965.
67. **Reeve, P. and Taverne, J.,** Strain differences in the behavior of TRIC agents in cell cultures, *Am. J. Ophthalmol.,* 63, 1167, 1967.
68. **Fan, V. S. C. and Jenkin, H.,** Glycogen metabolism in chlamydia-infected HeLa cells, *J. Bacteriol.,* 104, 608, 1970.
69. **Evans, A.,** The development of TRIC organisms in cell cultures during multiple infections, *J. Hyg.,* 70, 39, 1972.
70. **Jenkins, H. M. and Fan, V. S. C.,** Contrast of glycogenesis of *Chlamydia trachomatis* and *Chlamydia psittaci* in HeLa cells, in *Proc. Symp. on Trachoma and Related Disorders Caused by Chlamydial Agents,* Excerpta Medica International Congress No. 223, Nichols, R. L., Ed., Excerpta Medica, 1971, 52.
71. **Jones, H., Rake, G., and Stearns, B.,** Studies on lymphogranuloma venereum. III. The action of the sulfonomides on the agent of lymphogranuloma venereum, *J. Infect. Dis.,* 76, 55, 1945.
72. **Schachter, J.,** A bedsonia isolated from a patient with clinical lymphogranuloma venereum, *Am. J. Opthalmol.,* 63, 1049, 1967.
73. **Wong, S. Y., Gray, E. S., Finlayson, S., and Johnson, F. W. A.,** Acute placentitis and spontaneous abortion caused by *Chlamydia psittaci* of sheep origin: a histological and ultrastructural study, *J. Clin. Pathol.,* 38, 707, 1985.
74. **Johnson, F. W. A., Matheson, B. A., Williams, H., Laing, A. G., Jandial, V., Davidson-Lamb, R., Holliday, G., Hobson, D., Wong, S. Y., Hadley, K. M., Moffat, M. A., and Postlewaite R.,** Abortion due to infection with *Chlamydia psittaci* in a sheep farmer's wife, *Br. Med. J.,* 200, 592, 1985.
75. **Forsey, T. and Darougar, S.,** Acute conjunctivitis caused by an atypical chlamydial strain: *Chlamydia* IOL207, *Br. J. Opthalmol.,* 68, 409, 1984.
76. **Saikku, P., Wang, S.-P., Kleemala, M., Brander, E., and Grayston, J. T.,** An epidemic of mild pneumonia due to an unusual strain of *Chlamydia psittaci, J. Infect. Dis.,* 151, 832, 1985.
77. **Dwyer, R. S. C., Treharne, J. D., Jones, B. R., and Herring, J.,** Chlamydial infections. Results of micro-immunofluorescence tests for the detection of type-specific antibody in certain chlamydial infections, *Br. J. Vener. Dis.,* 48, 452, 1972.
78. **Weiss, E., Schramek, G., Wilson, N. N., and Newman, L. W.,** Deoxyribonucleic acid heterogeneity between human and murine strains of *Chlamydia trachomatis, Infect. Immun.,* 2, 24, 1970.
79. **Caldwell, H. G., Kuo, C-c., and Kenney, G. E.,** Antigenic analysis of chlamydiae by two-dimensional immunoelectrophoresis. II. A trachoma-LGV-specific antigen, *J. Immunol.,* 115, 969, 1975.
80. **Stephens, R. S., Tam, M. R., Kuo, C.-c., and Nowinski, R. C.,** Monoclonal antibodies to *Chlamydia trachomatis*: antibody specificities and antigen characterization, *J. Immunol.,* 128, 1083, 1982.
81. **Stephens, R. S. and Kuo, C.-c.,** *Chlamydia trachomatis* species-specific epitope detected on mouse biovar outer membrane protein, *Infect. Immun.,* 45, 790, 1984.
82. **Caldwell, H. D. and Schachter, J.,** Antigenic analysis of the major outer membrane protein of *Chlamydia* spp., *Infect. Immun.,* 35, 1024, 1982.
83. **Stephens, R. S., Kuo, C.-c., Newport, G., and Agabian, N.,** Molecular cloning and expression of *Chlamydia trachomatis* major outer membrane protein in *Escherichia coli, Infect. Immun.,* 47, 713, 1985.
84. **Caldwell, H. C. and Judd, R. C.,** Structural analysis of chlamydial outer membrane proteins, *Infect. Immun.,* 138, 960, 1982.
85. **Batteiger, B. E., Newhall, W. J. V., and Jones, R. B.,** Differences in outer membrane proteins of the lymphogranuloma venereum and trachoma biovars of *Chlamydia trachomatis, Infect. Immun.,* 50, 480, 1985.
86. **Grayston, J. T. and Wang, S.-P.,** New knowledge of chlamydiae and the diseases they cause, *J. Infect. Dis.,* 132, 87, 1975.
87. **Schachter, J. and Dawson, C. R.,** *Human Chlamydial Infections,* PSG Publishing, Littleton, Mass. 1978, Chaps. 3 and 4.

88. **Banks, J., Eddie, B., Schachter, J., and Meyer, K. F.,** Plaque formation by *Chlamydia* in L cells, *Infect. Immun.*, 1, 259, 1970.

89. **Rota, T. R. and Nichols, R. L.,** Infection of cell cultures by trachoma agent: enhancement by DEAE-dextran, *J. Infect. Dis.*, 124, 419, 1971.

90. **Lee, C. K.,** Interaction between a trachoma strain of *Chlamydia trachomatis* and mouse fibroblasts (L cells) in the absence of centrifugation, *Infect. Immun.*, 31, 584, 1981.

91. **O'Brien, T. L. and MacLeod, R.,** Description of chlamydia-like micro-organisms in *Hydra viridis* and an initial characterization of the relationship between microorganisms and hydra, *Cytobios*, 36, 141, 1983.

92. **Harshbarger, J. C., Chang, S. C., and Otto, S. V.,** Chlamydiae (with phages), mycoplasmas, and rickettsias in Chesapeake Bay bivalves, *Science*, 196, 666, 1977.

93. **Morrison, C. and Shum, G.,** Chlamydia-like organisms in the digestive diverticula of the bay scallop (*Argopecten irradians* Lmk), *J. Fish Dis.*, 5, 173, 1982.

94. **Buchanan, J. S.,** Cytological studies on a new species of rickettsia found in association with a phage in the digestive gland of the marine bivalve mollusc, *Tellina tenuis* (da Costa), *J. Fish Dis.*, 1, 27, 1978.

95. **Comps, M. and Dettrail, J. P.,** Un microorganisme de type rickettsien chez l'Huitre portugaise *Crassostrea angulata* Lmk, *C.R. Acad. Sci. Paris Ser. D*, 289, 169, 1979.

96. **Osaki, H.,** Electron microscopic observations of chlamydia-like organism in hepatopancreas cells of the spider, *Coelotes luctuosus*, *Acta Arachnol.*, 25, 23, 1973.

97. **Morel, G.,** Studies on *Parachlamydia buthi* g. n. sp. n., an intracellular pathogen of the scorpion *Buthus occitanus*, *J. Invert. Pathol.*, 28, 167, 1976.

98. **Shay, M. T., Bettica, A., Vernon, G. M., and Witkus, E. R.,** *Chlamydia isopodii* sp. n., an obligate intracellular parasite of *Porcellio scaber*, *Exp. Cell Biol.*, 53, 115, 1985.

99. **Sparks, A. K., Morado, S. F., and Hawkes, J. W.,** A systemic microbial disease in the Dungeness crab, *Cancer magister*, caused by a *Chlamydia*-like organism, *J. Invert. Pathol.*, 45, 204, 1985.

100. **Page, L. A. and Cutlip, R. C.,** Morphological and immunological confirmation of the presence of chlamydiae in the gut tissues of the Chesapeake Bay clam, *Mercenaria mercenaria*, *Ann. Microbiol.*, 7, 297, 1982.

101. **Meyers, T. R.,** Preliminary studies on a chlamydial agent in the digestive diverticular epithelium of hard clams *Mercenaria merceneria* (L.) from Great South Bay, New York, *J. Fish Dis.*, 2, 179, 1979.

102. **Schleifer, K. H. and Stackebrandt, E.,** Molecular systematics of procaryotes, *Annu. Rev. Microbiol.*, 37, 143, 1983.

103. **Rogers, H. J., Perkins, H. R., and Ward, J. B.,** *Microbial Cell Walls* and *Membranes*, Chapman and Hall, London, 1980, Chap. 1.

104. **Gerloff, R. K., Ritter, D. B., and Watson, R. O.,** DNA homology between the meningopneumonitis agent and related organisms, *J. Infect. Dis.*, 116, 197, 1966.

105. **Kingsbury, D. T. and Weiss, E.,** Lack of deoxyribonucleic acid homology between species of the genus *Chlamydia*, *J. Bacteriol.*, 96, 1421, 1968.

106. **Brenner, D. J., Steigerwalt, A. G., Gorman, G. W., Wilkinson, H. W., Bibb, W. F., Hackel, M., Tyndall, R. L., Campbell, J., Feeley, J. C., Thacker, W. L., Skaliy, P., Martin, W. T., Brake, B. J., Fields, B. S., McEachern, H. V., and Corcoran, L. K.,** Ten new species of *Legionella*, *Int. J. Syst. Bacteriol.*, 35, 50, 1985.

107. **Fox, G. E., Stackebrandt, E., Hespell, R. B., Gibson, J., Maniloff, J., Dyer, T. A., Wolfe, R. S., Balch, W. E., Tanner, R. S., Magrum, L. J., Zablen, L. B., Blakemore, R., Gupta, R., Bonen, L., Lewis, B. J., Stahl, D. A., Luehrsen, K. R., Ghen, K. N., and Woese, C. R.,** The phylogeny of procaryotes, *Science*, 209, 457, 1980.

108. **Weisburg, W., Hatch, T. P., and Woese, C. W.,** *J. Bacteriol.*, 167, 570, 1986.

109. **Weisburg, W., Woese, C. R., Dobson, M. E., and Weiss, E.,** A common origin of rickettsiae and certain plant pathogens, *Science*, 230, 556, 1985.

110. **Woese, C. R.,** Why study evolutionary relationships among bacteria? in *Evolution of Prokaryotes*, Schleifer, K. H. and Stackebrandt, E., Eds., Academic Press, London, 1985, 17.

111. **Staley, J. T.,** Budding bacteria of the *Pasteuria-Blastobacter* group, *Can. J. Microbiol.*, 19, 609, 1973.

112. **Moulder, J. W.,** The relation of the psittacosis group (Chlamydiae) to bacteria and viruses, *Annu. Rev. Microbiol.*, 20, 107, 1966.

113. **Richmond, S. J., Stirling, P., and Ashley, C. R.,** Virus infecting the reticulate bodies of an avian strain of *Chlamydia psittaci*, *FEMS Microbiol. Lett.*, 14, 31, 1982.

114. **Schachter, J., Banks, J., Sugg, N., Sung, M., Storz, J., and Meyer, K. F.,** Serotyping of *Chlamydia*. I. Isolates of ovine origin, *Infect. Immun.*, 9, 92, 1974.

115. **Schachter, J., Banks, J., Sugg, N., Sung, M., Storz, J., and Meyer, K. F.,** Serotyping of *Chlamydia*: isolates of bovine origin, *Infect. Immun.*, 11, 904, 1975.

116. **Staley, J. T. and Krieg, N. R.,** Classification of procaryotic organisms: an overview, in *Bergey's Manual of Systematic Bacteriology*, Krieg, N. R., Ed., Vol. 1, Williams & Wilkins, Baltimore, Md., 1984, 1.

117. **LaPage, S. P., Sneath, P. H. A., Lessel, E. F., Skerman, V. B. D., Seeliger, H. P. R., and Clark, W. A., Eds.,** *International Code of Nomenclature of Bacteria,* American Society of Microbiologists, Washington, D.C., 1976.

118. **Moulder, J. W.,** Order *Chlamydiales* and family *Chlamydiaceae,* in *Bergey's Manual of Systematic Bacteriology,* Krieg, N. R., Ed., Vol. 1, Williams & Wilkins, Baltimore, Md., 1984, 729.

119. **Southern, P. J., Blount, P., and Oldstone, M. B. A.,** Analysis of persistent virus infection by in vitro hybridization to whole-mouse sections, *Nature,* 312, 555, 1984.

120. **Perez-Martinez, J. A. and Storz, J.,** Antigenic diversity of *Chlamydia psittaci* of mammalian origin determined by immunofluorescence, *Infect. Immun.,* 50, 905, 1985.

Chapter 2

STRUCTURAL CHARACTERISTICS OF CHLAMYDIAL BODIES

Akira Matsumoto

TABLE OF CONTENTS

I. INTRODUCTION

In this chapter our current studies on the morphology of chlamydiae have been summarized. Its purpose is to emphasize and perhaps explain the significance of the structure to events in the multiplication of chlamydiae, as revealed mainly by electron microscopy. Several points should be made at the outset: (1) chlamydiae are obligate intracytoplasmic parasites; both species in the genus *Chlamydia* have a complicated, unique life cycle which has never been observed in any other microorganisms, (2) in the life cycle, there are two different cell types, one is the infectious elementary body (EB) and the other is the vegetative, noninfectious reticulate body (RB), and (3) the cell types differ from each other not only in morphology, but also in many respects, including strength against mechanical agitation, cell wall permeability, and chemical composition in the cell envelopes. Many of these facts have been learned from electron microscopic observations on several strains of *Chlamydia psittaci*, some of which are well controlled in cell culture in vitro. Therefore, we know more about these strains than we do about other strains of *C. psittaci* and of *C. trachomatis*. The information obtained from these strains are compared with the morphology of the other chlamydiae at many points.

Electron microscopy has had a great impact on our understanding of chlamydial morphology. The information obtained to date has been, on one hand, derived from electron microscopy of *in situ* chlamydiae, and on the other hand, from examinations of isolated chlamydial bodies and their structural components which were prepared by the various physicochemical techniques. A number of preparation techniques for electron microscopy have also been established. The specimen images viewed with the electron microscope are dependent on the specimen preparation technique. This means that the image, even though the same object is being examined, varies with the techniques. The reconstitution of an intrinsic morphology from the different images is thus more meaningful. It is from this view that chlamydial morphology is discussed.

II. GENERAL ASPECTS OF INTRACYTOPLASMIC INCLUSIONS AND *IN SITU* CHLAMYDIAL BODIES

From studies on chlamydial multiplication by conventional light microscopy, Bedson and Bland[1] demonstrated the complex sequence of events in an infected L cell from the first stage of infection to the release of progeny of *C. psittaci*. They found that chlamydial populations consisted of two very different kinds of bodies: one small and the other a little larger in size. Their findings were later extended by them and by other workers and recognized to apply to all members of the genus *Chlamydia* with minor modifications.[2,3] Using a one-step growth curve, Higashi[4] first studied the sequential morphogenesis of *C. psittaci* by light and electron microscopy and reported a description of its multiplication which is summarized as follows:

1. The particles which stain red with Macchiavello's stain enter the host cells by viropexis (phagocytosis) and then convert their forms into large, blue particles. The red and blue particles correspond to elementary bodies and reticulate bodies, respectively.
2. Although RBs rapidly multiply by fission from 10 to 20 hr after infection, no infectivity is detectable during this period; that is, chlamydial bodies lose their infectivity during the lag phase even though morphological forms are continuous throughout their intracytoplasmic multiplication.

Due to poor specimen preservation, which undoubtedly resulted from the use of methacrylate embedding medium, the inclusion membrane was not clarified in his early electron

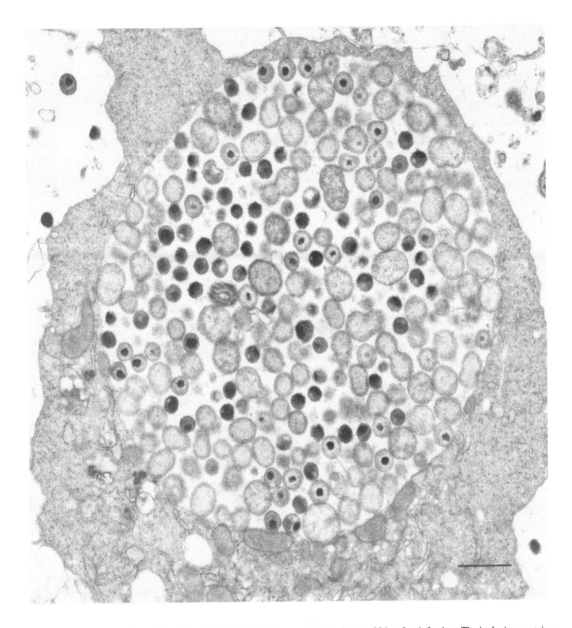

FIGURE 1. Inclusion of a strain of *C. psittaci*[14] isolated from a grass parakeet at 30 hr after infection. The inclusion contains three types of chlamydial bodies; EBs, RBs, and intermediate bodies. No glycogen particle is seen. Bar indicates 1 μm.

microscopic study. The introduction of epoxy or polyester resin for thin sectioning made a great improvement in the preservation of chlamydial bodies and host cells. Higashi[5] carried out further electron microscopic observations on the cells infected with EBs, which were purified by the method of Tamura and Higashi.[6] These studies were done in parallel with growth experiments and light microscopy, and confirmed the results mentioned above. It was also revealed that all events in chlamydial multiplication occurred within inclusions bound with a membrane, which was derived from the host plasma membrane as EBs of both species were phagocytized. Such a complicated morphogenesis was confirmed not only for other *C. psittaci* strains but also for strains of *C. trachomatis*.[7-13]

Figure 1 illustrates an inclusion at 30 hr in an L-929 cell inoculated with a *C. psittaci* strain which has been isolated from a pet grass parakeet kept by a psittacosis patient.[14] There are three types of chlamydial bodies; EBs, which have a high density and are small in size (0.3 to 0.35 μm in diameter); RBs, which consist of rather homogeneous internal material and are large in size (0.5 to 1.3 μm in diameter); and intermediate bodies which have a characteristic dense nucleus at the center of the body. Some RBs show the process of binary fission. As shown in Figure 1, inclusions of *C. psittaci* regularly contain a number of chlamydial bodies throughout the multiplication cycle. Consequently, the inclusions stained with Giemsa are seen as rather homogeneous, purple-blue, compact areas in the host cytoplasm under the light microscope. To compare the dimensions of some structures, such as membrane thickness and diameter of ribosomes, measurements have to be done within the same micrograph. By this method, the dimensional aberration which is caused by the mechanical aberration in the electron microscope is completely eliminated. At higher magnification, it is clear that all chlamydial bodies, EBs, RBs, and intermediate bodies, are surrounded by a trilaminar membrane (Figure 2). Measurement of the outer membrane or cell wall gives a thickness of 75 to 80 Å. Another trilaminar membrane, the cytoplasmic membrane, is seen just beneath the cell wall in RBs and intermediate bodies, while the cytoplasmic membrane cannot be seen in EBs. This is presumably because the electron opacity of the cytoplasmic membrane is identical with that of the internal material, which is composed of ribosomes and nuclei in amorphous material. A stalk, 30 to 33 nm in width, is occasionally seen between two intermediate bodies and may be formed by fission of an RB. The stalks, as far as examined, are detected only between the intermediate bodies, but are not seen between RB and EB. It is, therefore, likely that the nucleus formation at the central region occurs after the final fission of an RB.

From measurements on electron micrographs, the cell walls of *C. trachomatis* strains (L2/434/Bubo, A/G-17/OT, and J/UW-36/CX) show identical thickness to those of *C. psittaci* strains (meningopneumonitis, Izawa,[14] and Nose strains[15]), although the electron opacity of EB cell walls is higher than that of RB cell walls in each case.

In contrast to *C. psittaci*, inclusions of *C. trachomatis* strains are quite different. Figure 3, at the same magnification as Figure 1, shows an inclusion at 30 hr postinoculation in an L cell infected with *C. trachomatis* L2/434/Bubo strain. Although the dimensions of chlamydial bodies are different from those of *C. psittaci*, each body can be recognized as EB, RB, and intermediate bodies based on their morphology; that is, one can not differentiate chlamydial species by morphology. However, the appearance of *C. trachomatis* inclusions is quite different from that of *C. psittaci* at three points: (1) the cell population in a *C. trachomatis* inclusion is much lower than that of *C. psittaci*; (2) degradation of RBs and accumulation of highly dense particles in membrane-bounded bodies are regularly encountered (the membranes containing the dense particles may be derived from RBs because of the presence of two trilaminar membranes); and (3) the accumulation of a number of dense particles occurs in the intrainclusion space. Concerning (1) and (2), the morphology of *C. trachomatis* RB, together with the RB degradation and production of the dense particles in RBs, will be discussed later (see section IV).

The iodine stainable carbohydrate in the inclusions of *C. trachomatis* was first reported by Rice.[16] After her finding, many workers reported on the kinetics of the carbohydrate synthesis during chlamydial multiplication and the characterization of the carbohydrate from many strains of *C. trachomatis*.[17-24] The results are briefly summarized as follows:

1.   The iodine-stainable carbohydrate is accumulated in inclusions during multiplication of *C. trachomatis* strains, but not in inclusions of *C. psittaci* strains though *C. psittaci* does synthesize carbohydrate during multiplication.
2.   The carbohydrate is biochemically characterized as glycogen with an average chain length of 14 to 16 glucose units.

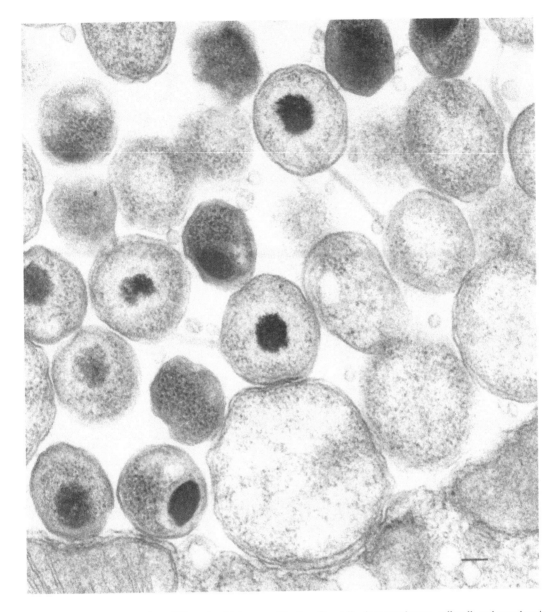

FIGURE 2. Part of the inclusion of Figure 1 at higher magnification. Two trilaminar membranes, cell wall, and cytoplasmic membrane are obvious in RBs and intermediate bodies, while the cytoplasmic membrane is not identified in EBs. A stalk is seen between two intermediate bodies. Bar indicates 100 nm.

3. As measured by the number of inclusions which are visualized with iodine stain, the accumulation of glycogen occurs at 20 to 24 hr after infection. Our study on the glycogen accumulation indicated that the number of iodine-positive inclusions increased from 18 hr and reached a maximum level at 24 hr postinoculation in L-929 cells when the cells were incubated at 37°C in 5% CO_2 and 95% air atmosphere. The maximum level was retained up to 60 hr and more postinoculation.[25]

4. Penicillin, which strongly inhibits the normal RB fission and maturation to form EB from RB,[26,27] effectively inhibits glycogen accumulation. The glycogen, however, does accumulate in the presence of 5-fluorouracil, although chlamydial development is effectively inhibited.[28] Thus, it is suggested that inhibition of glycogen accumulation with penicillin is a deceptive phenomenon and that the glycogen accumulation is not essential for the growth of *C. trachomatis*.

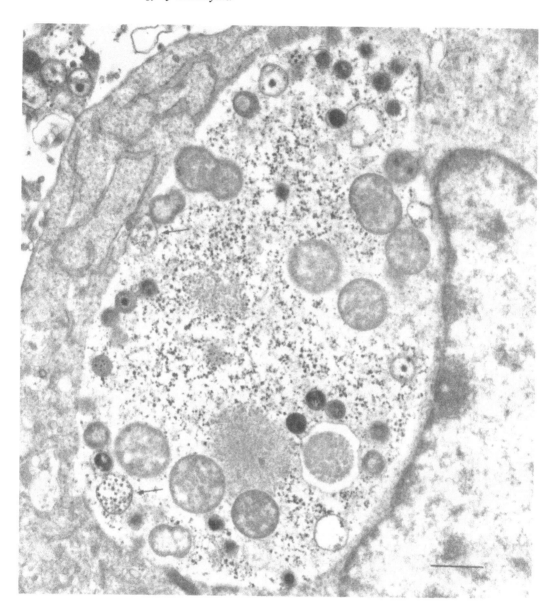

FIGURE 3.   Inclusion of *C. trachomatis* L2 strain at 30 hr after infection. EBs, RBs, and an intermediate body are seen. There are a number of densely stained glycogen particles in the intrainclusion space. Several degenerated RBs, some of which contain glycogen particles, are noted (arrows). Bar indicates 1 μm.

Regarding points (1) and (2), the glycogen accumulation, which is readily detectable with the iodine or lugol staining, is one of the important criteria for differentiation of chlamydial species, together with the compression of the host nucleus to the periphery with the expansion of inclusions formed with many *C. trachomatis* strains on the 2nd to 3rd day after infection. The dense particles scattered throughout the inclusion space (Figure 3) are glycogen particles, which readily disappear during specimen preparation, especially in the fixation procedures.

Chlamydial bodies of *C. trachomatis* strains show very active, rapid movement in their inclusions.[29] This movement is easily seen under the phase contrast microscope at a relatively low magnification. No effect is seen on the movement after UV irradiation, which completely inactivates the infectivity. Thus, the movement appears to be caused by Brownian motion. On the contrary, one cannot observe movement in the inclusions of *C. psittaci* strains. The

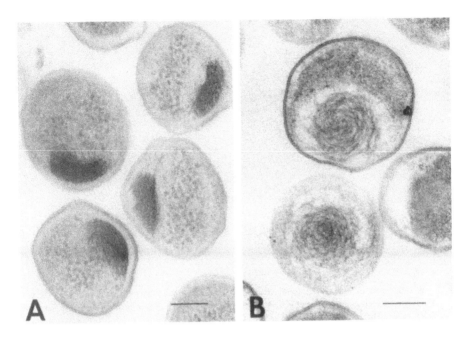

FIGURE 4. Purified EBs of the meningopneumonitis strain in thin sections. (A) EBs recovered with sucrose gradient centrifugation. The nucleus is located at an eccentric region and the cytoplasm is composed of many ribosomes in moderately dense, amorphous material. (B) EBs recovered with Renografin gradient centrifugation. Thin fibers, presumably DNA, tangle concentrically. No ribosome is differentiated in the cytoplasm. Bar indicates 100 nm.

movement seems to be dependent on the population density of the chlamydial bodies in the inclusions. Brownian movement in *C. psittaci* inclusions may be prevented with packed chlamydial bodies.

III. MORPHOLOGY OF EB

Studies on EB morphology were greatly improved not only by the development of techniques in electron microscopy, but also by the establishment of purification methods for EBs and their components, such as cell envelopes and cell walls. A method by which EBs are recovered in an extremely high purity was first established for *C. psittaci* meningo-pneumonitis strain by Tamura and Higashi.[6] The infectious EBs are recovered from the infected cell cultures at 48 hr postinoculation by a series of treatments consisting of differential centrifugation, sonic disruption to eliminate fragile RBs and cell debris, incubation with RNase, DNase, and trypsin, and sucrose gradient centrifugation.

The internal organization of purified EBs is identical to that of the *in situ* EBs, and no clear evidence showing a difference in morphology between the two species is revealed by thin sectioning. Each EB has a dense nucleus at an eccentric region and cytoplasm that is composed of a number of ribosomes and moderately dense, amorphous material (Figure 4A). The diameter of the ribosomes ranges from 12 to 15 nm in thin section. When EBs are recovered by sucrose density gradient centrifugation, the ribosomes in each EB tend to be clearly seen in thin sections prepared after double fixation with glutaraldehyde and osmium solutions. However, the ribosomes are not so clearly differentiated when EBs are purified by Renografin gradient centrifugation as described by Caldwell et al.[30] Alternatively, the cytoplasm appears to be composed of a mass of amorphous material, and a circular rolling arrangement of fine DNA fibers becomes evident in the nuclear region (Figure 4B). In such

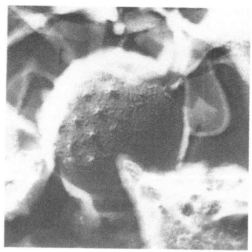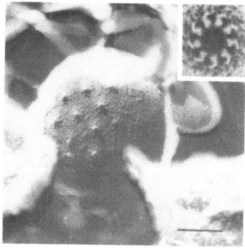

FIGURE 5.   A stereo pair of an EB viewed by freeze-deep-etching. Each projection emerged from the center of a "flower". The flower is seen as a radial arrangement of nine subunits (inset). Bar indicates 100 nm. (From Matsumoto, A., *J. Bacteriol.*, 151, 1040, 1982. With permission.)

preparations, parallel arrangement of the fibers is occasionally encountered. Thus, the arrangement may indicate the mode of DNA folding in the nuclear region in each EB.

Figure 5 illustrates a stereo pair of a purified EB prepared by freeze-deep-etching.[31] It is to be noted that there are unique surface projections which are arrayed hexagonally at a center-to-center spacing of approximately 50 nm. Each projection emerges from the center of a "flower" structure, about 30 nm in diameter, which appears to be a radial arrangement of several leaves. An identical morphology is observed on *in situ* EB surfaces when the deep-etching is applied to inclusions in the host cells at 40 hr postinoculation.[31] That the flower is composed of nine leaves in a nonagonal arrangement was determined by the application of the rotation technique of Markham et al.[32] (Figure 5, inset).

The surface projections can also be observed on the surface of EBs under the scanning electron microscope. However, the thickness of the metal coating reduces the resolution. Consequently, the detailed morphology of the projections is concealed and the projections are seen as hemispheres.[33,34] Nevertheless, scanning electron microscopy is a powerful tool for observing the projections. Based on a statistical analysis on EBs observed by scanning electron microscopy, Matsumoto and Higashi reported that each EB of the meningopneumonitis strain had only one surface area containing projections ranging from 10 to 23 with a peak at 18 (mean = 17.5/EB).[33,35] A recent study on EBs of a newly isolated *C. psittaci* strain revealed the projections ranged from 13 to 30 with a mean number of 22 per EB.[14] With similar examinations by scanning electron microscopy, Gregory et al. reported the presence of projections on EBs of many strains of *C. psittaci* and *C. trachomatis* and proposed the projections as a phenotypic marker for recognizing members of genus *Chlamydia*.[34]

The ordinary thin sectioning technique failed to identify the projections. This may be due to the inadequate electron opacity of the projections in thin sections. Application of ruthenium red to purified EBs of *C. psittaci* revealed that the projections were not distributed on the surface close to the nucleus, which is normally located in an eccentric region, but on the surface at the far side from the nucleus (Figure 6).[36] When EBs were collapsed gently by repeated freezing-thawing and then treated with RNase, the cytoplasmic membrane sites where the projections were located were bound to DNA fibers, which were detached from the projection sites after trypsin treatment. By successive treatments with DNase and washings, the internal materials were almost completely removed, and cell envelopes were ob-

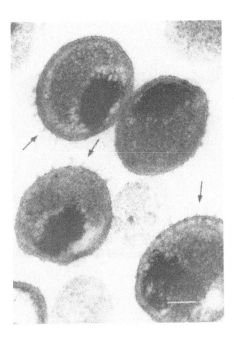

FIGURE 6.　EBs of the meningopneumonitis strain treated in
block with ruthenium red to enhance the electron opacity of the
projections. The projections are seen on the surface far from
the nucleus (arrows). Bar indicates 100 nm.

tained. In such a preparation, it was shown that one end of each projection was fitted into
the cytoplasmic membrane while the other end of the projection protruded beyond the cell
wall (Figure 7A, B).[36] The DNA-binding sites seem, therefore, to be bound to the cytoplasmic
membrane. These results strongly suggest that the DNA fibers extended from the nucleus
to the cytoplasmic membrane connecting with the projections at some points of the molecule.
It is likely that some regions of the DNA molecule are exposed directly to the cytoplasm
of EB, and that the contact of the DNA molecule to the cytoplasmic membrane is mediated
by a trypsin-sensitive binding component. Using the freeze-deep-etching method, the presence of the projections in purified C. psittaci EB envelopes was confirmed by Wyrick and
Davis.[37] Nichols et al.[38] demonstrated the projections in in situ C. trachomatis intermediate
forms which were fixed by unique fixation procedures consisting of prefixation with a mixture
of glutaraldehyde and acrolein, and postfixation with osmium tetroxide.

The presence of the projections on in vivo EB surfaces was first confirmed in thin sections
by Soloff et al.,[39] who prepared the specimens with an additional treatment of tannic acid
after double fixation of cervical tissue from guinea pig infected intravaginally with a C.
psittaci guinea pig inclusion conjunctivitis (GPIC) strain. The finding by Soloff et al. that
the projections on EBs appeared, in some cases, to impinge on adjacent RB caused them
to suggest that the projections may be involved in intrachlamydial communication.

IV. MORPHOLOGY OF RB

There is no difference in the essential organization between RBs of C. psittaci and C.
trachomatis strains as seen by electron microscopy. Each RB is surrounded with two trilaminar membranes, the cell wall, and cytoplasmic membranes, which possess an identical
thickness (approximately 80 Å) in thin sections. However, some difference is seen in internal
structure of RBs depending on the stages of the multiplication. Careful examinations indicate

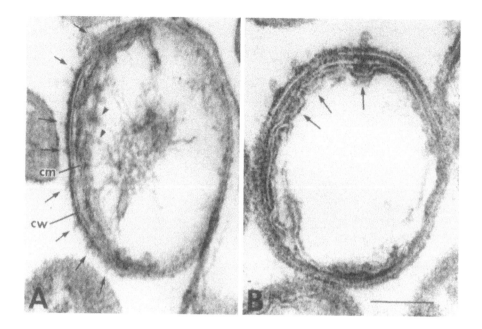

FIGURE 7. Envelopes of the meningopneumonitis strain treated with tannic acid to enhance the electron opacity. (A) An envelope prepared with repeated freeze-thawing followed by the treatment with RNase. Some fibers protruding from the nucleus bind to some sites of the cytoplasmic membrane (arrowheads) which roughly coincide with the sites of the projections (arrows). (B) A complete EB envelope prepared with successive treatments with trypsin and DNase. inside structures, such as cytoplasm and nucleus, are removed. Fitting of one end of each projection into the cytoplasmic membrane is seen (arrow). (cw) Cell wall, (cm) cytoplasmic membrane. Bar indicates 100 nm. (From Matsumoto, A., *J. Electron Microsc.*, 30, 315, 1981. With permission.)

that many RBs, between 10 to 18 hr postinoculation, tend to show a homogeneous distribution of internal components, such as ribosomes and DNA fibers. RBs at 8 to 10 hr and after 18 hr frequently show an uneven distribution of the marginal condensation of the internal material. Consequently, the central area appears to be less dense and one can see aggregation of fibrous material. At the early phase, the less dense area is rapidly replaced by the homogeneous distribution of the internal component. In contrast, at the late phase, the area rapidly increases its density for the condensation of fibrous material, presumably DNA, and then the RB converts into the intermediate form. These morphological changes may reflect the metabolic activity of RBs at different stages of multiplication.

Tamura et al.[40] established a purification method of RBs from L cells infected with *C. psittaci* meningopneumonitis strain at 18 hr postinoculation. At this stage, the infected cells do not contain any EB. The isolated RBs, which were examined in thin sections, showed identical morphology with that of *in situ* RBs. When the isolated RBs were shadowcast, many small particles about 15 nm in diameter were released from RBs. The particles were presumably ribosomes released from broken RBs, as their number was found to increase after mechanical agitation of RBs and because the RB envelopes were consequently fragile to mechanical agitation. To purify *C. trachomatis* EBs, a fraction containing EBs and RBs are frequently treated with sonication.[30] Thus, *C. trachomatis* RBs seem to be as fragile as the RBs of *C. psittaci*.

A study on the glycogen accumulation in the inclusions of a *C. trachomatis* L2 strain and a clinical isolate was recently carried out by the iodine and Giemsa staining methods and by electron microscopy.[25] Glycogen accumulated in the inclusions of the L2 strain after 18 hr. All inclusions became iodine-positive at 24 hr postinoculation. The results obtained by

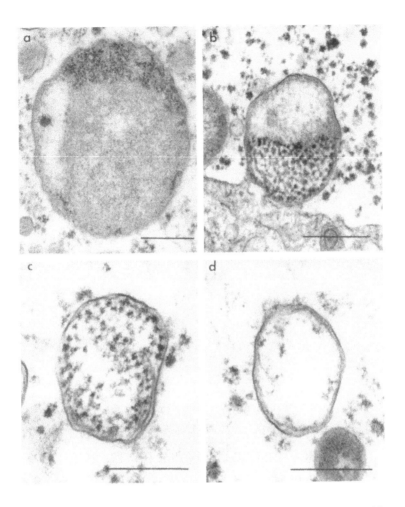

FIGURE 8.   A possible sequence (from a to d) of accumulation of glycogen particles within RBs of *C. trachomatis* L2 strain. (a) RB at 18 hr, (b) RB at 20 hr, (c) and (d) RB at 22 hr. Bars indicate 0.5 μm.

electron microscopy strongly suggested that the glycogen accumulation appeared to be initiated within some RBs from 18 hr postinoculation; the RBs were rapidly filled up with glycogen particles and then the glycogen particles were released into the intrainclusion space by the rupture of RB envelopes. Figure 8 shows the possible sequence of glycogen accumulation in RBs. Such glycogen accumulation in inclusions, as reported by many workers, is never encountered in RBs of *C. psittaci* strains, indicating one of the distinct differences in RB morphology between *C. trachomatis* and *C. psittaci*.

Another difference seen between *C. trachomatis* and *C. psittaci* concerns the degradation of RBs. As shown in Figure 9A, one could frequently observe the expansion of *C. trachomatis* RBs, followed by degradation during multiplication. The examinations indicated no correlation between such degradation and glycogen accumulation. The expansion of RB envelopes appeared to occur concurrently with the fading of internal materials and then, in many cases, homogeneous, less dense areas appeared to be formed (Figure 9B). This area seems to be identical with the translucent area which is frequently seen in inclusions stained with iodine under the light microscope.[29]

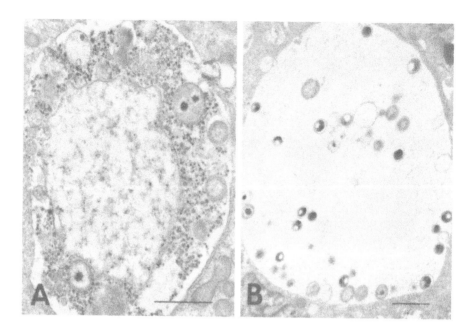

FIGURE 9.   Inclusions of *C. trachomatis* strains in thin sections. (A) Inclusion of L2 strain at 30 hr after infection. A largely expanded RB is seen at the center of inclusion. (B) Inclusion of a *C. trachomatis* strain isolated from a nongonoccocal urethritis patient at 30 hr after infection of 5th passage. By the prolonged fixation in glutaraldehyde and dehydration in ethanol, the less dense areas became homogeneous masses with moderate opacity, though the opacity of glycogen particles had been reduced. Bars indicate 1 μm.

V. MORPHOLOGY OF CELL ENVELOPES

Manire and Tamura[41] established an excellent purification method of *C. psittaci* EB cell walls which provided for great advances in determining the morphology of the cell walls. By electron microscopic examinations of purified EB cell walls of the meningopneumonitis strain, Manire[42] first demonstrated a regularly packed, hexagonal macromolecular structure about 100 Å in diameter located on the inner layer of the cell walls. He suggested that these structures are responsible for the rigidity of EB cell walls to mechanical agitation. The hexagonal arrangement is seen only on the partially exposed inner surface of cell walls when the specimens are prepared by the shadowcast technique (Figure 10A). However, the structures are evident all over the cell wall when stained negatively because of the penetration of the staining solution into the inside of the cell wall (Figure 10B).[43] In an early study, we reported an approximate 18 nm periodicity of the hexagonal arrangement as measured from a center-to-center spacing on the micrographs of shadowcast and negatively stained EB cell walls.[43] To determine the exact periodicity, optical diffraction and optical transformation by the use of the laser beam were directly applied to the cell walls stained negatively.[44] From the diffraction, it was shown that the cell wall was composed of three structural units with dimensions of 167, 90, and 60 Å, respectively (Figure 11). When an image was reconstituted by optical transformation using only spots showing the 167 Å periodicity, the image obtained was hexagonally arrayed pattern. This indicated that the hexagonal image corresponded to the hexagonal structure on the inside layer of the EB cell wall and that the periodicity of the units composing the hexagonal structures was 167 Å. In the negatively stained preparations, the specimens lie upon one another and this is a main obstacle in determining the outer surface of cell wall. This disadvantage is eliminated by the examination

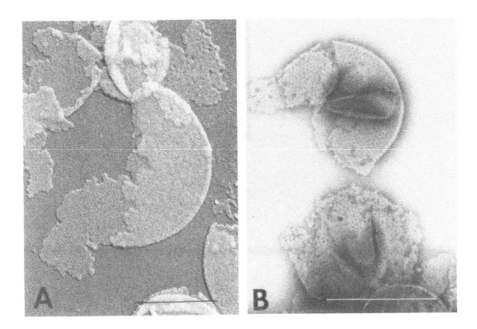

FIGURE 10. Purified EB cell walls. (A) EB cell walls shadowcast with Pt-palladium alloy. The hexagonal pattern is seen only on the inside surface. (B) EB cell walls stained negatively. The hexagonal arrayed structures are evident throughout cell wall. Bars indicate 0.5 μm.

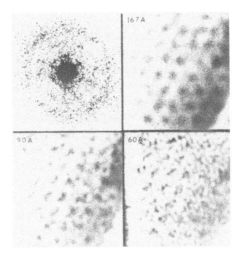

FIGURE 11. Diffraction pattern and images reconstituted by optical transformation. (167 Å) Image reconstituted by filtration with the spots at 167 Å, showing an evident hexagonal pattern. (90 Å) Image reconstituted by the filtration with the spots at 167 and 90 Å. The hexagonal pattern is modified, but the modification is not clearly differentiated. (60 Å) Image reconstituted by filtration using all indexes at 167, 90, and 60 Å. The fine particles laying on the hexagonal pattern are differentiated.

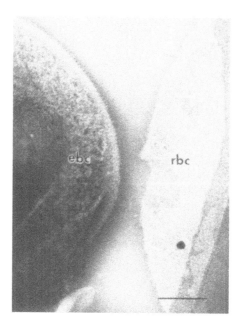

FIGURE 12.   Parts of EB and RB cell walls stained negatively. The protrusions of fine particles 50 to 60 Å in diameter are evidently seen at the edge of EB cell wall. The particles are partially distributed in RB cell wall. (ebc) EB cell wall, (rbc) RB cell wall. Bar indicates 100 nm.

of the folded edge of the cell wall. At the folded edge of EB cell walls one can see fine particles about 50 to 60 Å in diameter protruding from the surface with a center-to-center spacing of 70 to 80 Å (Figure 12).[44] These particles seem to be a major structural component of the outer most surface of the EB cell wall. Optical transformation using the spots of 60, 90, and 167 Å indicated an image composed of fine particles lying on the hexagonal structure. However, the image reconstituted with the 90 and 167 Å spots was only a deformed hexagon without fine particles. It is, therefore, likely that the units giving the 60 Å dimension are fine particles on the outer surface of the cell wall. The structure showing the units at a dimension of 90 Å has not been determined.

The only noticeable structures observed in EB cell walls are fine holes, referred to as "rosettes", which were roughly arrayed hexagonally in a group at a center-to-center spacing ranging from 35 to 49 nm (Figure 13A).[45] Each rosette, 19 to 20 nm in diameter, is composed of several subunits surrounding a central zone, 10 to 12 nm in diameter, into which the negative-stain penetrates. Results obtained by photographic enhancement of subunit contrast by the rotation technique clearly showed a nonagonal profile, indicating that each rosette has nine subunits (Figure 13A, inset). The number of rosette subunits is exactly the same as that of flower leaves seen in the freeze-replicas of the intact EB surface (Figure 5, inset). Therefore, it may be concluded that the rosettes and flowers are the same structures, and consequently that the rosette is a hole from which the projection protrudes beyond the cell wall. This was confirmed by the examinations of negatively stained EB envelopes in which the projections were occasionally split off from the rosettes during preparation (Figure 13B).[35] Therefore, it is very likely that the rosette is a morphological marker indicating the presence of the projection.

When EBs are examined by the freeze-replica technique, they appear as convex or concave faces (Figure 14).[45] The "button structures" (B structures) are observed only on the concave

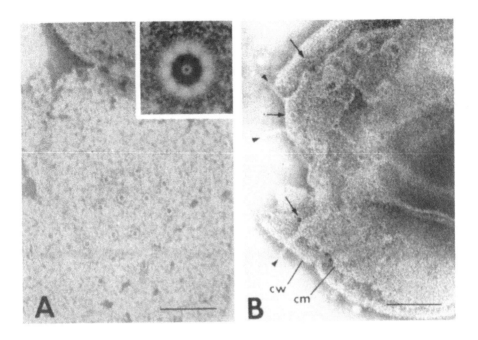

FIGURE 13. EB cell wall fragment and envelope stained negatively. (A) A fragment containing rosettes. Application of the rotation technique clearly shows nine subunits forming a rosette (inset). (B) Part of an envelope. The projections (arrowheads) and rosettes (arrows) are clearly seen. (cw) Cell wall, (cm) cytoplasmic membrane. Bars indicate 100 nm.

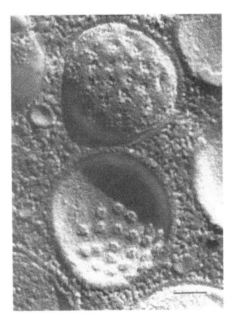

FIGURE 14. Freeze-replica of purified EBs. The B structures are seen on a limited area of the concavity. On the convexity, there are some particles in hexagonal arrangement with an identical center-to-center spacing with that of the B structures. The cell wall is regularly cut through its whole thickness. Bars indicate 100 nm. (From Matsumoto, A., *J. Electron Microsc.*, 28 (Suppl.), 57, 1979. With permission.)

face. Each B structure has a diameter of 27 nm and several B structures are grouped together in a hexagonal arrangement with a center-to-center spacing similar to that of the projections. On the convex face, one could see larger particles 16 nm in diameter. They too are arrayed hexagonally with a spacing also identical with the B structures. Therefore, the B structure and large particle seem to be originally complementary. The "craters" reported by Louis et al.[46] are undoubtedly the same as the B structures. It is noteworthy that the large particles referred to as the central granules by Louis et al. possessed a fine pit at the center. Based on these observations, it may be concluded that one end of each projection tightly connects with cytoplasmic membrane and the other end projects beyond the cell wall through a hole (rosette) composed of nine leaves and that the B structure or crater is one end of projection connecting with the cytoplasmic membrane viewed from the inside of the EB.

RB cell walls can also be recovered in high purity from purified RBs by an SDS treatment and differential centrifugation.[47,48] RB cell walls are normally composed of fine particles about 50 Å in diameter (Figure 12).[44] Optical diffraction of the RB cell wall indicated that the fine particles were randomly distributed.[44] Some RB cell walls were, however, seen with small areas showing hexagonal structures identical with those of the EB cell wall.[43] Such partial hexagonal structures were only occasionally encountered in shadowcast preparations if the inside surface was exposed,[49] but they were readily detected by negative staining. When RB cell walls were prepared from a single culture at different intervals after infection, and then examined, the number of cell walls with the structures decreased from 23% at 15 hr to only 2% at 24 hr, indicating that the hexagonal structures were not newly synthesized, but were carried over from the infected EB and were then diluted out during multiplication of RB. A parallel correlation between the presence of the hexagonal structures and the rigidity of the cell wall against mechanical agitation suggests that the hexagonal structures are responsible for the cell wall rigidity. However, the correlation between disulfide bondings within EB cell walls[41,50,51] and the hexagonal structures is not known.

RBs of the meningopneumonitis strain possess the surface projections with a morphology identical with that on EBs, one end of each projection is connected with the cytoplasmic membrane, but the other end of the projection protrudes beyond the cell wall through a rosette in the cell wall.[35] After removal of the cytoplasmic membrane with diluted SDS, cell walls with the rosettes are obtained. The rosettes in RB cell walls, as well as in EB cell walls, are morphological markers indicating the presence of the projections. By statistical analysis of the number of rosettes in RB cell walls at different intervals (10, 15, and 20 hr) after infection, it was demonstrated that the average number of projections in RB from the early (44.7 projections at 10 hr) to the late stages (20.4 projections at 20 hr) of the multiplication approached the number of projections on EB (17.5 projections).[35] In spite of these facts, no B structure or crater in RB was detected by the freeze-replica technique.[45] This seems to be due to the difference of cleavage-site; that is, frozen EB are normally cleaved at the gap between cell wall and the postulated intermediate layer (discussed later), but the cleavage in RB occurs in the cytoplasmic membrane intramembranously, resulting in exposing the outer surface of the inner leaflet as a convex face and the inner surface of the outer leaflet of the cytoplasmic membrane as a concave face. RB cell wall is transversely cut as well as EB cell wall.

Morphological and biochemical studies failed to detect peptidoglycan in EB envelopes.[30,41,52,53] These results led us to the consensus that chlamydiae have no peptidoglycan layer which is normally observed as a single layer between the cell wall or outer membrane and the cytoplasmic membrane in Gram negative bacteria. However, when EBs were freeze-fractured, multiple layers were occasionally exposed (Figure 15A).[33,44] The organization of this multilayered membrane system is hard to explain by a simple consideration for the presence of only two membranes, cell wall and cytoplasmic membrane, in the EB envelope, because the cell wall is never cleaved intramembranously (this is one of the reasons why

37

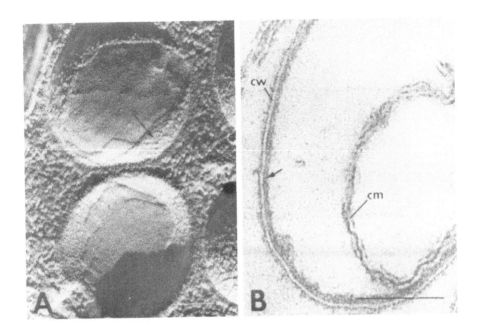

FIGURE 15. Freeze-replica of EBs and thin section of EB envelope. (A) The multiple layers are frequently exposed by freeze fracture. To explain such complex layers, it is necessary to suppose the face is covered with many particles as the outside surface of the intermediate layer (arrow) (B) EB envelope treated with tannic acid. An additional dense layer is visible very close to the inside surface of the cell wall (arrow). (cw) Cell wall, (cm) cytoplasmic membrane. Bars indicate 100 nm. (From Matsumoto, A., *J. Electron Microsc.*, 28 (Suppl.), 57, 1979. With permission.)

the inner surface of the cell wall is normally exposed as the concave face). Such multilayered cleavage might be explained if one supposes an additional intermediate layer whose outer surface is covered with a number of particles located between the cell wall and cytoplasmic membrane. If this is the case, the intermediate layer is not responsible for the rigidity of the cell wall, although the site of the intermediate layer is the same with that of peptidoglycan layer in Gram negative bacteria. When EB envelopes were treated with tannic acid to enhance their opacity, an additional single layer was seen at just beneath the cell wall (Figure 15B). It is, however, still not clear whether this layer is a true layer to which tannic acid is bound or is a pseudoimage being formed by the deposition of tannic acid on the inner surface of cell wall. Nichols et al.[38] demonstrated an elaborate proliferation of the cytoplasmic membrane, consisting of regularly spaced infoldings in an intermediate body. Such a configuration might explain the multilayered cleavage presented here. Even if the multilayered profiles resulted from proliferation of the cytoplasmic membrane, the nonetchable particles must be located not only on the uppermost face, but also on the other faces in the multilayered cleavage. However, such an image has never been encountered. Thus, the morphology of EB envelope is not completely clarified yet.

Dhir and Boatman[54] demonstrated double layered polysaccharides in the intact EB cell walls of the meningopneumonitis strain with silver-methanamine only after the treatment of periodate and suggested that the periodate-sensitive polysaccharides were a component of the group specific antigen. Using the same silver-methanamine staining, Tamura et al.[55] obtained similar results in the cell walls of purified intact EBs. The staining was, however, limited only to the inside layer of the EB cell walls, suggesting removal of the outer-most polysaccharide layer in intact EBs during the cell wall preparation. The subunits composing the hexagonal structures in the inside surface of EB cell walls were isolated from purified cell walls by the treatment with formamide at 160°C.[43] Tamura et al. demonstrated the

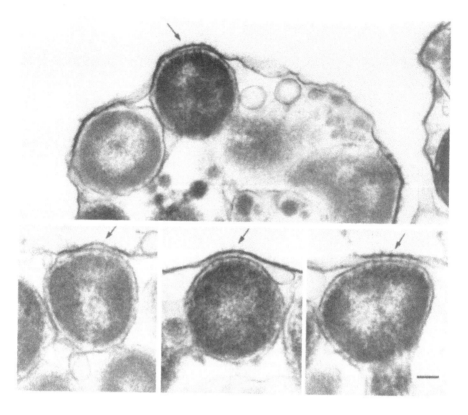

FIGURE 16.   Connection between RBs and inclusion membrane of isolated inclusions. The inclusions were treated with tannic acid to enhance the projection opacity. Arrows indicate the projections which appear to pierce the inclusion membrane. Bar indicates 100 nm.

presence of polysaccharide in the isolated subunits of the hexagonal structures by the polyacrylamide electrophoresis.[56] In intact RBs, the silver-methanamine failed to stain the double layers of the cell walls; instead, only one layer was stained. The results suggested that part of the polysaccharide in the EB cell wall is removed during conversion of EB to RB at the early stage of infection.[56]

VI. CONNECTION BETWEEN RB AND INCLUSION MEMBRANE

In the freeze-replica technique, inclusions in the host cell at 18 to 20 hr postinoculation are occasionally seen as very smooth concave or convex faces without the membrane particles which are commonly seen on the fractured faces of the host plasma membrane. It was, however, noted that several groups of fine particles were observed on the convex face of the inclusion membrane.[45] In each group, the arrangement of the particles was roughly hexagonal with the spacing ranging from 20 to 50 nm, and the number of the particles was varied from 10 to 40 in a group. In the concave face, these particles appeared to be fine depressions. The group of particles on the convex face was frequently encountered on the top of the RB in the inclusion. These facts suggest strongly that a connection exists between the inclusion membrane and RB located close to the inclusion membrane. To clarify such a correlation, inclusions containing *C. psittaci* RBs were isolated from the host cells and examined by electron microscopy.[57] In thin sections, some RB were closely connected to the inside surface of the inclusion membrane by means of fine projections, cylindrical in shape and 10 to 13 nm in diameter, which appeared to pierce the inclusion membrane (Figure 16). The arrangement of the particles on the surface of isolated inclusions was also seen by

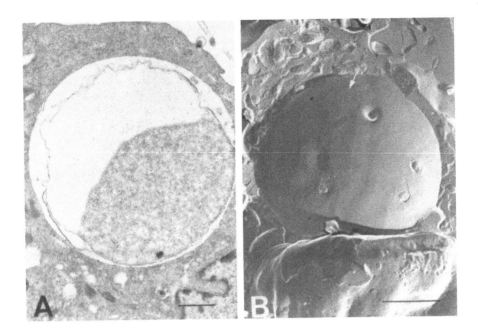

FIGURE 17. Inclusions of *C. trachomatis* L2 strain formed in the presence of penicillin. (A) The inclusion containing an abnormally large RB which cell wall can be traced throughout its outline. (B) Fractured surface of inclusion. Many projections are seen in a limited area (arrow). Bars indicate 1 μm.

the freeze-replication. These results indicate that the particles on the cleaved inclusion surface are profiles of fractured RB projections, which pierce the inclusion membrane, indicating that RBs are connected directly with the host cytoplasm through the canals of the projections. Identical particles were also detected on the convex face of *in situ* inclusions of *C. trachomatis* L2 and C strains. It is, therefore, likely that a direct connection between RBs and the inclusion membrane with the projections is a common phenomenon during the multiplication of genus *Chlamydia*. In this manner, RBs interchange information with the host cell through the projections. The movement of chlamydial bodies in the inclusion is supposed to play a role in the connection between the RB and inclusion membrane.

It was noted that many mitochondria were closely associated with the outer surface of the isolated inclusions, with a minimum spacing of 5 nm.[57] This fact is quite interesting for the chlamydiae are "energy parasites". However, morphological correlation between the location of mitochondria and the projections is not clear although the mitochondrial association seems to reflect the parasitism of chlamydiae.

In the presence of penicillin, the normal fission and maturation of *C. psittaci* RB to form EB is strongly inhibited. Consequently, an abnormal, giant RB is formed.[26,27] Such an effect on RB is also observed in the RB morphology of *C. trachomatis* L2 and C strains. The giant RB appears to be derived from one EB, because the continuity of the cell wall surrounding the giant RB can be followed in an inclusion which was seen in the host cell infected at less than one IFU (inclusion forming unit) per cell inoculum (Figure 17A).[58] When a portion of the same preparation was examined by the freeze-replica technique, many projections grouped together within an area were occasionally encountered on the convex face of the inclusion (Figure 17B). In the cases so far as examined, the area contained more than 70 projections, but the number of the projection containing areas was limited to only one in one inclusion. This observation strongly suggests that even when penicillin suppresses normal multiplication, the giant RB connects directly with the host cell through the projections. The increased number of projections in one area may be due to the inhibition of the normal RB fission, while the giant RB can assemble the projections.

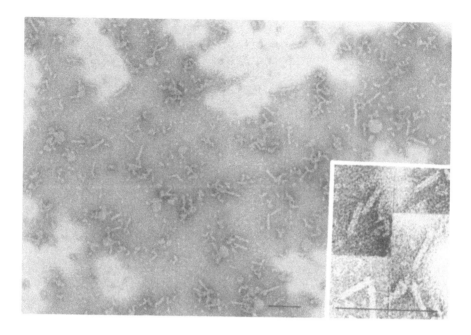

FIGURE 18. Isolated projections stained negatively. The projections normally show a nail-like shape. Insets indicate the penetration of the stain into the longitudinal axis of projections and composition of the projections with several subunits. Bars indicate 100 nm.

VII. MORPHOLOGY OF ISOLATED SURFACE PROJECTIONS

When envelopes in negative-stained preparations were examined, the projections were frequently split off from the envelope rosettes (Figure 13B).[35] Therefore, if the cytoplasmic membrane is solubilized without any damage to the morphology of the projections, the projections should be detached and recovered from the envelope fraction. Based on this idea, the isolation of the projections was carried out.[59,60] EB envelopes of the meningo-pneumonitis strain were treated with a 1% Triton X-100 solution to solubilize the cytoplasmic membrane and then centrifuged at $12,000 \times g$ to remove insoluble cell walls. The supernatant obtained was again centrifuged at $120,000 \times g$ for 120 min. The precipitate was suspended in a 0.03 M Tris-HCl buffer pH 7.3 and then centrifuged in a sucrose density column (10 to 50%, w/v) at $88,000 \times g$ for 120 min. A number of projections were recovered in the gradient, although the purity was not so high (Figure 18). By negative staining, the isolated projections were nail-like in shape possessing a pointed end with an average diameter of approximately 6 nm and a full length of 45 nm, of which 7 to 8 nm measured 8 to 10 nm in diameter. It was noted that when the projections were kept in the staining solution for 60 min or more, penetration of the stain into the longitudinal axis of the projections was occasionally observed (Figure 18, inset). In such preparations, one could see the projections as composed of fine subunits. These results, therefore, suggest strongly that the projections are tubular in shape and composed of helical arrangement of the subunits. The projections seem to be identical with the spikelike projections of *C. trachomatis* intermediate forms demonstrated by Nichols et al.[38] However, the spikelike projection measured 90 nm in length, suggesting a difference in dimension in chlamydial species and strains.

VIII. ABNORMALITIES OF RB AND EB

In thin sections of inclusions of chlamydial strains, EBs with stellate outlines are occasionally found. Most of them seem to be caused by artificial distortion during preparation,

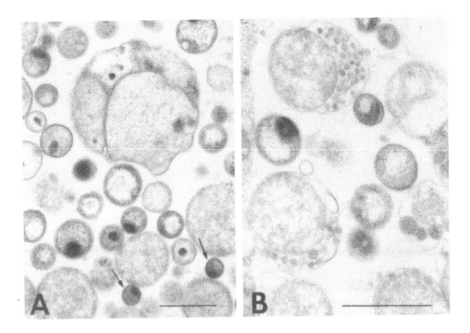

FIGURE 19.   Abnormal RBs of *C. psittaci*[64] strain isolated from an upper cervical lymphadenitis patient at 32 hours after infection of 3rd passage in L cell cultures. (A) A large RB subdivided into irregular areas is seen among RBs and EBs pleomorphic in size. Arrows indicate EBs normal in size. (B) RBs having many small particles in the periplasmic space. Bars indicate 1 μm.

because EBs are regularly seen as smooth spheres when the specimens are prepared by the freeze-replica[45] and freeze-substitution-fixation techniques,[61] in which a momentary freezing of the specimen is done at liquid nitrogen temperature. Contrarily, RBs in inclusions are regularly seen as smooth spheres or tumblers with smooth outline at the stage of binary fission. This may be due to the cell wall permeability of the fixative. Abnormal RBs are, however, frequently found in newly isolated *C. psittaci* strains when first adapted to cell cultures. There are two types of abnormal forms: one is extremely large RBs being subdivided into irregular small areas (Figure 19A) and the other is RBs having many small particles 40 to 180 nm in diameter within the periplasmic space (Figure 19B). The small particles, as described by Avakyan and Popov,[62] appear to be formed by the budding from the RB spheroplast and to be bound with the cytoplasmic membrane. The morphology of the former is similar to the abnormal RBs readily formed in the presence of penicillin[27] and the small particles in the latter RBs must be analogous to the miniature RBs of *C. psittaci* goat pneumonitis strain reported by Tanami and Yamada.[63] The small particles are readily induced in the in vitro L cells by the inoculation of the meningopneumonitis strain with a high multiplicity of infection (MOI) of more than five inclusion-forming units per cell without penicillin.[59] The mechanism by which the high MOI infection and penicillin promotes the proliferation of such miniature RBs is not known. As shown in Figure 19A, EBs are polymorphic in size. However, it is not clear whether these abnormally large EBs are infective or not.

IX. SUMMARY

Based on the results of the morphological studies on the EB of the meningopneumonitis strain, a diagram as shown in Figure 20 has been proposed,[60,65] although further study may be required to determine whether the intermediate layer is a true structure or whether another

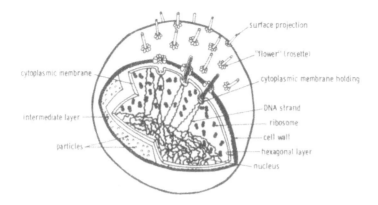

FIGURE 20. Diagrammatic representation of EB. (From Matsumoto, A., *Kawasaki Med. J.*, 8, 149, 1982. With permission.)

interpretation is required for the multiple layers exposed by the freeze-replica technique. Nevertheless, the intermediate layer is included in the diagram based on the observations of the replica and thin sections. At the same time, further study is required to determine whether such a diagram is applicable to all EBs of *C. psittaci* and *C. trachomatis* strains. On the morphology of RBs, such a diagram is not yet appropriate, because the mechanism of projection formation and the correlation between the projections and DNA molecules are obscure.

It is well known that chlamydial bodies share many antigens, such as genus specific, species specific, and strain specific antigens. They seem to be localized on the cell wall surface of the EB and the RB. At present, evidence on the correlation between the fine structures revealed by the electron microscopy and the components sharing the serological properties is still not clear enough to allow any firm generalizations.

The unique structures, especially projections, and their relation to the inclusion membrane cause us to make some assumptions as to their function in the obligate parasitism of chlamydiae. There is, however, no alternative but to wait for further study to clarify the structure and function.

ACKNOWLEDGMENT

The author thanks G. P. Manire and P. B. Wyrick for reading this manuscript.

REFERENCES

1. **Bedson, S. P. and Bland, J. O. W.,** A morphological study of psittacosis virus, with the description of a developmental cycle, *Br. J. Exp. Pathol.*, 13, 461, 1932.
2. **Weiss, E.,** The extracellular development of agents of the psittacosis-lymphogranuloma group. I. (Chlamydiozoaceae), *J. Infect. Dis.*, 84, 125, 1949.
3. **Weiss, E.,** The nature of the psittacosis-lymphogranuloma group of microorganisms, *Annu. Rev. Microbiol.*, 9, 227, 1955.
4. **Higashi, N.,** Some fundamental differences of the mechanism of multiplication between psittacosis group viruses and poxviruses, *Annu. Rep. Inst. Virus Res. Kyoto Univ., Ser. B*, 2, 1, 1959.
5. **Higashi, N.,** Electron microscopic studies on the mode of reproduction of trachoma virus and psittacosis virus in cell cultures, *Exp. Mol. Pathol.*, 4, 24, 1965.
6. **Tamura, A. and Higashi, N.,** Purification and chemical composition of meningopneumonitis virus, *Virology*, 20, 596, 1963.

7. **Anderson, D. R., Hopps, H. E., Barile, M. F., and Bernheim, B. C.,** Comparison of the ultrastructure of several rickettsiae, ornithosis virus, and mycoplasma in tissue culture, *J. Bacteriol.*, 90, 1387, 1965.

8. **Moulder, J. W.,** The relation of the psittacosis group (Chlamydiae) to bacteria and viruses, *Annu. Rev. Microbiol.*, 20, 107, 1966.

9. **Armstrong, J. A. and Reed, S. E.,** Fine structure of lymphogranuloma venereum agent and the effects of penicillin and 5-fluorouracil, *J. Gen. Microbiol.*, 46, 435, 1967.

10. **Moulder, J. W.,** A model for studying the biology of parasitism: *Chlamydia psittaci* and mouse fibroblasts (L cells), *Biol. Sci.*, 19, 875, 1969.

11. **Kramer, M. J. and Gordon, F. B.,** Ultrastructural analysis of the effects of penicillin and chlortetracycline on the development of a genital tract chlamydia, *Infect. Immun.*, 3, 333, 1971.

12. **Lepinay, A., Robineaux, R., Orfila, J., Orme-Rosselli, L. et Boutry, J. M.,** Ultrastructure et cytochimie ultrastructurale des membranes de *Chlamydia psittaci, Arch. ges. Virusforsch.*, 33, 271, 1971.

13. **Costerton, J. W., Poffenroth, L., Wilt, J. C., and Kordova, N.,** Ultrastructural studies of *Chlamydia psittaci* 6BC *in situ* in yolk sac explants and L cells: a comparison with gram-negative bacteria, *Can. J. Microbiol.*, 21, 1433, 1975.

14. **Matsumoto, A., Bessho, H., Soejima, R., and Hino, J.,** Biological properties of a chlamydia strain isolated from a pet bird, budgerigar which was kept by a psittacosis patient, *Kawasaki Med. J.*, 10, 77, 1984.

15. **Hino, J., Watanabe, M., Fujiwara, T., Matsushima, T., Soejima, R., Matsumoto, A., and Sato, M.,** An outbreak of psittacosis in a family and isolation of *C. psittaci* from a pet grass parakeet, presented at 50th Annu. Meet. Jap. Soc. Internal Med., Chugoku-Shikoku Branch, Hiroshima, May, 1984, 16.

16. **Rice, C. E.,** Carbohydrate matrix of the epithelial cell inclusion in trachoma, *Proc. Soc. Exp. Biol. Med.*, 33, 317, 1935.

17. **Gordon, F. B. and Quan, A. L.,** Occurrence of glycogen in inclusions of the psittacosis-lymphogranuloma venereum-trachoma agents, *J. Infect. Dis.*, 115, 186, 1965.

18. **Gordon, F. B. and Quan, A. L.,** Isolation of the trachoma agent in cell culture, *Proc. Soc. Exp. Biol. Med.*, 118, 354, 1965.

19. **Bernkopf, H., Mashiah, P., and Becker, Y.,** Correlation between morphological and biochemical changes and appearance of infectivity in FL cell cultures infected with trachoma agent, *Ann. N.Y. Acad. Sci.*, 98, 62, 1962.

20. **Fan, V. S. C. and Jenkin, H. M.,** Glycogen metabolism in chlamydia infected HeLa cells, *J. Bacteriol.*, 104, 608, 1970.

21. **Evans, A.,** The development of TRIC organisms in cell cultures during multiple infection, *J. Hyg.*, 70, 39, 1972.

22. **Weigent, D. A. and Jenkin, H. M.,** Contrast of glycogenesis and protein synthesis in monkey cells and HeLa cells infected with *Chlamydia trachomatis* lymphogranuloma venereum, *Infect. Immun.*, 20, 632, 1978.

23. **Garrett, A. J.,** Some properties of the polysaccharide from cell cultures infected with TRIC agent *(Chlamydia trachomatis), J. Gen. Microbiol.*, 90, 33, 1975.

24. **Page, L. A.,** Influence of temperature on glutamate catabolism and glycogen production by organisms of the genus *Chlamydia, Proc. Soc. Exp. Biol. Med.*, 143, 73, 1973.

25. **Matsumoto, A. and Bessho, H.,** Glycogen accumulation in the intracytoplasmic inclusions of *Chlamydia trachomatis*, presented at 41st Annu. Meet. Jap. Soc. Electron Microsc., Sapporo, June, 1985, 27-II-8.

26. **Tamura, A. and Manire, G. P.,** Effect of penicillin on the multiplication of meningopneumonitis organisms *(Chlamydia psittaci), J. Bacteriol.*, 96, 875, 1968.

27. **Matsumoto, A. and Manire, G. P.,** Electron microscopic observations on the effects of penicillin on the morphology of *Chlamydia psittaci, J. Bacteriol.*, 101, 278, 1970.

28. **Becker, Y.,** Effect of fluorouracil on glycogen synthesis and degradation, in *The Agent of Trachoma*, Melnick, J. L., Ed., S. Karger, Basel, 1974, 53.

29. **Matsumoto, A.,** *Chlamydia* — in special reference to biological properties, antigenicity and isolation procedures —, *Modern. Media*, 31, 227, 1985 (in Japanese).

30. **Caldwell, H. D., Kromhout, J., and Schachter, J.,** Purification and partial characterization of the major outer membrane protein of *Chlamydia trachomatis, Infect. Immun.*, 31, 1161, 1981.

31. **Matsumoto, A.,** Surface projections of *Chlamydia psittaci* elementary bodies as revealed by freeze-deep-etching, *J. Bacteriol.*, 151, 1040, 1982.

32. **Markham, R., Frey, S., and Hills, G. J.,** Method for enhancement of image detail and accentuation of structure in electron microscopy, *Virology*, 20, 88, 1963.

33. **Matsumoto, A. and Higashi, N.,** Morphology of the envelopes of *Chlamydia* organisms as revealed by freeze-etching technique and scanning electron microscopy, *Annu. Rep. Inst. Virus Res. Kyoto Univ.*, 18, 51, 1975.

34. **Gregory, W. W., Gardner, M., Byrne, G. I., and Moulder, J. W.,** Arrays of hemispheric surface projections on *Chlamydia psittaci* and *Chlamydia trachomatis* observed by scanning electron microscopy, *J. Bacteriol.,* 138, 241, 1979.

35. **Matsumoto, A.,** Electron microscopic observations of surface projections on *Chlamydia psittaci* reticulate bodies, *J. Bacteriol.,* 150, 358, 1982.

36. **Matsumoto, A.,** Electron microscopic observations of surface projections and related intracellular structures of *Chlamydia* organisms, *J. Electron Microsc.,* 30, 315, 1981.

37. **Wyrick, P. B. and Davis, C. H.,** Elementary body envelopes from *Chlamydia psittaci* can induce immediate cytotoxicity in resident mouse macrophages and L cells, *Infect. Immun.,* 45, 297, 1984.

38. **Nichols, B. A., Setzer, P. Y., Pang, F., and Dawson, C. R.,** New view of the surface projections of *Chlamydia trachomatis, J. Bacteriol.,* 164, 344, 1985.

39. **Soloff, B., Rank, R. G., and Barron, A. L.,** Ultrastructural studies of chlamydial infection in guinea-pig urogenital tract, *J. Comp. Pathol.,* 92, 547, 1982.

40. **Tamura, A., Matsumoto, A., and Higashi, N.,** Purification and chemical composition of reticulate bodies of the meningopneumonitis organisms, *J. Bacteriol.,* 93, 2003, 1967.

41. **Manire, G. P. and Tamura, A.,** Purification and chemical composition of the cell walls of mature infectious dense forms of meningopneumonitis organisms, *J. Bacteriol.,* 94, 1178, 1967.

42. **Manire, G. P.,** Structure of purified cell walls of dense forms of meningopneumonitis organisms, *J. Bacteriol.,* 91, 409, 1966.

43. **Matsumoto, A. and Manire, G. P.,** Electron microscopic observations on the fine structure of cell walls of *Chlamydia psittaci, J. Bacteriol.,* 104, 1332, 1970.

44. **Matsumoto, A.,** Recent progress of electron microscopy in microbiology and its development in future: from a study of the obligate intracellular parasites, *Chlamydia* organisms, *J. Electron Microsc.,* 28 (Suppl.), s57, 1979.

45. **Matsumoto, A.,** Fine structures of cell envelopes of *Chlamydia* organisms as revealed by freeze-etching and negative staining techniques, *J. Bacteriol.,* 116, 1355, 1973.

46. **Louis, C., Nicolas, H., Eb, F., Lefebvre, J.-E., and Orfila, J.,** Modification of the envelope of *Chlamydia psittaci* during its developmental cycle: freeze-fraction study of complementary replica, *J. Bacteriol.,* 141, 868, 1980.

47. **Tamura, A. and Manire, G. P.,** Preparation and chemical composition of the cell membranes of developmental reticulate forms of meningopneumonitis organisms, *J. Bacteriol.,* 94, 1184, 1967.

48. **Tamura, A., Matsumoto, A., Manire, G. P., and Higashi, N.,** Electron microscopic observations on the structure of the envelopes of mature elementary bodies and developmental reticulate forms of *Chlamydia psittaci, J. Bacteriol.,* 105, 355, 1971.

49. **Matsumoto, A.,** Technique for high resolution electron microscopy — taking a morphological analysis of cell walls of obligate intracellular parasite, *Chlamydia* organisms as an example, *Saibo,* 11, 549, 1979 (in Japanese).

50. **Baviol, P., Ohlin, A., and Schachter, J.,** Role of disulfide bonding in outer membrane structure and permeability in *Chlamydia trachomatis, Infect. Immun.,* 44, 479, 1984.

51. **Hatch, T., Allan, I., and Pearce, J. H.,** Structural and polypeptide differences between envelopes of infective and reproductive life cycle forms of *Chlamydia* spp., *J. Bacteriol.,* 157, 13, 1984.

52. **Ganett, A. J., Harrison, M. J., and Manire, G. P.,** A search for the bacterial mucopeptide component, muramic acid, in *Chlamydia, J. Gen. Microbiol.,* 80, 315, 1974.

53. **Barbour, A. G., Amano, K.-I., Hackstadt, T., Perry, L., and Caldwell, H. D.,** *Chlamydia trachomatis* has penicillin-binding proteins but not detectable muramic acid, *J. Bacteriol.,* 151, 420, 1982.

54. **Dhir, S. P. and Boatman, E. S.,** Location of polysaccharide on *Chlamydia psittaci* by silver-methanamine staining and electron microscopy, *J. Bacteriol.,* 111, 267, 1972.

55. **Tamura, A., Higashi, N., and Fujiwara, E.,** Electron microscopic study of polysaccharide localization on *Chlamydia psittaci* by silver-methanamine staining, *Annu. Rep. Inst. Virus Res. Kyoto Univ.,* 16, 30, 1973.

56. **Tamura, A., Tanaka, A., and Manire, G. P.,** Separation of the polypeptides of *Chlamydia* and its cell walls by polyacrylamide gel electrophoresis, *J. Bacteriol.,* 118, 139, 1974.

57. **Matsumoto, A.,** Isolation and electron microscopic observations of intracytoplasmic inclusions containing *Chlamydia psittaci., J. Bacteriol.,* 145, 605, 1981.

58. **Matsumoto, A.,** unpublished data, 1985.

59. **Matsumoto, A.,** Isolation and electron microscopy of surface projections of *Chlamydia psittaci* elementary bodies, presented at 40th Annu. Meet. Jap. Soc. Electron Microsc., Sendai, June, 1984, 28-B-II-2.

60. **Matsumoto, A.,** Morphology of elementary body of obligate intracellular parasite *Chlamydia psittaci,* in *Proc. 3rd Asia-Pacific Conf. Electron Microsc.,* M. F. Chung, Ed., Applied Research Corp., Singapore, 1984, 432.

61. **Uehira, K., Matsumoto, A., and Suda, T.,** Morphology of *Chlamydia* organisms as revealed by freeze-substitution-fixation technique, *Kawasaki Med. J.,* 9, 135, 1983.

62. **Avakyan, A. A. and Popov, V. L.,** Rickettsiaceae and chlamydiaceae; Comparative electron microscopic studies, *Acta Virol.,* 28, 158, 1984.
63. **Tanami, Y. and Yamada, Y.,** Miniature cell formation in *Chlamydia psittaci, J. Bacteriol.,* 114, 408, 1973.
64. **Mukai, A., Mukai, C., Asaoka, K., and Suzuki, I.,** Chlamydial infection with main symptoms of upper cervical lymphadenitis (UCLA) — isolation of *Chlamydiae, Jap. J. Otol. Tokyo,* 88, 1200, 1985 (in Japanese).
65. **Matsumoto, A.,** Morphology of *Chlamydia psittaci* elementary bodies as revealed by electron microscopy, *Kawasaki Med. J.,* 8, 149, 1982.

Chapter 3

MACROMOLECULAR AND ANTIGENIC COMPOSITION OF CHLAMYDIAE

Wilbert J. Newhall, V

TABLE OF CONTENTS

I. INTRODUCTION

Chlamydiae are very similar to other Gram-negative bacteria in that they possess both an inner and an outer membrane. The outer membrane is perhaps the most studied component of the bacterium since its surface represents the point of interaction with the host. Such interaction is vital for chlamydial survival since the organism must initially attach to the host cell and be ingested, and then be able to compete with the host for intracellular nutrients and precursors to grow and develop. Many of these functions, as well as other pathogenic mechanisms including the ability to prevent phagosome-lysosome fusion, are attributable to the composition and structure of the chlamydial cell envelope. The present chapter will explore the macromolecular composition of chlamydiae from a descriptive, structural/functional, and immunological perspective to provide a background with which to better understand chlamydial virulence, elementary body - reticulate body - elementary body transformation during the developmental cycle, and the immune response to chlamydial infections. The discussion will begin with chlamydial proteins partly because more is presently known about them than other components, and because they appear to be intimately involved in every aspect of chlamydial biology.

II. CHLAMYDIAL PROTEINS

A. Biochemical Characterization

The total number of different proteins that chlamydia can make is unknown. However, with a genome of about 7×10^8 molecular weight,[1] several hundred different proteins are possible. Analysis of the polypeptide composition of purified elementary bodies (EB) of a representative *Chlamydia trachomatis* strain by SDS-PAGE (Figure 1) shows that whole chlamydia do indeed possess many proteins and that a small number of these account for the majority of the total protein mass. When whole chlamydia are treated with the ionic detergent Sarkosyl using a procedure originally adapted for preparing outer membranes (OM) of *Escherichia coli*,[2] outer membrane complexes are generated that sediment upon centrifugation and which appear as cell wall ghosts under the electron microscope.[3] The protein profile of these OMs contains a single predominant protein (about 60% of the OM protein mass) with a molecular mass of about 42,000 daltons (42K).[3] Hence this protein has been termed the major outer membrane protein (MOMP). Other proteins are also present in the Sarkosyl insoluble material including those having estimated masses of 60K, 96K, and several minor proteins. Many chlamydial proteins are notably soluble in Sarkosyl, such as the 62K and 45K proteins. Although not easily distinguishable in Coomassie blue-stained gels such a Figure 1, the OM also possesses a 12K protein that was originally identified by Hatch et al.[4] in *C. psittaci*. This protein is easily labeled with [^{35}S]cysteine and has been termed a "cysteine-rich" protein. An analogous protein has also been identified in each of the 12 *C. trachomatis* serovars tested.[5] Analysis of Sarkosyl-insoluble material derived from organisms grown in the presence of [^{35}S]cysteine is shown in Figure 2 for six different serovars. Interestingly, all trachoma biovar strains have a "cysteine-rich" protein of about 12K, whereas the LGV biovar strains have a slightly larger (12.5K) protein.

Surface radioiodination of chlamydial proteins by a method employing lactoperoxidase has shown that MOMP and a number of other proteins may be exposed on the bacterial surface.[3,6] A potential drawback of this type of study, however, may be the inability to obtain a preparation of purified EBs that are all intact. Studies in our laboratory indicate that both the Renografin and Percoll density gradient methods that are commonly used to purify EBs yield as much as 5% damaged EBs and ghosts. Thus, the available data likely result in a liberal estimate of the number of surface-exposed proteins. Alternative methods such as immune electron microscopy (described below) may yield more convincing information.

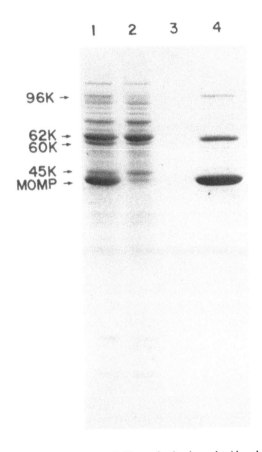

FIGURE 1. Coomassie blue-stained polyacrylamide gel after SDS-PAGE of a typical *C. trachomatis* strain and the fractions derived during an extraction with Sarkosyl. (Lane 1) Intact elementary bodies, (Lanes 2 and 3) 100,000 × *g* supernatants after first and second extraction in 2% Sarkosyl, respectively, and (Lane 4) 100,000 × *g* pellet after Sarkosyl extraction (outer membrane complexes).

There are currently 15 different serovars of *C. trachomatis* that have been recognized on the basis of complex antigenic cross-reactivities using the micro-immunofluorescence (micro-IF) test developed by Wang and Grayston.[7,8] The protein profiles of representative strains of each of these serovars have been compared by SDS-PAGE.[9] In general, these profiles are quite similar with proteins of the same masses present in each serovar (Figure 3). Similar findings have been described by Salari and Ward[6] and to a limited extent by Caldwell et al.[10] All strains possess a MOMP, however this protein varies by 3K in molecular mass from about 42K for the G serovar to about 45K for serovars K and L_3.[9] A 62K protein (Sarkosyl soluble) is present in each of the serovars, and the 60K outer membrane protein appears to be more predominant in the trachoma biovar (serovars A to K) than in the LGV biovar strains. While other studies have shown that LGV strains possess a 60K doublet,[4] this is not readily apparent in Coomassie blue-stained gels of whole EBs. Another apparently common peptide has a molecular mass of about 45K in most stains and about 45.5K in serovars E, L_1, L_2, and L_3. Although this protein appears to be Sarkosyl-soluble, it can be radioiodinated during surface labeling experiments[3,11] suggesting that perhaps this is a loosely associated, extrinsic outer membrane protein.

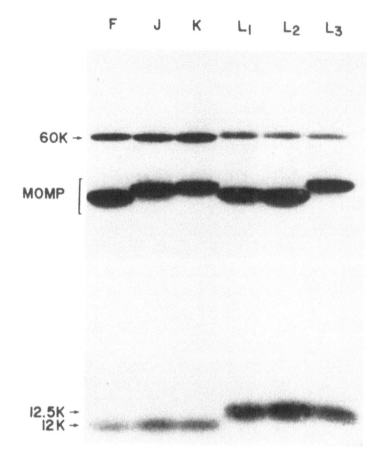

FIGURE 2. Fluorogram of polyacrylamide gel after SDS-PAGE of Sarkosyl-insoluble outer membrane complexes derived from EBs after growth in the presence of [³⁵S]cysteine. Serovar designations are given above each lane.

An interesting set of proteins consists of the 20K, 26K, and 29K polypeptides. The 20K protein is observed in the A, B, and Ba serovars which are mainly associated with endemic trachoma.[12] The 26K protein is present in C, D, F, and G to K; whereas, the 29K protein is found in E and the three LGV serovars. Most of the serovars of the former group, with the main exception being C, are responsible for genital tract infections while the LGV serovars are mostly associated with lymphogranuloma venereum. Recent studies with monoclonal antibodies suggest that these proteins are antigenically related to each other and to a 15K polypeptide that is common to all the serovars.[13] In addition, these proteins are analogous to the putative host cell binding proteins identified by Hackstadt[14] and by Wenman and Meuser[15] based on the reactivity of those proteins with the monoclonal antibody described above.[16] Attempts to demonstrate surface exposure of these proteins by surface radioiodination have proven unsuccessful. In addition, these proteins are not associated with Sarkosyl derived outer membrane complexes. However, these proteins are resistant to radioiodination even when they are in a soluble form, perhaps suggesting a lack of tyrosine residues, and they appear to be present in outer membranes that are prepared by mechanical disruption or with other detergents.[13] Additional studies are needed to define more clearly the significance of these proteins in pathogenesis, as well as their membrane location and surface exposure.

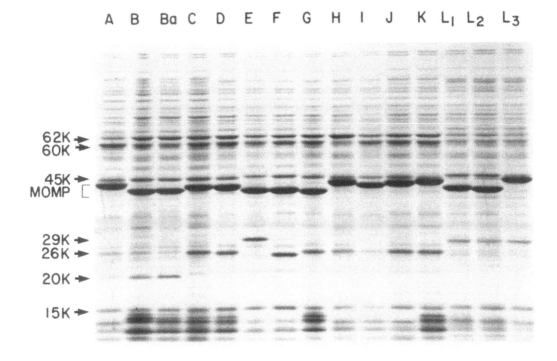

A B Ba C D E F G H I J K L₁ L₂ L₃

62K
60K

45K
MOMP

29K
26K

20K

15K

FIGURE 3.   Coomassie blue-stained polyacrylamide gel of the 15 serovars of *C. trachomatis* and the *C. psittaci* CAL-10 strain (MN). Serovar designations are given at the top of each lane.

B. Disulfide Bonds and Protein Crosslinking

Tamura and Manire[17] were the first to speculate that disulfide bonds might, through an unknown mechanism, be related to the rigid structure of the EB cell wall. Evidence for a relationship between disulfide bonds and cell wall rigidity was obtained by Hatch et al.[18] who observed that the solubilization of *C. psittaci* envelope proteins required the presence of a reducing agent such as 2-mercaptoethanol. Subsequent studies by Newhall and Jones[19] demonstrated that MOMP and the 60K outer membrane protein are extensively crosslinked by disulfide bonds into dimers and trimers as well as multimeric complexes having masses greater than 5×10^6. Other investigators have confirmed these observations[4,20,21] and also shown that the 12K ''cysteine-rich'' outer membrane protein is present in crosslinked complexes. Based on these studies a model of chlamydial cell-wall structure has evolved in which the covalent crosslinking of these outer membrane proteins by disulfide bonds is the primary structural determinant, rather than peptidoglycan or a related structure.

Assuming that crosslinked outer membrane proteins provide the rigid cell wall in the EB, then the fragile reticulate body (RB) would be predicted to lack such extensive crosslinking. Analyses of the proteins present in intact RBs compared to EBs of both *C. psittaci*[4] and *C. trachomatis*[22] indicate that the protein compositions are very similar. However, when the Sarkosyl-insoluble proteins of a *C. psittaci* strain are analyzed, the RBs appear to be deficient in the 60K doublet (59K and 62K) and the 12K ''cysteine-rich'' proteins compared to EBs.[4] In contrast, MOMP appears to be present in nearly the same amount in outer membranes of both EBs and RBs. The extent of crosslinking of the outer membrane proteins is also different between EBs and RBs. Newhall and Jones[19] noted that the MOMP of RBs was much less crosslinked than the MOMP of EBs. As shown in Figure 4, most of the MOMP in RBs is present in monomeric form and only a trace in the dimeric form. In addition, essentially no 60K or 12K proteins were detected in RBs confirming the observation of Hatch et al.[4] In contrast, the MOMP in EBs is present in high molecular mass aggregates along with 60K and 12K which are often located at the top of the stacking gel.

Recent evidence suggests that in *C. psittaci* the 60K doublet and 12K protein become crosslinked as soon as they are synthesized, whereas MOMP remains in a monomeric form until the chlamydia are released from their intracellular environment.[23] In contrast, study of an L_2 serovar strain of *C. trachomatis* suggests that while the crosslinking of the 60K doublet proteins appear to be extensive through most of the growth cycle (60 to 80% crosslinked), both the MOMP and the 12K protein are predominantly in their monomeric forms during the first 24 hr postinfection and become gradually more crosslinked during the final 24 hr of the growth cycle to a maximum of about 70%.[24] The latter data might suggest an enzymatic mechanism for the crosslinking of outer membrane proteins in *C. trachomatis* while in *C. psittaci* the crosslinking of MOMP may occur mainly by autooxidation.

These relatively unique properties of chlamydia are probably very important in terms of the maturation of the cell envelope structure during the formation of infectious EBs. The precise role that disulfide bonds play in the modulation of the in vivo porin function[21] and in the virulence determinants of this organism remain to be defined. In addition, studies to discover the regulation of protein biosynthesis and protein crosslinking may provide new clues as to the control of differentiation and acquisition of pathogenic properties.

C. Isoelectric Points and Charge

In addition to the characterization of chlamydial proteins by their molecular masses many have been characterized by their intrinsic charge. When proteins of EBs are resolved by two-dimensional isoelectric focusing and SDS-PAGE[5] most chlamydial proteins are found to have isoelectric points that are neutral to slightly acidic. Testing of 13 of the *C. trachomatis* serovars revealed that the MOMPs all had acidic isoelectric points (pI) within a very narrow range (pI of 5.3 to 5.5) in spite of differences of as much as 3K in the molecular masses of these proteins. This conservation of electric charge among MOMPs may be necessary for the specific functions of this protein which are thought to include roles as a porin[21] and as a major structural determinant as discussed above.

A similar analysis of the 60K protein and protein doublets indicated that in the trachoma biovar strains these proteins have apparent pIs between 7.4 and 7.7, and thus are nearly neutral. The analogous 60K doublet proteins of the LGV biovar strains were found to be considerably more basic with pIs for both proteins of greater than 8.5. Thus, the 60K doublet proteins presumably bear a net positive charge at neutral pH. Variation between the trachoma and LGV biovar also occurs for the low molecular mass "cysteine-rich" proteins. The 12K protein of the trachoma biovar strains has a pI of about 5.4, whereas the 12.5K protein of the LGV strains has an approximate pI of 6.9. Thus, the LGV biovar strains possess a neutral 12.5K and a positively charged 60K doublet while the trachoma strains possess a negatively charged 12K and a neutral 60K. These differences in charge characteristics are the first correlates at the molecular level that are consistent with the separation of the different chlamydia serovars into distinct biovars. Recent work by Newhall and Basinski[25] further suggests that for the 60K proteins there are also significant differences in amino acid sequence between the biovars as determined by limited proteolytic mapping, and epitope mapping with anti-60K monoclonal antibodies. Similar studies have not yet been performed for the 12K and 12.5K proteins.

It is currently not known if the charge or mass differences between these outer membrane proteins are in any way responsible for the documented differences in biological behavior between biovars, for example in the sensitivity of attachment and inclusion formation to

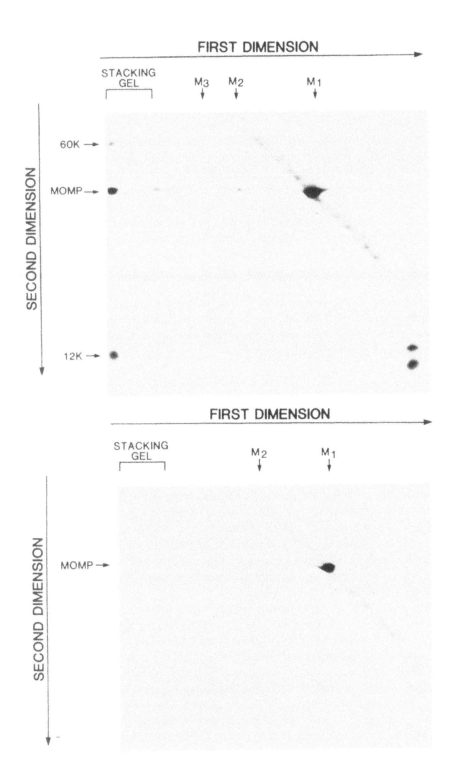

FIGURE 4.   Fluorograms of chlamydial proteins resolved by two-dimensional SDS-PAGE. The fluorogram at the top was obtained using purified EBs, and that at the bottom using purified RBs. The first dimension for both was run under nonreducing conditions to avoid reduction of disulfide bonds. Such bonds were then cleaved prior to electrophoresis in the second dimension. M_1, M_2, and M_3 correspond to the positions of monomeric, dimeric, and trimeric MOMP in the first dimension.

positively charged compounds such as DEAE-dextran or poly-l-lysine.[26,27] However, further study of these intriguing proteins may provide data relevant to the processes of attachment and ingestion in addition to defining their role in the structure and function of the outer membrane.

D. Glycoproteins

Evidence to suggest the possible glycosylation of outer membrane proteins has been largely indirect. Using a fairly sensitive staining method, Hatch et al.[18] obtained evidence that the MOMP of a *C. psittaci* strain contains carbohydrate. In contrast, Caldwell et al.,[3] using a periodic acid-Schiff staining method, did not detect carbohydrate associated with any of the proteins of various *C. trachomatis* strains. This difference could be due to the use of different organisms or may reflect the different sensitivities of the methods used. In an attempt to obtain more direct evidence, Gibson et al.[28] attempted to label chlamydial proteins of an F and L$_2$ serovar strain with [^{3}H]glucosamine. When samples were assessed by SDS-PAGE, a major peak of radioactivity was associated with MOMP. Analysis of MOMP purified by HPLC methods[25] and subjected to acid hydrolysis indicated that as much as 50% of the radioactivity in MOMP was present as the radiolabeled sugar. Before concluding that these data indicate glycosylation of MOMP, further studies are required to demonstrate the nature of the covalent linkages of the sugars to the protein and that such linkages are not formed by nonspecific, nonenzymatic mechanisms.[29]

III. LIPOPOLYSACCHARIDE

The genus-reactive complement fixation (CF) test of Bedson,[30] originally used to diagnose psittacosis, contains a reactive component that is soluble in ethyl ether[31] and inactivated by periodate[32] suggesting a glycolipid composition. An analogous glycolipid antigen was extracted in ethyl ether from three *C. trachomatis* strains by Dhir and co-workers.[33] From this was derived a water-soluble polysaccharide antigen that did not fix complement in the presence of antibody, but that competitively inhibited complement fixation by the antibody with the extracted glycolipid antigen. This suggests that the genus-reactive CF antibody is directed against the carbohydrate portion of the molecule. Additional studies indicated that the reactive epitopes present on this antigen are related, but not identical, to 2-keto-3-deoxyoctonoic acid (KDO).[34] These observations provided the first indication that the chlamydial genus-reactive antigen had biochemical properties similar to those ascribed to bacterial lipopolysaccharide (LPS).

An LPS-like molecule has been extracted from purified EBs using two different standard LPS extraction methods.[35,36] Nurminen and co-workers,[36] using the extraction method of Galanos et al.,[37] extracted the LPS from a Re mutant of *Salmonella typhimurium* that produces an LPS that lacks the O-polysaccharide side chain and part of the core polysaccharide. They compared this Re LPS with a similar preparation derived from EBs of an L$_2$ serovar strain of *C. trachomatis*. Both preparations contained KDO and migrated similarly in SDS-PAGE. In addition, both preparations were very similar antigenically when tested with antisera to Re LPS and to various chlamydia and salmonella strains. Similar cross-reactions were also observed with Re LPS from mutants of *E. coli* and *Proteus mirabilis*, and with an LPS extract prepared from an *Acinetobacter* strain. Subsequently, these workers compared extracts from chlamydia with those of an *Acinetobacter calcoaceticus* strain and Re and Rb$_2$ mutants of *Salmonella* by SDS-PAGE and immunoblotting.[38] When gels were silver-stained the chlamydial preparation was found to possess a single band that migrated more slowly and hence with a higher apparent molecular mass than the Re LPS of *Salmonella*, but more quickly than RB$_2$ LPS of *Salmonella*. When probed with antiserum raised against a Triton X-100 extract of chlamydial EBs, reactions were observed with the Re LPS, the chlamydial component, and one of the three silver-staining bands of the *Acinetobacter* preparation. While these studies indicated antigenic relatedness among these molecules a precise chemical description of the chlamydial antigen was lacking prior to the work of Nurminen et al.[39] Their chemical analysis indicated the presence of D-glucosamine, 3-hydroxy fatty acids, KDO, and phosphate in a molar ratio similar to that of Re LPS. These data provide the most compelling chemical evidence to date that this antigenic material is indeed LPS.

Using the LPS extraction procedure of Westphal and Jann,[40] Caldwell and Hitchcock[35] derived a preparation from chlamydial EBs that had gross biochemical characteristics similar to that of Re LPS of *S. typhimurium*. The preparation contained about 9% KDO by weight, gave a positive reaction with *Limulus* lysate, and reacted with antisera raised against Re LPS and against lipid A. When analyzed by SDS-PAGE and stained with silver this preparation possesses a single entity that migrates below about 10,000 daltons. Thus, this material is probably equivalent to that described by Nurminen et al.[36]

Analysis with monoclonal antibodies showed that the putative chlamydial LPS possesses a genus-specific epitope(s) since one antibody, L2I-6, binds to the purified component and cross-reacts in the micro-IF test with all *C. trachomatis* serovars and a number of *C. psittaci* strains.[35] Interestingly, this antibody did not react in immunoblot with any antigens present in *Neisseria gonorrhoeae*, nor with various LPS mutants of *S. typhimurium* and *E. coli* suggesting a specificity for chlamydia spp. Based on these analyses, and those of Brade et al.,[41] the chlamydial LPS appears to have at least three sets of epitopes. Two of these are expressed by the intact LPS molecule including one that is chlamydia genus-specific and another that is shared with the Re LPS of various enterobacteria. The third epitope appears to be located within the lipid A and is shared with other bacterial LPS molecules. There is, however, some question whether this latter epitope is expressed in the native LPS or only after acid hydrolysis. Regardless, the available evidence clearly indicates that chlamydia possess an LPS-like molecule that possesses genus-specificity.

IV. LIPIDS

Membranes form a continuous barrier around all cells and thus help protect the cell interior from adverse fluctuations in its environment. Membrane lipids also provide the proper biophysical environment for optimal functioning of membrane proteins, whether they are structural or enzymatic. In addition, lipids form, in part, the surface of the cell which is a major determinant of the outcome of any cell-cell interaction. Lipid composition is carefully controlled to ensure successful completion of these roles in cytoplasmic protection, membrane fluidization, and cell surface modulation.

The generalizations presented above may be subdivided into more specific processes. Of particular interest are those which could potentially influence the pathogenesis of chlamydial disease, although such processes have been best studied in other systems. Those processes that are known to require specific degrees of membrane fluidity for optimal efficiency include antibody-dependent activation of the first component of complement,[42] activity of the terminal components of complement,[43] antibody-dependent binding of haptenated phospholipid vesicles to some phagocytic cells,[44] adhesion,[45] permeability and the activities of certain membrane enzymes,[46] membrane fusion,[47] and killing of target cells by natural killer cells and cytotoxic T cells.[48,49] Biophysical and biochemical characterization of the membranes of chlamydia will help determine the applicability of any of these processes to the pathogenesis of chlamydial disease.

Relative to the many reports on chlamydial proteins, there are few studies of chlamydial lipid composition or synthesis. Early morphological and mechanical data indicated that the EB form is more structurally stable than the RB form[17,50-52] and led the authors to speculate that a highly permeable membrane structure is necessary for the growth of the RB forms.[17]

All other studies to date have focused on the EB cell form, presumably because the RB form is difficult to purify. An attempt to compare the lipid composition of psittacosis and trachoma agents led to the suggestion that it was possible to distinguish members of the psittacosis group from host material and that a branched-chain fatty acid was associated with the bacteria.[53] Confirmation of these results followed when a detailed study of *C. psittaci* strain 6BC lipids revealed several odd-numbered, branched-chain fatty acids and the presence of phosphatidylglycerol (PG), one of the lipids uniquely absent in the host membranes.[54] Also detected was a lower ratio of phosphatidylcholine (PC) to phosphatidylethanolamine (PE) in the bacteria when compared to the host membranes,[54] a change now known to be one of several causes of lowered membrane fluidity.[55] Branched chain fatty acids were also found in the CAL-10 strain of *C. psittaci*.[56] In the same study an apparent host-independent ability to incorporate radioactive carbon from asparate, isoleucine, and glucose-6-phosphate into cell lipids was detected. This synthetic activity was limited to the fatty acids of phospholipids, principally PE.[56]

The lipid metabolism of the host monkey kidney cells during chlamydial infection with an LGV strain of *C. trachomatis* appeared to be somewhat inhibited as judged by incorporation of radioactive acetate into total lipids[57] and phospholipids.[58] In the latter study a burst of PE synthetic activity was noted at 24 to 36 hr and was attributed to the infecting organism. The presence of a branched-chain fatty acid in the trachoma strain was also reported.[57] This result was confirmed and extended in a qualitative comparison of five *C. trachomatis* strains using capillary gas chromatography.[59] Chlamydial lipid metabolism has also been studied more directly in mouse L cells treated with cycloheximide to inhibit host lipid and protein synthesis.[60] As measured by the incorporation of radioactive isoleucine into total lipids, chlamydial lipid synthesis was first detectable about 6 to 12 hr after infection and increased until a constant rate of synthesis was reached at 24 to 30 hr. This time course suggests that lipid synthesis is associated with chlamydial multiplication and not necessarily linked to chlamydial maturation.[60]

A. Phospholipid Composition of Whole Chlamydia

Together with Richard Haak and Mark Halvorson, we have begun comprehensive studies on the lipid composition of isolated *C. trachomatis* EBs and RBs, and are attempting to correlate lipid composition with membrane biophysical properties and chlamydial pathogenesis. To maximize the validity of comparisons, all *C. trachomatis* strains were cultured by established methods in HeLa 229 cells.[61] After purification of the EBs and RBs, lipids were extracted and phospholipids were separated using 2-dimensional thin layer chromatography. Individual spots were analyzed for phosphate content and the results were expressed as a percentage of the total (Table 1).

Elementary body phospholipids of the F and L$_2$ serovars, chosen as representatives of the trachoma and LGV biovars, were compared. In addition, EBs and RBs of L$_2$ were compared to gain information about the chlamydial developmental cycle. In all cases PE was the major phospholipid found, which perhaps explains the burst of PE syntheses previously noted by Fan and Jenkin.[58] PC was the next most abundant, although in far less quantity than in the host lipids. As in *C. psittaci*,[54] the PC/PE ratio was much less in *C. trachomatis* membranes than in the host cell membranes. Diphosphatidylglycerol (DPG) and PG were found in significant amounts in the chlamydial lipids, although they could not be detected in the host HeLa cells. Other phospholipids present in small quantities included phosphatidylinositol (PI), sphingomyelin (SPH), and phosphatidylserine (PS). Lysophosphatidylcholine (LPC), which tends to destabilize lipid bilayers, was not detected in the bacterial membranes.

B. Fatty Acid Composition of Chlamydial Phospholipids

Halvorson et al.[62] have analyzed the fatty acids in the individual phospholipids using capillary column gas chromatography. Greater than 45% of the EB fatty acids contained

Table 1
PHOSPHOLIPID COMPOSITION OF *C.*
TRACHOMATIS AND HeLa 229 MEMBRANES

| | *Chlamydia trachomatis* | | | |
| | F/UW6/Cx | L$_2$/434/Bu | | |
	EB	EB	RB	Hela 229[a]
PE	46.6[b] ± 3.3[c]	43.3 ± 0.8	41.5 ± 0.6	25.0
PC	19.2 ± 0.7	21.8 ± 1.3	27.5 ± 1.0	45.7
DPG	15.2 ± 2.7	13.0 ± 1.7	12.7 ± 0.6	ND
PG	13.1 ± 1.0	12.1 ± 1.2	7.6 ± 0.8	ND
PI	4.1 ± 0.8	4.7 ± 1.5	6.1 ± 1.3	7.7
Sph	1.0 ± 0.4	3.0 ± 1.0	2.2 ± 0.1	3.6
PS	4.1 ± 0.8	1.3 ± 0.7	1.1 ± 0.3	5.4
LPC	ND[d]	ND	trace[e]	5.0
origin	0.9 ± 0.6	0.8 ± 0.7	1.1 ± 0.2	1.0

[a] Contained significant amounts of plasmalogens.
[b] Percent of total phospholipid phosphorus.
[c] Standard deviation.
[d] ND, Not detected.
[e] Trace, less than 1%.

branched chains, a result which confirms and extends the previous studies by many others.[53,54,56-58] Noticeable by their relative absence were the unsaturated fatty acids. Evidently the methyl anteiso-branched chain fatty acids are compensating for the lack of fluidizing, unsaturated bonds in maintaining optimal membrane properties, as has been reported for the genus *Bacillus*.[63]

In all cases, the fatty acid composition of PE, DPG, and PG were similar and differed from that of PC. The major fatty acids found in PE, DPG, and PG were 12-methyltetradecanoate (15:0 ba) and hexadecanoate (16:0), while for PC the major ones were 9-octadecanoate (18:1) and (16:0). Since the major fatty acids of HeLa cells are also 18:1 and 16:0,[64] it is possible that some host PC is incorporated into bacterial membranes. However, branched chain fatty acids are not found in HeLa cells,[64] whereas the chlamydial PC does. Thus, chlamydia must either synthesize at least some of their own PC or be capable of modifying the fatty acids of PC derived from the host.

C. Fluidity of Chlamydial Membranes

To complement the compositional data discussed above, the membrane fluidity of whole chlamydia was characterized using spin label (SL) electron spin resonance (ESR) methods. A fatty acid analog, 5-doxyl stearic acid, was used as the SL and the order parameter calculated.[65] The order parameter is a measure of SL order and motion, and thus provides a measure of membrane fluidity. The results are listed in Table 2 along with compositional data averaged from Table 1 and the fatty acid results. Parameters known to be partial determinants of membrane fluidity are the PC/PE ratio, fatty acid chain length, degree of unsaturation, and percent branched chain fatty acids.[46,55,63]

Differences in fluidity of the chlamydia membranes were detected at a statistically significant level of $p < 0.01$ when the null hypothesis was tested in a t-test. The RBs were the most fluid (the least ordered), followed by L$_2$ EBs, and then F EBs. Although these differences appear small on a numerical basis, one must remember that the order parameter is constrained by definition to a value between zero and one and that differences of the order of 1% are biologically significant and not unusual.[46,66,67]

Table 2

COMPARISON OF CHLAMYDIAL PHOSPHOLIPIDS, AVERAGE PROPERTIES OF THE PHOSPHOLIPID FATTY ACIDS, AND MEMBRANCE FLUIDITY AS MEASURED BY THE ORDER PARAMETER

Serotype	Cell form	PC/PE[a]	$\bar{c}$[b]	# of =[c]	Y[d]	Order parameter
L2	RB	0.662	16.6	0.56	36.9	0.644
L2	EB	0.503	16.3	0.31	44.7	0.656
F	EB	0.412	16.2	0.23	48.4	0.661

[a] Phosphatidylcholine to phosphatidylethanolamine ratio.
[b] Average number of carbons in the phospholipid fatty acids.
[c] Average number of unsaturated carbon-carbon bonds in the phospholipid fatty acids.
[d] Average percent of phospholipid fatty acids that have branched chains.

In general the determinants of fluidity act together to exert a net effect and a shift in one determinant is often counterbalanced by the others to maintain the optimal fluidity characteristic of that membrane. This also appears to be true for chlamydial membranes. For example, the decreases in the PC/PE ratio and number of unsaturated bonds should produce increases in the order parameter. These effects are apparently only partially offset by the decrease in chain length and increase in branched-chain fatty acids which should lower the order parameter. The result of these opposing factors is to leave small net differences among the chlamydial membrane types, perhaps revealing that they have different fluidity optima. One might speculate, for example, that the metabolically active RB forms have the most fluid membranes to meet their increased demand for permeation of nutrients and membrane enzymatic activity.

V. PEPTIDOGLYCAN

When thin-sections of chlamydia are viewed under the electron microscope a clearly defined periplasmic space or peptidoglycan layer is not identified.[68] Yet, since the EBs are capable of resisting various methods of chemical and mechanical disruption, some type of stability-conferring structure must be present. Many mechanisms have been postulated to explain the structural stability of the EB cell envelope, as well as the relative fragility of the RB cell envelope. One possible mechanism is that chlamydia possess peptidoglycan, which is the primary structural determinant of most bacteria including Gram-negative bacteria. However, attempts to verify biochemically the presence of peptidoglycan in chlamydia have not yet provided convincing evidence. Most investigations have attempted to identify muramic acid in acid hydrolysates of purified or enriched preparations of chlamydia and in infected tissue culture cells. Perkins and Allison[69] detected muramic acid in two *C. psittaci* and two *C. trachomatis* strains, as did Jenkin[70] for the CAL-10 strain of *C. psittaci*. However, in an analysis of purified EBs and cell walls of the CAL-10 strain by Manire and Tamura[52] muramic acid was not detected. In a reevaluation of this question by Garrett et al.[71] using a more sensitive, radiochemical method of detection, no muramic acid was identified in either whole EBs or in cell walls of the CAL-10 strain. Based on their detection limit they concluded that no more than 0.04% of the wall is muramic acid. More recently Barbour et al.[22] analyzed whole cells and SDS-insoluble material (classically contains peptidoglycan) from an L$_2$ serovar strain of *C. trachomatis* and also did not detect muramic acid (less than 0.02% by mass).

In unpublished studies performed by Gibson et al.[28] attempts were made to label carbohydrate-containing macromolecules of an F and L$_2$ serovar of *C. trachomatis* with [^{3}H]glucosamine. In other bacteria, exogenous glucosamine is converted into N-acetyl-glucosamine and into N-acetyl-muramic acid. These are incorporated into peptidoglycan thereby providing an efficient means of labeling this molecule. When purified, radiolabeled chlamydial EBs were mixed with intact cells of *Neisseria sica*, solubilized in SDS, and subjected to conventional peptidoglycan purification methods;[72] no radioactivity copurified with the *N. sica* peptidoglycan suggesting that the EBs either lack a significant amount of peptidoglycan or that the peptidoglycan cannot be labeled with glucosamine. In addition, in purified EBs as well as in whole infected HeLa cells, no glucosamine was found to be converted into muramic acid. Thus, taken together, these different studies indicate that peptidoglycan is either absent or is present in such small amounts as to be structurally insignificant.

In spite of the inability to demonstrate peptidoglycan, chlamydia are clearly sensitive to antibiotics that are known to interfere with peptidoglycan biosynthesis such as penicillin,[73-77] D-cycloserine,[77,78] and bacitracin,[79] and they also possess three penicillin binding proteins.[22] However, other antibiotics that interfere with peptidoglycan biosynthesis, including phosphomycin and vancomycin show no antichlamydial effect.[79] The mechanism by which penicillin inhibits peptidoglycan biosynthesis is thought to be through inhibition of the crosslinking between specific peptide side chains on adjacent glycan backbones by transpeptidation.[80] To explain the effect of penicillin on chlamydia Garrett et al.[71] speculated that the peptides involved in transpeptidation in chlamydia may be located on macromolecular structures other than the classical polysaccharide backbones composed of alternating residues of N-acetyl-glucosamine and N-acetyl-muramic acid. Perhaps, as suggested by Barbour et al.,[22] the peptide side chain typically associated with peptidoglycan is involved in crosslinkages within a protein or lipoprotein matrix. Direct evidence for such hypotheses has not been obtained. Regardless of the presence of a penicillin-sensitive mechanism of macromolecular crosslinking, penicillin clearly affects chlamydial growth. This effect is likely mediated through the complex signaling and feedback mechanisms that are responsible for cell division and RB to EB transformation.

VI. NUCLEIC ACIDS

A nucleic acid analysis of purified *C. psittaci* EBs by Tamura and Higashi[81] demonstrated the presence of both DNA and RNA in these organisms. Studies by Sarov and Becker[82] of DNA from a *C. trachomatis* strain showed that the genome consists of double standard DNA with an average length of 343 μm. A similar value was found for a *C. psittaci* strain.[83] This length corresponds to a molecular mass of about 6.6 × 10^8. DNA is also present in RBs in which it is actively being replicated. In addition to the chlamydial genome both species of chlamydia also possess a 4.4 Mdalton DNA plasmid.[84,85] The function of these plasmids is unknown. The plasmids from different *C. trachomatis* strains give similar restriction profiles. These profiles are different, however, from that of the *C. psittaci* plasmid.[84-86] This difference was confirmed by Joseph et al.[87] who also analyzed the proteins encoded by these plasmids and found both similarities and differences between the two chlamydia species in the polypeptides synthesized by in vitro transcription-translation, as well as in vivo in maxicells and minicells. There is currently no evidence that these plasmids are involved in the acquisition of antibiotic resistance such as that described by Johnson and Spencer.[88]

Electron micrographs have long suggested the presence of ribosomes within chlamydia, particularly in RBs, and the chemical demonstration of rRNA was first described by Tamura and Higashi.[81] Subsequently, three species were identified including 21s, 16s, and 4s in both EBs and RBs.[89] For the EB most of the RNA was present in the 4s form. Similar rRNA species have been identified in *C. trachomatis*.[90] It is interesting to note that for EBs, which are metabolically inert, the ratio of RNA to DNA is roughly 1:1; whereas, in the metabolically active RB the ratio is increased to about 3:1.[51] This probably reflects the greatly increased protein biosynthetic activity of these forms. Since the amount of rRNA decreases during condensation of RBs into EBs, presumably RNA as well as ribosomal proteins are turned over at the end of the growth cycle. Studies to document this possibility have not yet been performed.

In addition to rRNA, RBs also possess mRNA; however, methods to analyze this short-lived species have not been adequately developed. Once available it should be possible to determine, for example, if certain genes are turned on and off at different times in the growth cycle. This will be of particular importance in attempting to understand the cycle-regulated biosynthesis of the 60K and 12K outer membrane proteins and the mechanism and signals for RB to EB transformation.

VII. CHLAMYDIAL ANTIGENS

The rapid accumulation of information in the last few years concerning chlamydial antigens, particularly outer membrane antigens, is partly due to two observations. The first is that the serovar-specific epitopes on the chlamydial surface are heat labile and trypsin-sensitive, and thus are presumably located on surface-exposed proteins. The second is that protective immunity in a limited number of studies in humans and some animals appears to be serovar-specific. While the search for the serovar-specific antigens is not over, knowledge of chlamydial antigens has increased tremendously in recent years. Some of this information has been reviewed by others.[91-93]

The current classification of *C. trachomatis* serovars is based on the antigenic cross-reactivities observed using micro-IF[7,8] in which mouse antisera raised against purified or partially purified chlamydia are used to probe the surface reactivity of a panel of different strains. Based on the reactivity of individual antigens with different prototype antisera and on the cross-reactivity of antisera with different antigens, most strains of *C. trachomatis* can be classified into 1 of 15 distinguishable serovars which are designated by letters: A to K, Ba, L_1, L_2, and L_3.

A tremendous amount of experience with the micro-IF test suggests that some antigens yield antisera that are monospecific (e.g., anti-A and anti-H), bispecific (e.g., anti-F, G, and K), some that cross-react with a few different serovars (e.g., anti-B and anti-C), and some that are broadly cross-reactive (e.g., anti-D and anti-E).[7] These different reaction patterns indicate that the chlamydial surface consists of different classes of antigens. These include serovar-specific, various subspecies-specific antigens including the B and C-complex antigens, species-specific, and genus- (or group) specific antigens.

Much of what is now known about the location of various epitopes on specific macromolecules has come from an analysis of serum from humans or animals with chlamydial disease, serum from immunized animals, and monoclonal antibodies by immunoblotting methods. While immunoblotting has certain limitations particularly in the use of protein antigens that have been denatured by heating in SDS, a wealth of knowledge about chlamydial antigens has been obtained.

As an illustration of the resolution of the immunoblotting method for identifying reactions with different antigens, Figure 5 represents the proteins of the 15 serovars of *C. trachomatis* probed with serum from a mouse immunized and boosted twice with purified EBs of strain I/UW-12. A number of distinct reactivities were observed. Interestingly this serum reacted with each of the 15 different MOMPs, but slightly more intensely with the homologous I MOMP. Thus, this serum appears to recognize a species as well as a serovar-specific epitope on MOMP. Many other species-specific reactions were also observed, with the 60K outer membrane proteins, and proteins having apparent molecular masses of 90K, 70K, 55K, 25K, 15K, and 12K. A trachoma biovar-specific reaction was observed with a 32K protein.

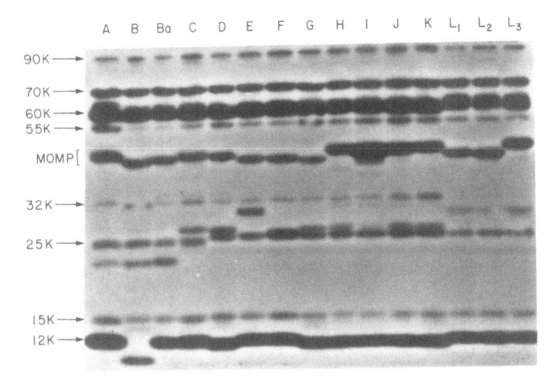

FIGURE 5. Autoradiogram of nitrocellulose membrane bearing the proteins of the 15 *C. trachomatis* serovars after probing with a mouse antiserum against serovar I EBs, followed by probing with ^{125}I-labeled protein A. Serovar designations are given above each lane and estimated molecular masses are indicated by arrows.

Since *C. psittaci* proteins were not included in this analysis, it is not known if any of these reactions may actually be genus-specific. However, these data indicate that chlamydia are antigenically complex and that while most studies have tended to concentrate on MOMP, some of these other antigens may be important in chlamydial biology and immunology. The following sections will briefly summarize the antigenic nature of various chlamydial components.

A. Glycolipid/Lipopolysaccharide

Chlamydia appear to have a LPS component that is present in both RBs and EBs (discussed above). Based on its chemical composition and biochemical behavior this LPS appears to be equivalent to the heat-stable CF antigen described by Bedson[30] and other workers.[31-33] Antigenically the LPS or CF antigen is genus-reactive and this reactivity is destroyed by treatment with periodate, suggesting that the reactive epitopes are located in the carbohydrate region of the molecule.[32] Monoclonal antibodies specific for a low molecular mass (< 10,000 daltons) heat-stable, protease-resistant antigen that is presumably LPS have been characterized by various laboratories.[35,94,95] Each of these antibodies is genus-specific. Analysis of one of these monoclonal antibodies, as well as others with analogous specificities, by immune electron microscopy suggested that the reactive epitopes are not accessible on the surface of EBs, but can easily be demonstrated on the RB surface[96] (see Figure 6). This finding is consistent with the observations of Yong et al.[97] that serum antibodies reactive with RBs tend to be genus-specific while those recognized by EBs are serovar and subspecies-specific. Similarly, Stephens et al.[95] observed that genus-specific monoclonal antibodies that probably

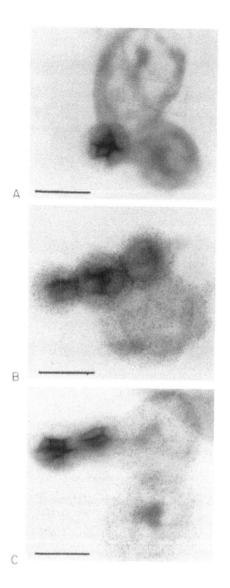

FIGURE 6.   Transmission electron micrograph of whole chlamydia adsorbed to a plastic-coated grid and probed with (A)
an irrelevant monoclonal antibody, (B) a species-specific anti-
MOMP monoclonal antibody, and (C) a genus-specific anti-
LPS monoclonal antibody, followed by protein A-gold for localization of bound antibody. The bar in each micrograph represents 500 nm.

recognize LPS epitopes react preferentially with RBs. This suggests that the genus-reactive surface epitopes recognized by micro-IF are associated mainly with the RB outer membrane and are perhaps covered up in the EB. Further studies are needed to decide this issue. Nevertheless, studies on neutralization of infectivity and inhibition of attachment and ingestion should be approached with this possibility in mind.

In addition to LPS, a glycolipid antigen possessing genus-specific epitopes has been described by Stuart and MacDonald[98,99] to be secreted by chlamydia-infected cell monolayers. This macromolecule is made up of a lipid portion containing C_{17} and $C_{18:1}$ fatty acids, as well as a polysaccharide portion containing galactose, gulose, and mannose.[100] This component appears to be distinct from LPS, however antigenic cross-reactivities have not yet been assessed.

B. MOMP

The variation in molecular mass that has been noted for MOMP, its surface exposure, and its predominance, led to early speculation that this protein may be responsible for the antigenic variation among chlamydia, including serovar specificity. The first evidence to this effect was derived by Caldwell and coworkers[3,10] using rabbit antisera raised against MOMP purified by preparative SDS-PAGE. While the anti-MOMP sera were species-specific when assessed by micro-IF, they also showed subspecies reactivity, particularly within an expanded B-complex (B, Ba, D, E, F, G, K, L_1, L_2, and L_3). In addition, a serovar-specific reaction was observed with antiserum against H-MOMP.

More recently, monoclonal antibodies have been used to study the antigenic heterogeneity of MOMP. The monoclonal antibodies developed by Stephens et al.[95] showed that MOMP possesses both species and subspecies epitopes. The subspecies reactivities included B, Ba, D, E, L_1, and L_2 which has been defined as the B-complex[7], and B, Ba, G, F, K, L_2, and L_3 with weak reactions with C and J. Monoclonal antibodies that recognize serovar-specific epitopes on MOMP were first identified by Matikainen and Terho.[94] These antibodies were specific for L_1 MOMP. Subsequent work by Newhall et al.[61] has identified serovar-specific monoclonal antibodies recognizing epitopes on the MOMPs of serovars A, B, C, D, E, F, G, K, L_1, and L_2. Each of these antibodies appears to react with a surface-exposed epitope of MOMP in intact EBs and RBs as well as with SDS-denatured MOMP in immunoblotting. Presumably these antibodies recognize epitopes determined by amino acid sequence. However, since the extent of renaturation of secondary and tertiary MOMP structure during immunoblotting is unknown, some of these epitopes may also be conformational. Interestingly, monoclonal antibodies specific for the H, I, and C/J serovars do not react with any antigen in immunoblot experiments.[61] Perhaps these antibodies recognize heat or SDS-labile MOMP epitopes.

A number of other epitopes have also been demonstrated to reside on MOMP. These epitopes are summarized by the serovar cross-reactivity of corresponding monoclonal antibodies in Table 3. The reactivities of these antibodies range from serovar, bispecific, and trispecific, to B- and C-complex, species, and genus. Thus, MOMP appears to contain many of the antigens that are probably recognized in the micro-IF test and that are responsible for the current definitions of serovars and subgroups. In addition, MOMP possesses epitopes that are shared across species, but these genus-specific epitopes do not appear to be surface-exposed and are presumably buried in the membrane.[96] Perhaps these are located on buried regions of MOMP that are structurally conserved among all chlamydial MOMPs.

Work has just begun to assess the functional relevance of many of these MOMP epitopes. Neutralization experiments in tissue culture systems suggest that antibody to some of the MOMP epitopes can prevent chlamydial infection.[101,102] Further studies are needed to evaluate these antibodies in animal models before convincing roles for the corresponding epitopes in infection and pathogenesis can be fully defined.

Table 3
ANTI-MOMP MONOCLONAL ANTIBODY
SPECIFICITIES

Contribution	Specificity
Stephens et al.[95]	Species
	B-complex (B, Ba, D, E, L$_1$, L$_2$)
	B, Ba, G, F, K, L$_2$, L$_3$ (C, J)
Matikainen and Terho[94]	L$_1$
Newhall et al.[61]	Serovar specific: A, B, C, D, E, F,
	G, K, L$_1$, L$_2$
	Bispecific: B/Ba, C/J, C/L$_3$
Caldwell and Hitchcock[35]	L$_2$
	Ba, D, E, F, G, K, L$_1$, L$_2$, L$_3$
Lucero and Kuo[101]	Serovar specific: B, H, I, K, L$_2$
	Trispecific: B/Ba/E, B/Ba/L$_2$
	B-complex
	B-complex + F, G
	C-complex (A, C, H, I, J)
	Species
Batteiger et al.[110]	Bispecific: F/G, G/L$_3$
	Trispecific: D/F/L$_1$, B/Ba/L$_2$
	B, Ba, D, F, L$_1$
	C-complex + K, L$_3$
	B-complex
	B-complex + F, G
	B-complex + F, G, K, L$_3$
	Species
	Genus

C. 155K Protein

In an analysis of a number of antigens that had been identified by crossed immunoelectrophoresis,[103] one was selected and used to produce a monospecific antiserum that was reactive only with *C. trachomatis* organisms (species-specific).[104] Subsequent purification of the antigen by immuno-affinity chromatography and biochemical characterization indicated that this antigen was a protein with an apparent mass of 155K.[105] Antigenicity was sensitive to protease and heat treatment, but was resistant to periodate oxidation. A protein with a similar mass has been identified as being surface-exposed by Salari and Ward.[6] Whether these two polypeptides are the same is not known.

D. 60K Outer Membrane Protein

As discussed above, all chlamydia appear to have either a 60K protein or a doublet of proteins of about 60K in their outer membranes. Analysis of sera from chlamydia-infected humans indicates that most possess antibodies against this protein as well as against the 62K Sarkosyl-soluble (non-outer membrane) protein. Evaluation of the cross-reactivity of selected sera indicated that some epitopes on the 60K proteins are shared among the *C. trachomatis* serovars and not with a *C. psittaci* strain, while others are apparently shared by both species.[9] Thus, the 60K protein has both species- and genus-specific epitopes. The identification of a species-specific monoclonal antibody that recognizes only the 60K outer membrane proteins has confirmed these observations.[5] In addition, monoclonal antibodies that cross-react with the 60K doublets of *C. psittaci* outer membranes have also been identified.[106] Interestingly, none of these antibodies appears to react with the chlamydial surface as assessed by micro-IF or immune electron microscopy.[96]

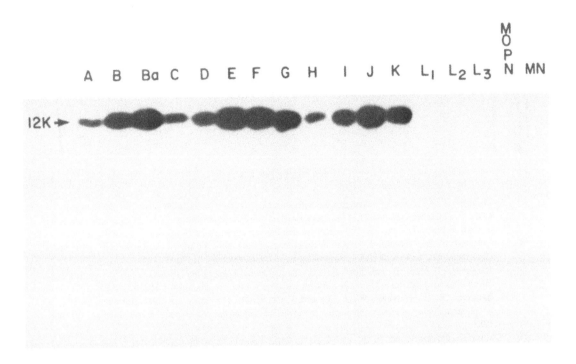

FIGURE 7.   Autoradiogram of nitrocellulose membrane bearing the proteins of the 15 serovars and the mouse pneumonitis (MoPn) strains of *C. trachomatis* and the MN strain of *C. psittaci* after probing with a monoclonal antibody specific for the 12K "cysteine-rich" outer membrane protein.

E. 12K Outer Membrane Protein

The antigenic relatedness of the 12K "cysteine-rich" proteins has been assessed with a monoclonal antibody derived from a fusion with a mouse immunized with 12K of serovar F that was partially purified by HPLC.[25] This monoclonal antibody reacted only with the 12K proteins of the trachoma biovar strain and not with the 12.5K proteins of the LGV strains (Figure 7). This trachoma biovar-specific epitope does not appear to be accessible at the bacterial surface.[96]

F. 30—32K Proteins

In addition to the serovar-specific epitopes on MOMP, serovar-specific antigens that are not MOMP have also been identified.[107-108] Sachs et al.[107] identified a Triton X-100 soluble component from a serovar B strain that reacted more strongly with the homologous rabbit antisera than with antisera against serovar A or C. Antigenicity was sensitive to heat and protease treatment. The antigen has a rough average molecular mass of 30K. This antigen is probably analogous to a serovar-specific antigen from the A serovar that was purified by SDS-PAGE following extraction with Nonidet P-40.[109] Hourihan et al.[108] purified an antigen from a serovar A strain with Triton X-100 that possessed serovar specificity when tested with adsorbed hyperimmune sera against serovars A, B, and C. This antigen is presumably identical to that extracted with Nonidet P-40.[109] The possibility that these antigens may represent breakdown products of larger macromolecules has been noted by MacDonald.[93] It should be possible to determine if they are derived from MOMP by testing reactivity with anti-MOMP monoclonal antibodies.

ACKNOWLEDGMENT

I thank Richard A. Haak for his invaluable contribution in the writing of the section on chlamydial lipids, as well as Mark Halvorson, Bobbie A. Collett, and Margret Basinski for their contributions of relevant data.

REFERENCES

1. **Moulder, J. W.,** Looking at chlamydiae without looking at their hosts, *ASM News,* 50, 353, 1984.
2. **Filip, C., Fletcher, G., Wulff, J. L., and Earhart, C. F.,** Solubilization of the cytoplasmic membrane of *Escherichia coli* by the ionic detergent sodium lauryl sarcosinate, *J. Bacteriol.,* 115, 717, 1973.
3. **Caldwell, H. D., Kromhout, J., and Schachter, J.,** Purification and partial characterization of the major outer membrane protein of *Chlamydia trachomatis, Infect. Immun.,* 31, 1161, 1981.
4. **Hatch, T. P., Allan, I., and Pearce, J. H.,** Structural and polypeptide differences between envelopes of infective and reproductive life cycle forms of *Chlamydia* supp., *J. Bacteriol.,* 157, 13, 1984.
5. **Batteiger, B. E., Newhall, W. J., V, and Jones, R. B.,** Differences in outer membrane proteins of lymphogranuloma venereum and trachoma biovars of *Chlamydia trachomatis, Infect. Immun.,* 50, 488, 1985.
6. **Salari, S. H. and Ward, M. E.,** Polypeptide composition of *Chlamydia trachomatis, J. Gen. Microbiol.,* 123, 197, 1981.
7. **Wang, S.-P. and Grayston, J. T.,** Microimmunofluorescence antibody responses in *Chlamydia trachomatis* infection, a review, in *Chlamydial Infections,* Mardh, P. A., Holmes, K. K., Oriel, J. D., Piot, P., and Schachter, J., Eds., Elsevier, Amsterdam, 1982, 301.
8. **Wang, S.-P. and Grayston, J. T.,** Classification of TRIC and isolated strains with micro-immunoflu-orescence, in *Trachoma and Related Disorders,* Nichols, R. L., Ed., Excepta Medica, Amsterdam, 1971, 305.
9. **Newhall, W. J., V, Batteiger, B., and Jones, R. B.,** Analysis of the human serological response to proteins of *Chlamydia trachomatis, Infect. Immun.,* 38, 1181, 1982.
10. **Caldwell, H. D. and Schachter, J.,** Antigenic analysis of the major outer membrane protein of *Chlamydia* spp., *Infect. Immun.,* 35, 1024, 1982.
11. **Batteiger, B. E. and Newhall, W. J.,** unpublished data, 1981.
12. **Wang, S.-P., Grayston, J. T., Kuo, C.-C., Alexander, E. R., and Holmes, K. K.,** Serodiagnosis of *Chlamydia trachomatis* infection with the micro-immunofluorescence test, in *Nongonococcal Urethritis and Related Infections,* Hobson, D. and Holmes, K. K., Eds., American Society for Microbiology, Washington, D.C., 1977, 237.
13. **Newhall, W. J.,** unpublished data, 1985.
14. **Hackstadt, T.,** Identification and properties of chlamydial polypeptides that bind eucaryotic cell surface components, *J. Bacteriol.,* 165, 13, 1986.
15. **Wenman, W. M. and Meuser, R. U.,** *Chlamydia trachomatis* elementary bodies possess proteins which bind to eucaryotic cell membranes, *J. Bacteriol.,* 165, 602, 1986.
16. **Newhall, W. J. and Hackstadt, T.,** unpublished data, 1986.
17. **Tamura, A. and Manire, G. P.,** Preparation and chemical composition of the cell membrane of devel-opmental reticulate forms of meningopneumonitis organisms, *J. Bacteriol.,* 94, 1184, 1967.
18. **Hatch, T. P., Vance, D. W., Jr., and Al-Hossainy, E.,** Identification of a major envelope protein in *Chlamydia* spp., *J. Bacteriol.,* 146, 426, 1981.
19. **Newhall, W. J., V and Jones, R. B.,** Disulfide-linked oligomers of the major outer membrane protein of chlamydiae, *J. Bacteriol.,* 154, 998, 1983.
20. **Hackstadt, T., Todd, W. J., and Caldwell, H. D.,** Disulfide-mediated interactions of the chlamydial major outer membrane protein: role in the differentiation of chlamydiae?, *J. Bacteriol.,* 161, 25, 1985.
21. **Bavoil, P., Ohlin, A., and Schachter, J.,** Role of disulfide bonding in outer membrane structure and permeability in *Chlamydia trachomatis, Infect. Immun.,* 44, 479, 1984.
22. **Barbour, A. G., Amano, K.-I., Hackstadt, T., Perry, L., and Caldwell, H. D.,** *Chlamydia trachomatis* has penicillin-binding proteins but not detectable muramic acid, *J. Bacteriol.,* 151, 420, 1982.
23. **Hatch, T. P., Miceli, M., and Sublett, J. E.,** Synthesis of disulfide-bonded outer membrane proteins during the developmental cycle of *Chlamydia psittace* and *Chlamydia trachomatis, J. Bacteriol.,* 165, 379, 1986.

24. **Newhall, W. J. V.**, Biosynthesis and disulfide crosslinking of outer membrane components during the growth cycle of *Chlamydia trachomatis, Infect. Immun.*, 55, 1987.

25. **Newhall, W. J., V and Basinski, M. B.**, Purification and structural characterization of chlamydial outer membrane proteins, in *Chlamydial Infections*, Oriel, D., Ridgway, G., Schachter, J., Taylor-Robinsin, D., and Ward, M., Eds., Cambridge University Press, Cambridge, 1986, 93.

26. **Kuo, C.-C., Wang, S.-P., and Grayston, J. T.**, Differentiation of TRIC and LGV organisms based on enhancement of infectivity by DEAE-dextran in cell culture, *J. Infect. Dis.*, 125, 313, 1972.

27. **Kuo, C.-C., Wang, S.-P., and Grayston, J. T.**, Effect of polycations, polyanions, and neuraminidase on the infectivity of trachoma-inclusion conjunctivitis and lymphogranuloma venereum organisms in HeLa cells: sialic acid residues as possible receptors for trachoma inclusion conjunctivitis, *Infect. Immun.*, 8, 74, 1973.

28. **Gibson, J. L., Newhall, W. J., V, and Rosenthal, R. S.**, Glycosylation of outer membrane proteins of *Chlamydia trachomatis*, Annu. Meet. Am. Soc. Microbiol., New Orleans, La., Abstract B70, 1983.

29. **Watkins, N. G., Thorpe, S. R., and Baynes, J. W.**, Glycation of amino groups in protein. Studies on the specificity of modification of RNase by glucose, *J. Biol. Chem.*, 260, 10629, 1985.

30. **Bedson, S. P.**, Immunological studies with virus of psittacosis, *Br. J. Exp. Pathol.*, 14, 126, 1937.

31. **Hilleman, M. R. and Nigg, C.**, Studies on lymphogranuloma venereum complement fixing antigens. III. The solubility in ether of an active fraction, *J. Immunol.*, 53, 201, 1946.

32. **Barwell, C. F.**, Some observations on the antigenic structure of psittacosis and lymphogranuloma venereum viruses. II. Treatment of virus suspensions by various reagents and the specificity of acid extracts, *Br. J. Exp. Pathol.*, 33, 268, 1952.

33. **Dhir, S. P., Kenny, G. E., and Grayston, J. T.**, Characterization of the group antigen of *Chlamydia trachomatis, Infect. Immun.*, 4, 725, 1971.

34. **Dhir, S. P., Hakomori, S., Kenny, G. E., and Grayston, J. T.**, Immunological studies on chlamydial group antigen (presence of a 2-keto-3-deoxy-carbohydrate as immunodominant group), *J. Immunol.*, 109, 116, 1972.

35. **Caldwell, H. D. and Hitchcock, P. J.**, Monoclonal antibody against a genus-specific antigen of *Chlamydia* species: location of the epitope on chlamydial lipopolysaccharide, *Infect. Immun.*, 44, 306, 1984.

36. **Nurminen, M., Leinonen, M., Saikku, P., and Makela, P. H.**, The genus-specific antigen of *Chlamydia*: resemblance to the lipopolysaccharide of enteric bacteria, *Science*, 220, 1279, 1983.

37. **Galanos, C., Luderitz, O., and Westphal, O.**, A new method for the extraction of R lipopolysaccharides, *Eur. J. Biochem.*, 9, 245, 1969.

38. **Nurminen, M., Wahlstrom, E., Lleemola, M., Leinonen, M., Saikku, P., and Makela, P. H.**, Immunologically related ketodeoxyoctonate-containing structures in *Chlamydia trachomatis*, Re mutants of *Salmonella* species, and *Acinetobacter calcoaceticus* var. *anitratus, Infect. Immun.*, 44, 609, 1984.

39. **Nurminen, M., Rietschel, E. T., and Brade, H.**, Chemical characterization of *Chlamydia trachomatis* lipopolysaccharide, *Infect. Immun.*, 48, 573, 1985.

40. **Westphal, O. and Jann, K.**, Bacterial lipopolysaccharide. Extraction with phenol-water and further applications of the procedure, *Meth. Carbohyd. Chem.*, 5, 83, 1965.

41. **Brade, L., Nurminen, M., Makela, P. H., and Brade, H.**, Antigenic properties of *Chlamydia trachomatis* lipopolysaccharide, *Infect. Immun.*, 48, 569, 1985.

42. **Esser, A. F., Bartholomew, R. M., Parce, J. W., and McConnell, H. M.**, The physical state of membrane lipids modulates the activation of the first component of complement, *J. Biol. Chem.*, 254, 1768, 1979.

43. **Kato, K. and Bito, Y.**, Relationship between bactericidal action of complement and fluidity of cellular membranes, *Infect. Immun.*, 19, 12, 1978.

44. **Lewis, J. T., Hafeman, D. G., and McConnell, H. M.**, Kinetics of antibody-dependent binding of haptenated phospholipid vesicles to a macrophage-related cell line, *Biochemistry*, 19, 5376, 1980.

45. **Schaeffer, B. E. and Curtis, A. S. G.**, Effects on cell adhesion and membrane fluidity of changes in plasmalemmal lipids in mouse L929 cells, *J. Cell Sci.*, 26, 47, 1977.

46. **Schinitzky, M.**, Membrane fluidity and cellular function, in *Physiology of Membrane Fluidity*, Vol. 1, Shinitzky, M., Ed., CRC Press, Boca Raton, Fla., 1984, 1.

47. **Lucy, J. A.**, The membrane of the hen erythrocyte as a model for studies on membrane fusion, in *Structure of Biological Membranes*, Abrahamsson, S. and Pascher, I., Eds., Plenum Publishing, New York, 1977, 275.

48. **Schlager, S. I. and Ohanian, S. H.**, Role of membrane lipids in immunological killing of tumor cells. I. Target cell lipids, *Lipids*, 18, 475, 1983.

49. **Schlager, S. I., Meltzer, M. S., and Madden, L. D.**, Role of membrane lipids in immunological killing of tumor cells. II. Effector cell lipids, *Lipids*, 18, 483, 1983.

50. **Tamura, A. and Higashi, N.**, Purification and chemical composition of meningopneumonitis virus, *Virology*, 20, 596, 1967.

51. **Tamura, A., Matsumoto, A., and Higashi, N.**, Purification and chemical composition of reticulate bodies of the meningopneumonitis organisms, *J. Bacteriol.*, 93, 2003, 1967.

52. **Manire, G. P. and Tamura, A.,** Preparation and chemical composition of the cell walls of mature infectious dense forms of meningopneumonitis organisms, *J. Bacteriol.*, 94, 1178, 1967.

53. **Jenkin, H. M.,** Comparative lipid composition of psittacosis and trachoma agents, *Am. J. Ophthal.*, 63, 1087, 1967.

54. **Makino, S., Jenkin, H. M., Yu, H. M., and Townsend, D.,** Lipid composition of *Chlamydia psittaci* grown in monkey kidney cells in defined medium, *J. Bacteriol.*, 103, 62, 1970.

55. **Hirata, F. and Axelrod, J.,** Enzymatic methylation of phosphatidylethanolamine increases erythrocyte membrane fluidity, *Nature (London)*, 275, 219, 1978.

56. **Gaugler, R. W., Neptune, E. M., Adams, G. M., Sallee, T. L., Weiss, E., and Wilson, N. N.,** Lipid synthesis by isolated *Chlamydia psittaci*, *J. Bacteriol.*, 100, 823, 1969.

57. **Fan, V. S. C. and Jenkin, H. M.,** Lipid metabolism of monkey kidney cells (LLC-MK-2) infected with *Chlamydia trachomatis* strain lymphogranuloma venereum, *Infect. Immun.*, 10, 464, 1974.

58. **Fan, V. S. C. and Jenkin, H. M.,** Biosynthesis of phospholipids and neutral lipids of monkey kidney cells (LLC-MK-2) infected with *Chlamydia trachomatis* strain lymphogranuloma venereum, *Proc. Soc. Exp. Biol. Med.*, 148, 351, 1975.

59. **Larsson, L., Jimenez, J., Odham, G., Westerdahl, G., and Mardh, P.,** Preliminary studies on cellular lipids of *Chlamydia trachomatis* using capillary gas chromatography, in *Chlamydial Infections*, Mardh, P., Holmes, K., Oriel, J., Piot, P., and Schachter, J., Eds., Elsevier, Amsterdam, 1982, 37.

60. **Reed, S. I., Anderson, L. E., and Jenkin, H. M.,** Use of cycloheximide to study independent lipid metabolism of *Chlamydia trachomatis* cultivated in mouse L cells grown in serum-free medium, *Infect. Immun.*, 31, 668, 1981.

61. **Newhall, W. J., V, Terho, P., Wilde, C. E., III, Batteiger, B. E., and Jones, R. B.,** Serovar determination of *Chlamydia trachomatis* isolates by using type-specific monoclonal antibodies, *J. Clin. Microbiol.*, 23, 333, 1986.

62. **Halvorson, M. J., Newhall, W. J., and Haak, R. A.,** Characterization of the phospholipids of *Chlamydia trachomatis*, Annu. Meet. Am. Soc. Microbiol., Washington, D.C., Abstract No. D-155, 1986.

63. **Kaneda, T.,** Fatty acids of the genus *Bacillus*: an example of branched-chain preference, *Bacteriol. Rev.*, 41, 391, 1977.

64. **Johnston, D., Matthews, E. R., and Melnykovych, G.,** Glucocorticoid effects on lipid metabolism in HeLa cells: inhibition of cholesterol synthesis and increased sphingomyelin synthesis, *Endocrinology*, 107, 1482, 1980.

65. **Gaffney, B. J.,** Practical considerations for the calculation of order parameters for fatty acid or phospholipid spin labels in membranes, in *Spin Labeling: Theory and Applications*, Berliner, L. J., Ed., Academic Press, New York, 1976, 567.

66. **Thompson, G. A., Jr.,** *The Regulation of Membrane Lipid Metabolism*, CRC Press, Boca Raton, Fla., 1980.

67. **Shinitzky, M., Ed.,** *Physiology of Membrane Fluidity*, Vol. 2, CRC Press, Boca Raton, Fla., 1984.

68. **Costerton, J. W., Poffenroth, L., Wilt, J. C., and Kordova, N.,** Ultrastructural studies of *Chlamydia psittaci* 6BC *in situ* in yolk sac explants and L cells: a comparison with gram-negative bacteria, *Can. J. Microbiol.*, 21, 1433, 1975.

69. **Perkins, H. R. and Allison, A. C.,** Cell wall constituents of rickettsiae and psittacosis-lymphogranuloma organisms, *J. Gen. Microbiol.*, 30, 469, 1963.

70. **Jenkin, H. M.,** Preparation and properties of cell walls of the agent of meningopneumonitis, *J. Bacteriol.*, 80, 639, 1960.

71. **Garrett, A. J., Harrison, M. J., and Manire, G. P.,** A search for the bacterial mucopeptide component, muramic acid, in *Chlamydiae*, *J. Gen. Microbiol.*, 80, 315, 1974.

72. **Rosenthal, R. S.,** Release of soluble peptidoglycan from growing gonococci: hexaminidase and amidase activities, *Infect. Immun.*, 24, 869, 1979.

73. **Tamura, A. and Manire, G. P.,** Effect of penicillin on the multiplication of meningopneumonitis organisms *(Chlamydia psittaci)*, *J. Bacteriol.*, 96, 875, 1968.

74. **Kramer, M. J. and Gordon, F. B.,** Ultrastructural analysis of the effects of penicillin and chlortetracycline on the development of a genital tract *Chlamydia*, *Infect. Immun.*, 3, 333, 1971.

75. **Matsumoto, A. and Manire, G. P.,** Electron microscopic observations on the effects of penicillin on the morphology of *Chlamydia psittaci*, *J. Bacteriol.*, 101, 278, 1970.

76. **Clark, R. B., Schatzki, P. F., and Dalton, H. P.,** Ultrastructural effect of penicillin and cycloheximide on *Chlamydia trachomatis* strain HAR-13, *Med. Microbiol. Immunol.*, 171, 151, 1982.

77. **Tribby, I. I. E.,** Cell wall synthesis by *Chlamydia psittaci* growing in L cells, *J. Bacteriol.*, 104, 1176, 1970.

78. **Moulder, J. W., Novosel, D. L., and Officer, J. E.,** Inhibition of the growth of agents of the psittacosis group by D-cycloserine and its specific reversal by D-alanine, *J. Bacteriol.*, 85, 707, 1963.

79. **How, S. J., Hobson, D., and Hart, C. A.,** Studies in vitro of the nature and synthesis of the cell wall of *Chlamydia trachomatis*, *Curr. Microbiol.*, 10, 269, 1984.

80. **Strominger, J. L., Izaki, K., Matsuhashi, M., and Tipper, D. J.**, Peptidoglycan transpeptidase and D-alanine carboxypeptidase: penicillin-sensitive enzymatic reactions, *Fed. Proc.*, 26, 9, 1967.

81. **Tamura, A. and Higashi, N.**, Purification and chemical composition of meningopneumonitis virus, *Virology*, 20, 596, 1963.

82. **Sarov, I. and Becker, Y.**, Trachoma agent DNA, *J. Mol. Biol.*, 42, 581, 1969.

83. **Matsumoto, A. and Higashi, N.**, Electron microscopic observations of DNA molecules of the mature elementary bodies of *Chlamydia psittaci*, *Annu. Rep. Inst. Virus Res. Kyoto Univ.*, 16, 33, 1973.

84. **Lovett, M., Kuo, C.-C., Holmes, K., and Falkow, S.**, Plasmids of the genus *Chlamydia*, in *Current Chemotherapy and Infectious Disease*, Vol. 2, Nelson, J. D. and Grassi, C., Eds., American Society for Microbiology, Washington, D.C., 1980, 1250.

85. **Peterson, E. M. and de la Maza, L. M.**, Characterization of *Chlamydia* DNA by restriction endonuclease cleavage, *Infect. Immun.*, 41, 604, 1983.

86. **Hyypia, T., Larsen, S. H., Stahlberg, T., and Terho, P.**, Analysis and detection of chlamydial DNA, *J. Gen. Microbiol.*, 130, 3159, 1984.

87. **Joseph, T., Nana, F. E., Garon, C. F., and Caldwell, H. D.**, Molecular characterization of *Chlamydia trachomatis* and *Chlamydia psittaci* plasmids, *Infect. Immun.*, 51, 699, 1986.

88. **Johnson, F. W. A. and Spencer, W. N.**, Multiantibiotic resistance in *Chlamydia psittaci* from ducks, *Vet. Rec.*, 112, 208, 1983.

89. **Tamura, A. and Iwanaga, M.**, RNA synthesis in cells infected with meningopneumonitis agent, *J. Mol. Biol.*, 11, 97, 1965.

90. **Sarov, I. and Becker, Y.**, RNA in the elementary bodies of trachoma agent, *Nature (London)*, 217, 849, 1968.

91. **Ward, M. E.**, Chlamydial classification, development and structure, *Br. Med. Bull.*, 39, 109, 1983.

92. **Schachter, J. and Caldwell, H. D.**, Chlamydiae, *Annu. Rev. Microbiol.*, 34, 285, 1980.

93. **MacDonald, A. B.**, Antigens of *Chlamydia trachomatis*, *Rev. Infect. Dis.*, 7, 731, 1985.

94. **Matikainen, M.-T. and Terho, P.**, Immunochemical analysis of antigenis determinants of *Chlamydia trachomatis* by monoclonal antibodies, *J. Gen. Microbiol.*, 129, 2343, 1983.

95. **Stephens, R. S., Tam, M. R., Kuo, C.-C., and Nowinski, R. C.**, Monoclonal antibodies to *Chlamydia trachomatis*: antibody specificities and antigen characterization, *J. Immunol.*, 128, 1083, 1982.

96. **Collett, B. A., Newhall, W. J., V., Jersild, R. A., and Jones, R. B.**, Detection of surface exposed epitopes on *Chlamydia trachomatis* by immune electron microscopy, Annu. Meet. Am. Soc. Microbiol., Washington, D.C., Abstract No. D-159, 1986.

97. **Yong, E. C., Chinin, J.-S., Caldwell, H. D., and Kuo, C.-C.**, Reticulate bodies as single antigen in *Chlamydia trachomatis* serology with microimmunofluorescence, *J. Clin. Microbiol.*, 10, 351, 1979.

98. **Stuart, E. S. and MacDonald, A. B.**, Identification of two fatty acids in a group determinant of *Chlamydia trachomatis*, *Curr. Microbiol.*, 11, 123, 1984.

99. **Stuart, E. S. and MacDonald, A. B.**, Isolation of possible group antigenic determinant of *Chlamydia trachomatis*, in *Chlamydial Infections*, Mardh, P.-A., Holmes, K. K., Oriel, J. D., Piot, P., and Schachter, J., Eds., Elsevier, Amsterdam, 1984, 57.

100. **Stuart, E. S. and MacDonald, A. B.**, Identification of a hexose in the determinant of *Chlamydia trachomatis* group antigen, Annu. Meet. Am. Soc. Microbiol., New Orleans, La., Abstract No. K-93, 1983.

101. **Lucero, M. E. and Kuo, C.-C.**, Neutralization of *Chlamydia trachomatis* cell culture infection by serovar-specific monoclonal antibodies, *Infect. Immun.*, 50, 595, 1985.

102. **Peeling, R., MacLean, I. W., and Brunham, R. C.**, In vitro neutralization of *Chlamydia trachomatis* with monoclonal antibody to an epitope on the major outer membrane protein, *Infect. Immun.*, 46, 484, 1984.

103. **Caldwell, H. D., Kuo, C.-C., and Kenny, G. E.**, Antigenic analysis of chlamydiae by two-dimensional immunoelectrophoresis. I. Antigenic heterogeneity between *C. trachomatis* and *C. psittaci*, *J. Immunol.*, 115, 963, 1975.

104. **Caldwell, H. D., Kuo, C.-C., and Kenny, G. E.**, Antigenic analysis of chlamydiae by two-dimensional immunoelectrophoresis. II. A trachoma-LGV-specific antigen, *J. Immunol.*, 115, 969, 1975.

105. **Caldwell, H. D. and Kuo, C.-C.**, Purification of a *Chlamydia trachomatis*-specific antigen by immunoadsorption with monospecific antibody, *J. Immunol.*, 118, 437, 1977.

106. **Newhall, W. J.**, unpublished data, 1986.

107. **Sacks, D. L., Rota, T. R., and MacDonald, A. B.**, Separation and partial characterization of a type-specific antigen from *Chlamydia trachomatis*, *J. Immunol.*, 121, 204, 1978.

108. **Hourihan, J. T., Rota, T. R., and MacDonald, A. B.**, Isolation and purification of a type-specific antigen from *Chlamydia trachomatis* propagated in cell culture utilizing molecular shift chromatography, *J. Immunol.*, 124, 2399, 1980.

109. **Sacks, D. L. and MacDonald, A. B.,** Isolation of a type-specific antigen from *Chlamydia trachomatis* by sodium dodecyl sulfate-polyacrylamide gel electrophoresis, *J. Immunol.,* 122, 136, 1979.
110. **Batteiger, B. E., Newhall, W. J., V, Terho, P., Wilde, C. E., III, and Jones, R. B.,** Antigenic analysis of the major outer membrane protein of *Chlamydia trachomatis* using murine monoclonal antibodies, *Infect. Immun.,* 53, 530, 1986.

Chapter 4

THE CHLAMYDIAL DEVELOPMENTAL CYCLE

Michael E. Ward

TABLE OF CONTENTS

I. INTRODUCTION

The *Chlamydia* are amongst the most successful of animal pathogens. *C. trachomatis* is responsible for some 350 million cases of trachoma in the developing world, as well as being a major cause of sexually transmitted disease. Infection with or carriage of *C. psittaci* has been reported amongst arthropods, molluscs, some 130 species of birds, and a wide range of mammals; indeed no other group of pathogenic organisms has been considered more widespread in nature.[1] Both chlamydiae and rickettsiae have had to solve such fundamental problems of obligate parasitism[2] within cells as entry into the host cell, survival and replication after endocytosis, and transit between cells in the hostile extracellular environment (see Chapter 1). However, rickettsiae are restricted by their requirement for arthropod vectors to ensure transit between cells whereas chlamydiae evolved specialized developmental forms adapted for extracellular and intracellular survival.[3] Interestingly, *Coxiella burnetii*, a rickettsia which does not require arthropod vectors, also produces a specialized developmental form for extracellular survival.[4]

At their simplest *Chlamydia* can be regarded as specialized Gram negative bacteria with a growth cycle consisting of two major developmental forms, the infectious elementary body (EB) and the metabolically active but noninfectious reticulate body (RB) responsible for chlamydial replication within the cell (see Chapter 1). Given suitable conditions each RB produces one or more EB capable, on release, of infecting a new host cell. If conditions are adverse maturation of the RB may be delayed until a more favorable time. This inherent flexibility of the chlamydial growth cycle in adapting to the condition of the host cell goes some way in explaining the ability of chlamydiae to cause long-term and often inapparent clinical infections.

This review describes the process of chlamydial development in sequence, commencing with the attachment of EB to the host cell and culminating with the release of new infectious particles. In spite of recent advances, knowledge of chlamydial development is still essentially descriptive. Little is known of the metabolic regulation of the chlamydial developmental cycle, or of the genetic controls which must operate in an ordered manner at the levels of transcription and translation. Thus reference is made to a much better understood example of prokaryotic differentiation and development, endospore formation in *Bacillus subtilis*. Inevitably this review must overlap with the chapters on the general characteristics of chlamydiae (Chapter 1) and on chlamydial-host cell interactions (Chapter 7). It is hoped where this occurs the differing perspectives of the authors will better enable the reader to form his own evaluation.

II. PRODUCTIVE CHLAMYDIAL DEVELOPMENT

A. Attachment

The chlamydial growth cycle (Figure 1) commences with the attachment of the infectious EB to the host cell surface, a process critical to the whole developmental cycle. It is generally assumed that chlamydiae must have evolved high avidity attachment mechanisms to ensure their attachment to the host cell; such high avidity binding mechanisms have yet to be identified.

Chlamydial attachment has two components: nonspecific physical forces governing interactions of any particle with cell surfaces, and specific binding to host cell receptors. Both the chlamydial and host cell surfaces carry a net negative electric charge at physiological pH which gives rise to long range electrostatic repulsion between them. Given close approach to the host cell surface, attractive forces resulting from London-van der Waals interactions or hydrophobic interaction will come into play. In this respect chlamydiae, like other negatively charged particles, have to obey the laws of colloid physics, whose application to

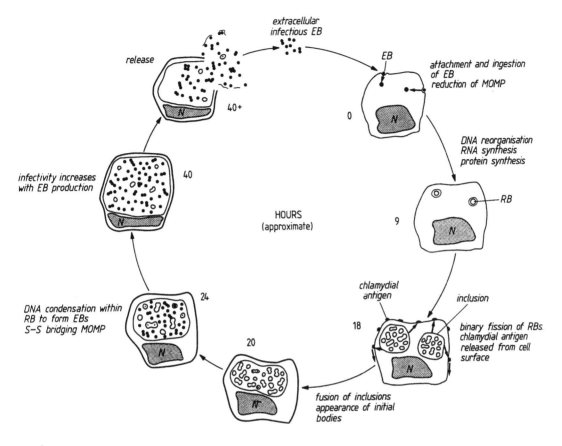

FIGURE 1. Diagram of chlamydial development.

microbial adhesion have been reviewed in detail elsewhere.[5] A notable prediction of these laws is that chlamydial penetration through the electrostatic repulsive barrier will be facilitated if the radius of contact of one of the interacting surfaces is minimized. Thus, initial chlamydial contact with the host cell is likely to be with filopodia[6] or other microvillous projections of the host cell surface (Figure 2).

This physically determined preference for projecting regions of the host cell surface will be overcome in the laboratory when chlamydiae are centrifuged onto the host cell. The real importance of electrostatic interactions in chlamydial attachment in vitro is shown by the fact that polycationic compounds, like DEAE-dextran or poly-L-lysine, enhance the susceptibility of HeLa-229 cells to infection by organisms of the trachoma biovar centrifuged onto the host cell surface.[7] Individual chlamydial serovars or biovars of C. trachomatis require different amounts of DEAE-dextran or poly-L-lysine for optimal infection of HeLa 229 cells reflecting likely differences in their surface properties.[8,9] Indeed strains of the LGV biovar will efficiently infect cells without either DEAE-dextran or centrifugation.[7] A simple explanation would be that trachoma biovar strains are more negatively charged then LGV strains, thus generating a greater electrostatic barrier to adhesion. Söderlund and Kihlström, using partitioning in aqueous two-phase systems and hydrophobic interaction chromatography, observed that a single trachoma strain was more negatively charged than a single LGV strain; both strains showed an equal limited surface hydrophobicity. However, isoelectric focusing data suggests that both an LGV serovar L2 and a trachoma serovar D strain had an identical net negative charge with an isoelectric point (pI) of 4.64.[1] As in other bacteria, carboxyl groups are likely to be the main negatively charged ionized groups on

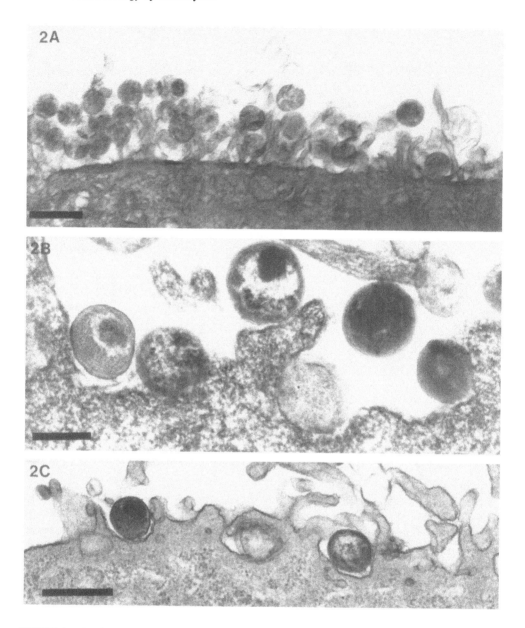

FIGURE 2. Attachment and uptake of *C. trachomatis* L2. (A) Initial chlamydial attachment is with microvilli projecting from the host cell surface. In this thick (approx. 1.0 μm) section for transmission electron microscopy at 125 kV, microvilli enfold both sides of the adherent chlamydiae. (B) Chlamydia make contact with the cell surface at the base of the microvilli. (C) internalization 5 min after inducing uptake of attached chlamydiae by warming to 36°C. Organisms are deeply enfolded by the host cell surface or within endosomes. In these electron micrographs the bars represent 1, 0.25, and 0.5 μm respectively.

chlamydial proteins at physiological pH. Acidic proteins which might contribute to the negative chlamydial surface charge include the major outer membrane protein (MOMP),[9-11] thought to comprise some 60% of the surface of the infectious EB, and two EB proteins, with apparent molecular mass of 18 and 32 kdaltons,[12] which bind the polyanionic compound heparin. Ultimately the most important factor may be the overall match of the charge mosaic on the chlamydial and host cell surfaces, rather than net charge.

Identification of specific chlamydial attachment mechanisms is of major importance. Local antibody directed against relevant chlamydial surface antigens might abort chlamydial development by preventing attachment to the host. Most studies of chlamydial adhesion have used HeLa, McCoy, or mouse L fibroblast as the model host cell simply because these cells are available in quantity and are susceptible in vitro to chlamydial infection. These cells probably lack receptors or other factors used by chlamydiae in vivo, as most clinically important isolates of *C. trachomatis* have to be centrifuged onto their surface to achieve adequate infectivity. Red cells might provide an alternative model for chlamydial adhesion in view of the weak haemagglutinating activity shown by chlamydial EB. Their advantage is ready availability, well described surface chemistry, and the enormous range of potential host cell receptors represented amongst the different human blood groups. A technically demanding, but more relevant approach, is the use of cells derived from explants of cervical tissue.[13] Unfortunately, even in these relevant cells or in fallopian tube organ culture,[14] chlamydial infection is still inefficient in the absence of centrifugation. Perhaps suitable host cell receptors are expressed only by certain epithelial cells, at critical stages in the cell cycle, or under restricted hormonal or environmental control.

The host cell receptors utilized by the different biovars of *C. trachomatis* in transformed cell lines differ considerably. LGV organisms, unlike those of the trachoma biovar, readily infect HeLa cells without the need for centrifugation and do not require neuraminidase-sensitive sites for their attachment, suggesting that sialic acid, which contains *N*-acetyl-mannosamine, is not involved.[7] Even within the LGV biovar differences have been reported.[15,16] Thus, treatment of host cells with wheat germ agglutinin, a plant lectin with affinity for *N*-acetylglucosamine residues, prevented infection with *C. trachomatis* LGV L1, but not LGV L2 or L3. Adhesion of LGV L1 to McCoy cells involves different host cell surface determinants, some sensitive and others resistant to protease-K digestion.[17] Association of LGV L1 with McCoy cells was inhibited by millimolar amounts of chitobiose or chitotriose, β 1-4 linked oligomers of *N*-acetyl-*D*-glucosamine. Cellobiose, a β 1-4 linked disaccharide of glucose was ineffective, suggesting that the *N*-acetyl amino group was critical in this adhesion.[18] Different strains of chlamydiae probably bind to a variety of glycoproteins or glycolipids on the host cell's surface as yet unidentified. Possible approaches include the use of a much wider range of oligosaccharides as inhibitors and the use of monoclonal antibodies to specific host cell surface components.

The chlamydial surface components mediating adhesion have also not been definitely identified. Adhesion probably involves a complex spectrum of determinants on a given organism and different strains may utilize different components. Thus a strain of *C. psittaci* required a heat-labile and trypsin-sensitive site on the EB for adhesion.[19,20] By contrast, extensive proteolytic degradation of exposed proteins on the surface of purified *C. trachomatis* L2 EBs failed to block their infectivity for HeLa 229 cells,[21] suggesting that if proteins were involved in chlamydial adhesion or uptake then their critical sites must be protease resistant. This proteolysis resulted in the loss of a species-specific epitope on MOMP, but the type specific epitope remained bound to the EB outer envelope as a heavily S-S cross-linked oligomeric complex. Proteolytic processing of the chlamydial surface may be important for both attachment and uptake (see Section V). Chlamydial polypeptides capable of binding surface-radioiodinated and Triton X-100-solubilized HeLa cell extracts have recently been identified by ligand-electroblotting.[12] In *C. trachomatis* two such proteins were identified, one with a constant molecular weight of 18 kdaltons, the other with an approximate molecular weight of 32 kdaltons depending upon the serotype. In *C. psittaci* a single binding protein with a molecular weight of 17 to 19 kdaltons was identified. These proteins were found in greater amounts on EBs rather than on noninfectious RBs, stained distinctively on silver-stained SDS-PAGE gels, and showed poor antigenicity in immunoblots. Attempts to confirm the surface location of these proteins on chlamydial EB by lactoperoxidase-mediated radioiodination were unsuccessful. This might indicate a lack of exposed tyrosin, trytophan, or histidine residues available for iodination. Alternatively, the polypeptides might not become surface-exposed until after close association with the host, perhaps as a result of protease cleavage. The polypeptides were acidic as shown by their ability to bind radio-active heparin. Thus their observed interaction with host cell components may be based on charge rather than specific interactions. Any specific interactions might be obscured by denaturation on SDS-PAGE. Nevertheless, ligand-electroblotting, preferably under mild conditions, offers a powerful technique for investigating chlamydial adhesion. Moreover the same technique using radio-active chlamydial proteins as probes could be reversed to investigate host cell receptors. Such experiments would need to take account of the fact that chlamydial lipopolysaccharide (LPS) or exoglycolipid might also be involved in adhesion of chlamydiae to host cells. Thus clones of *Escherichia coli* expressing a gene for a chlamydial enzyme capable of adding a chlamydial-specific epitope to *E. coli* LPS biosynthesis showed enhanced adhesion to eukaryotic cells.[22] This may have been a non-specific effect, since chlamydial LPS has hydrophobic moieties and might effect hydrophobic interaction between the surfaces. One suggestion is that LPS might promote chlamydial adhesion whilst a recently identified 30 kdalton serovar-specific chlamydial surface protein with esterase activity might promote uptake by increasing membrane fluidity.[23] This intriguing concept remains to be explored.

Overall, the relative importance of specific and nonspecific factors in chlamydial adhesion remains unclear. A practical problem is that bulky probes such as lectin or monoclonal antibodies targetted at specific cell surface determinants might sterically hinder adjacent receptors or alter nonspecific characteristics of the cell surface. Ligand-electroblotting is a powerful technique but poses the general problem of membrane biochemistry; solubilization vs. denaturation of membrane components. Clearly the intracellular character of chlamydial development and its specificity for restricted hosts and cell types all argue for the existence of specific adhesion mechanisms. However, *high avidity* adhesion mechanisms have yet to be identified; the half-life for chlamydial adhesion is measured in hours rather than minutes. Perhaps low avidity mechanisms are all that is necessary given the huge ratio of host cell surface to chlamydial surface in vivo. Such low avidity mechanisms might have positive advantages, permitting chlamydiae to bind reversibly to a wider variety of cells facilitating their eventual release.

B. Endocytosis

Ingestion of the adherent chlamydiae is essential for chlamydial development. Attachment and uptake are distinct processes.[20,24] Thus human leukocyte cell lines have been identified to which *C. trachomatis* L2 can attach, but in which ingestion does not occur (e.g. F4, Daudi), or in which ingestion may occur with (Rienta) or without (Nalm 6) subsequent inclusion development.[25] Uptake, unlike attachment, is blocked at low temperatures (4°C to 8°C).[24] There is no general stimulation of host cell endocytosis following attachment. On the contrary, chlamydial uptake involves a local segmental response of the host cell membrane adjacent to the adherent organism,[26] a process which Byrne and Moulder termed "parasite specified" uptake.[27]

Two main mechanisms have been postulated for chlamydial uptake; the microfilament and energy-dependent process of phagocytosis[28] and the microfilament independent uptake of chlamydiae into clathrin-coated vesicles by some variant of receptor-mediated endocytosis.[17,29,30,31] The latter mechanism envisages host cell receptors for chlamydiae diffusing laterally in the fluid cell membrane, interacting frequently with clathrin-coated pits ultimately destined to become coated vesicles (Figure 3A). Ligand binding to these receptors produces allosteric changes which are recognized by a trapping mechanism in the pit.[29] This type of clathrin-dependent mechanism is constitutively used by the host cell for the micro-pinocytic uptake of receptor-bound hormones and of molecules such as β-lipoprotein or interferon. Small viruses such as Semliki Forest Virus may "hitch-hike" their way into the cell by this

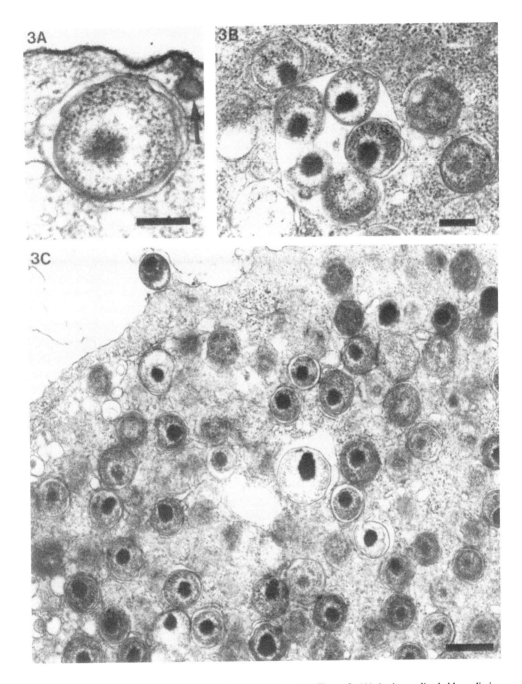

FIGURE 3. Uptake of *C. trachomatis* L2 initiated as described in Figure 2. (A) An internalized chlamydia in an endosome. Note the relative size of the clathrin-coated vesicle (arrow) which still retains its connection with the cell surface; (B). When many organisms are internalized, endosomes containing multiple chlamydiae are observed, presumably derived from fusion of individual endosomes. Staining of the host cell glycocalyx with colloidal thorium demonstrated that these chlamydiae were not lying in an invagination of the host cell surface. (C) The host cell was challenged at a ratio of 2000 EB per cell; 30 min after stimulating uptake numerous chlamydiae can be seen in tight-fitting endosomes. The bars represent 0.25 μm respectively.

mechanism, characteristically displacing some of the fluid otherwise pinocytosed. In reality phagocytosis and receptor mediated endocytosis may not be completely distinct; electron microscopy of carbon replicas of the cytosolic face of macrophage phagosomes shows the characteristic net-like meshwork of clathrin.[28] Both forms of endocytosis require clathrin for membrane recycling as well as enzymes regulated by the calcium-binding protein, calmodulin.[32-34]

The route of chlamydial endocytosis is probably dependent on the experimental system used. When chlamydiae are permitted to randomly associate with host cells, clear electron microscopic evidence was obtained for the entry of *C. psittaci* into clathrin-coated vesicles.[35] Monodansyl cadaverine or methylamine, classic inhibitors of receptor mediated endocytosis, prevented entry of *C. trachomatis* serovars E or L1 into "loosely confluent but not confluent" host cells.[17] Taken together this data suggests that following random association chlamydiae enter host cells by a clathrin-dependent mechanism similar to receptor mediated endocytosis. However, whilst calmodulin is considered important in the clathrin recycling essential for receptor mediated endocytosis, the calmodulin inhibitor trifluoperazine did not block chlamydial endocytosis.[36] By contrast, when EB of *C. trachomatis* L1 were centrifuged onto HeLa cells in the cold to facilitate adhesion and then rapidly warmed to promote entry, chlamydial endocytosis was inhibited by trifluoperazine in a dose-dependent manner.[37] In this latter system chlamydial entry into cells was extensive 5 min after stimulating uptake (Figure 3A, B and C). Electron microscopy showed chlamydiae apparently in process of ingestion, partly engulfed in microvillous processes of the host cell surface; there was no evidence of chlamydial entry into clathrin coated vesicles despite diligent searching and appropriate fixation.[26] Pinocytic uptake of extracellular fluid labeled with nonpermeant [^{3}H]-sucrose was not reduced by chlamydial endocytosis, showing that chlamydiae were not utilizing this constitutive, clathrin-dependent mechanism. Chlamydial ingestion was impaired by the microfilament inhibitor cytochalasin D, or the microtubule disruptors, vincristin and vinblastin, but was not impaired by inhibitors of receptor mediated endocytosis like monodansyl cadaverine or amantadine. However, hyperimmune rabbit antibody distal to the site of chlamydial attachment to the host cell did prevent ingestion, suggesting that chlamydial endocytosis requires the circumferential binding of chlamydial and host cell surface ligands by a microfilament-dependent, zipper-like mechanism.[26] This process was calmodulin-dependent since in this system trifluoperazine prevented endocytosis.[37] Given the observation that binding of chlamydiae to the host cell causes a rapid flux of Ca^{2+} across the host cell membrane[38] it was suggested that Ca^{2+} might interact with calmodulin to regulate phospholipase A_2 or other enzymes concerned with the generation of membrane fluidity, as well as myosin light chain kinase concerned with microfilament function. Such a mechanism would explain why treatment of HeLa cells with the Ca^{2+} ionophore A23187 increased their susceptibility to chlamydial infection.[38] However, the chlamydial surface also has esterase and protein kinase-like activity[23,39] which may also produce local changes in the host cell membrane triggering ingestion.

It seems likely that chlamydiae might use more than one route of endocytosis, dependent on the mode of presentation to the host, the chlamydial strain, and the host cell itself. Moreover, it should not be assumed that chlamydial uptake into clathrin-coated vesicles is the same mechanism as that used by small viruses; chlamydiae, unlike these viruses, are grossly larger than clathrin-coated vesicles. Indeed, chlamydial versatility might rewrite some of our basic understanding of endocytic mechanisms. Finally, one should be careful about inferring too much from experiments with "specific" inhibitors of host cell function.[17,18,26,40,41] Such inhibitors may have effects other than those predicted. Thus, cytochalasin B, an inhibitor of microfilament function, is a misleading probe of endocytic mechanisms in relation to chlamydiae because of its failure to inhibit microfilament dependent uptake of particles less than 600 nm.[42]

C. The Endosome and Inhibition of Endolysosomal Fusion

Throughout their intracellular life chlamydiae develop within an expanding endosomal vacuole. Viable *C. psittaci* are capable of blocking the normal process of lysosomal degranulation into the endosome.[43-46] Purified cell envelopes of *C. psittaci* are ingested by mouse L cells and also prevent endolysosomal fusion.[47] By contrast, heat-killed or antibody-coated *C. psittaci* are also ingested, but endolysosomal fusion is not prevented. Thus, it is the unmodified cell surface of *C. psittaci* which prevents endolysosomal fusion. Zeichner, in 1983, isolated endosomes containing *C. psittaci* from mouse L cells or macrophages.[48] The membrane of endosomes containing *C. psittaci* had fewer surface proteins by SDS-PAGE than the plasma membrane indicating that the endosome probably incorporates a specialized population of membrane proteins. Minor differences in protein composition were observed between endosomes containing viable chlamydiae and those containing heat-inactivated *C. psittaci*. It was postulated that *C. psittaci* might prevent the incorporation of host-specified markers into the endosomal membrane which normally designate the endosome for fusion with lysosomes. One should be wary of extrapolating this hypothesis too uncritically to *C. trachomatis* infection. In man, *C. psittaci* infection, unlike *C. trachomatis*, is commonly accompanied by bacteremia suggesting that psittacosis agents might have specifically evolved a mechanism for blocking lysosomal degranulation in immune effector cells. The extent to which *C. trachomatis* biovars block phagolysosomal fusion and its importance for cellular defence in the epithelia they commonly infect is unknown. Preliminary evidence suggests that some strains of *C. psittaci* or *C. trachomatis* might fail to grow in certain host cells following random attachment because they fail to prevent endolysosomal fusion.[49] The potentiating effect of cortisol on infection of cells by certain chlamydial strains may be due to its ability to stabilize endosomal membranes thereby reducing lysosomal degranulation. Endosomes containing *Chlamydia* are processed quite differently from those containing *Coxiella burnetii*. Like *Chlamydia*, *C. burnetii* enters cells primarily by endocytosis and remains within endocytic vacuoles throughout its life cycle. However, lysosomal fusion occurs and the resulting acidic conditions stimulate the transport and metabolism of both glucose and glutamate by *C. burnetii*; lysosomotropic amines prevent the intracellular multiplication of this organism.[50] Typhus rickettsiae are different again. Phagocytosis of these organisms requires the active expenditure of energy by the bacterium itself. Only those organisms which escape the phagosome prior to lysosomal fusion survive. Thus, *Chlamydia*, *Coxiella*, and *Rickettsia typhi*, obligately intracellular bacterial pathogens, interact quite differently with cell organelles;[50,51] there might be considerable differences within the genus *Chlamydia* itself.

D. Initiation of EB Maturation

Within a few hours of entering the cell the EB shows signs of morphological differentiation into RB (Figure 4B and C). The EB loses its prominent electro-dense DNA core, the cell envelope loses its rigidity, the size increases from 0.3 to 1.0 μm, and the cytoplasm becomes granular due to the production of ribosomes. Purified RBs of *C. psittaci* have approximately 3 times the RNA:DNA ratio of EBs. RBs of *C. psittaci* contain mainly 21S, 16S, and 4S RNA, whereas the EBs comprise mainly 4s RNA. Details of the synthesis and maturation of ribosomal RNA during the development of *C. trachomatis* have been published.[6,52] Clearly selective mRNA transcription is necessary at different stages of chlamydial development, but the species involved have not been identified.

E. Maturation of RB

By some 9 hr after infection, the original EB will have differentiated into the much larger RB. The latter morphologically resemble typical Gram-negative cocci, except that they lack the classical peptido-glycan layer in the outer wall. A detailed description of the ultra-

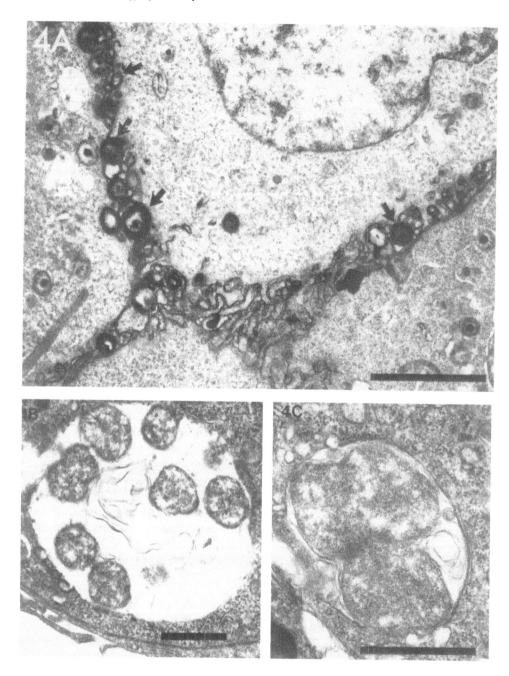

FIGURE 4. (A) HeLa cell monolayer 1 hr after initiating uptake of *C. trachomatis* L1. The cells were fixed and embedded *in situ*, then sectioned parallel to the cell surface approximately at the midline. Note the presence of intercellular chlamydiae (arrows) deep within the monolayer despite the close cell junctions. (B) Vesicles and blebs derived from the pleomorphic RB outer envelope are often found in inclusions around 16 hr after infection. Genus specific chlamydial glycolipid antigen is incorporated into the host cell surface and exported from the infected cell. (C) By 9 hr after infection the EB differentiate into RB and commence multiplication by binary fission. The bars represent 2.0, 1.0, and 1.0 μm respectively.

structure of chlamydial EB and RB is given in Chapter 2. The RB divide by binary fission in a typical bacterial manner but lack cytochromes and other components of the respiratory electron chain and are thus unable to synthesize compounds for high energy storage and utilization. ATP from the host cell is utilized to generate the proton motive force necessary to transport lysine and other essential nutrients across the cytoplasmic membrane.[53,54] Thus chlamydial development is kept in step with the metabolic health of the host cell,[3] chlamydial biosynthesis relying to a large extent on the host cell for vitamins, amino acids, nucleotides, and other essential nutrients. Thus chlamydiae normally compete with the host cell for nutrients and when their supply falls below a critical threshold value chlamydial development will be in abeyance.[55] When cycloheximide, an inhibitor of eukaryotic cell protein synthesis, is used to reduce host cell competition for the intracellular amino acid pool, high threshold levels of certain amino acids are needed for intracellular growth of *C. psittaci*. Thus amino acids probably play a regulatory role in chlamydial development. Differences in amino acid requirements exist between different chlamydial strains[56] and might be exploited to auxotype chlamydial isolates for epidemiological purposes. The metabolic requirements of replicating RBs have been recently reviewed[54,57] and are discussed in detail in Chapter 5.

The RB cell envelope is pleomorphic due to the absence of peptidoglycan and other forms of structural stabilization (Figures 4B and C).[58,59] Thus inclusions full of dividing RB are often filled with vesicles derived from the outer membrane (Figure 4B). Moreover, genus-specific antigen can be demonstrated by fluorescence microscopy or immunoelectron microscopy at the surface of infected host cells at this time, spreading to adjacent uninfected cells.[60] This material is an exoglycolipid derived from the RB and ultimately exported in quantity from the host cell where it forms the basis of some of the current commercial diagnostic kits for the immunoassay of chlamydial antigen in patient secretions. This exoglycolipid and its route of secretion have yet to be characterized in detail. The fact that preliminary airdrying of cells[60] or the brief use of detergent-containing fixatives (Ward, M.E., unpublished) is necessary for the immunoelectron microscopic demonstration of this antigen, suggests that this hydrophobic material may not be freely accessible at the host cell surface for interaction with antibody, but may be exported in a membrane-bound form. This might explain why attempts to demonstrate cytotoxic T cell attack on chlamydial infected cells were unsuccessful.[61] Further studies using monoclonal antibodies as immunocytochemical probes are necessary to resolve this problem.

A feature of chlamydial infected cells is a reduction in membrane fluidity at the host cell surface,[62] perhaps following the incorporation of exoglycolipid (see Chapter 3). Reduced membrane fluidity might decrease the susceptibility of cells expressing chlamydial antigen at their surface to immune attack, or alternatively might prevent superinfection of host cells already infected by interfering with endocytosis.[63] Incorporation of chlamydial glycolipid into the host cell surface might alter the transport properties of the plasma membrane permitting the entry of nutrients essential for chlamydial replication into the host cell. At present it is not clear whether the expression of chlamydial glycolipid at the surface of the host cell is merely fortuitous or is of vital importance to chlamydial development.

In transformed cell lines maturation of RB generally occurs for at least 30 hr after challenge by which time chlamydial ribosomal RNA and, in the case of *C. trachomatis*, glycogen synthesis from ADP-glucose are at their peak[64] (see Chapters 2 and 5, Figure 6). Strains of *C. trachomatis* have much higher rates of glycogen synthesis and accumulation than *C. psittaci*;[65] this glycogen can be demonstrated by iodine staining and forms a simple practical procedure for distinguishing *C. trachomatis* from *C. psittaci*. Developing RB of *C. trachomatis*, unlike most *C. psittaci* strains, are susceptible to growth inhibition by sulfonamides implying that *C. trachomatis* RB synthesize their own folate whereas *C. psittaci* RB do not.

F. Differentiation of EB from RB

By 18 to 20 hr after infection some of the RBs commence reorganization within the

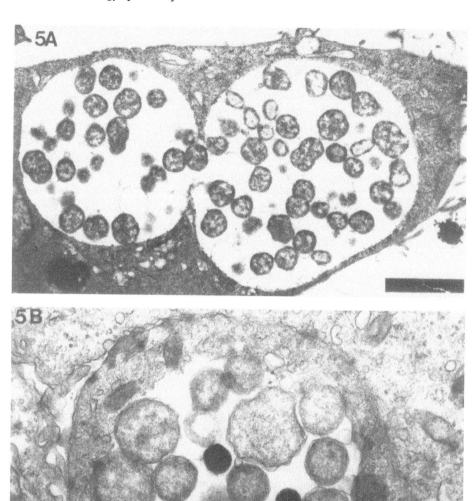

FIGURE 5.   (A) Fusion of two distinct RB-containing inclusions which have each presumably developed independently from an internalized EB of *C. trachomatis* L1. (B) In HeLa cells EB (arrowed) of *C. trachomatis* L2 begin to develop from RB around 20 hr after challenge. The bars represent 2.0 and 1.0 µm respectively.

expanding host cell inclusion to form infectious EB (Figures 5 and 6). These reorganizing RB are characterized by the reappearance of a single or multiple central core of DNA (Figure 6). The fact that, at the beginning of the developmental cycle, a single EB gives rise to a single RB has led to the widespread assumption that by the reverse process a single RB must also give rise to a single EB. Certainly, the most common form of reorganizing RB is that having only a single site of DNA condensation within its cytoplasm. However, simple

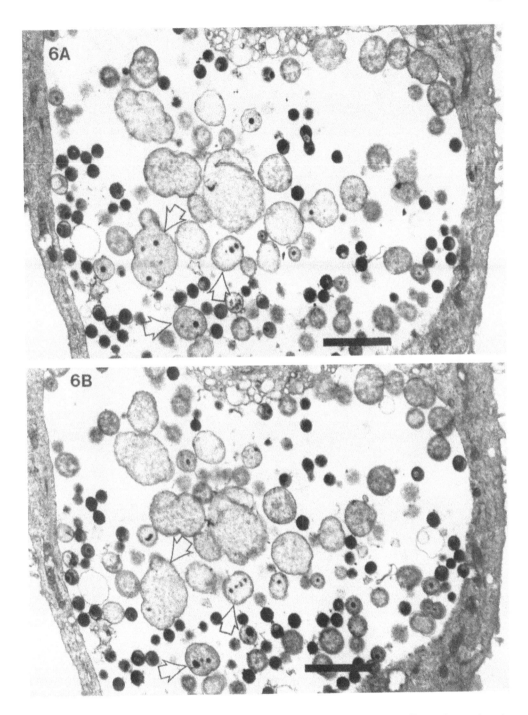

FIGURE 6. (A and B) Serial sections of the same inclusion of *C. trachomatis* serovar C spaced approximately 150 nm apart. Comparison of the two micrographs shows that RB containing multiple sites of DNA condensation (arrowed) are only observed if the section runs fortuitously along an intersecting axis. The bars in both micrographs represent 2.0 μm.

reorganization of DNA by unknown mechanisms does not account for the considerable DNA synthesis which is necessary during the transition of RB to EB.[66] The accompanying electron micrographs (Figures 6 and 7) show that RB not infrequently contain multiple sites of DNA condensation capable of cell division to give rise to multiple EBs. Indeed structures can be observed in which four EBs have apparently been produced within a single RB (Figure 7F).

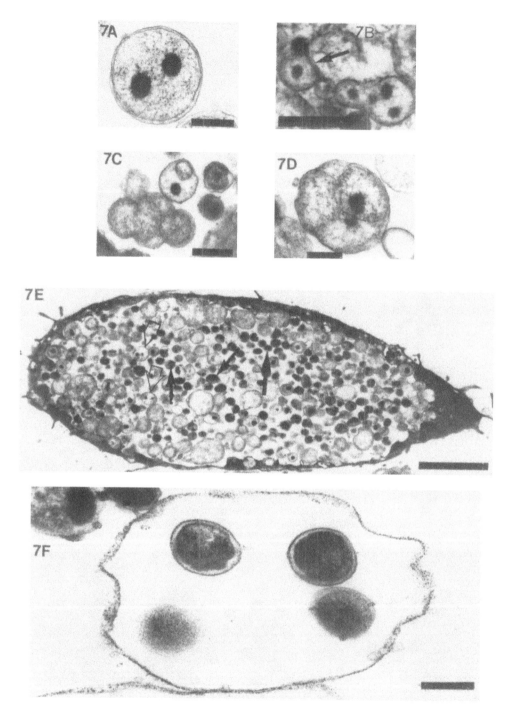

FIGURE 7. Multiple development of EB from a single reticulate/intermediate body. (A) Intermediate body with two distinct sites of DNA condensation. (B) Budding intermediate body with four sites of DNA condensation. One intermediate body (arrow) has already separated. (C and D) Reorganization and compartmentalization of membranes within RBs. Each compartment may subsequently be a site of DNA condensation as in (D). (E) Densely packed inclusion of *C. trachomatis* L2 containing EB (solid arrows) as well as RB with both single and multiple (hollow arrows) sites of DNA condensation. (F) Four mature EBs of *C. trachomatis* serovar A contained within an envelope possibly derived from the parental intermediate body following internal compartmentalization. The bars represent 0.25, 1.0, 0.5, 0.25, 5.0 and 0.5 μm respectively.

Multiple production of EB from a single RB might depend on an adequate supply of nucleotide precursors for necessary DNA synthesis and would explain why EBs often appear to be spatially clustered together within the inclusion (Figures 8A and B). Clearly, morphological stages in the production of EB from RB may be more complex than has hitherto been realized. Hydroxyurea and other reversible DNA inhibitors provide a means of synchronizing the chlamydial developmental cycle,[66] permitting more detailed investigation of these events.

At the end of productive chlamydial development, the original tight endosome membrane may have expanded to form an inclusion occupying over three-quarters of the cell volume (Figures 8A and B). Similar mature inclusions have been seen in electron micrographs of epithelia from patients with genital tract infections.[67] The frequent proximity of mitochondria to the inclusion membrane has led to the suggestion that this is a strategic location for the provision of energy to chlamydiae. Given the expanding nature of the chlamydial inclusion such a fortuitous arrangement of the mitochondria is probably inevitable. It has been postulated that "spear-like" projections on the surface of RB adjacent to the inclusion membrane might somehow penetrate the membrane and aid chlamydial nutrition.[68,69] Such a mechanism might not benefit chlamydiae deep within the inclusion. It is unclear how the inclusion membrane expands its size ahead of chlamydial development. One possibility is that this is a result of chlamydial-directed lipid synthesis (see Chapter 5) or that the chlamydial glycolipid itself becomes incorporated into the endosomal membrane. Such a mechanism might explain the apparent "dilution" of host plasma membrane proteins in the inclusion membrane[48] and lead to altered nutrient transport into the inclusion. These problems require careful electron microscopic immuno- and cyto-chemical investigation.

G. Release Mechanisms

Massive productive chlamydial development of the type described inevitably destroys the host cell. Starting from 20 to 30 hr after infection cellular organelles show progressive degenerative and necrotic changes. These include loss of host cell ribosomes and polysomes, dilation and vesiculation of the endoplasmic reticulum, vesiculation, and loss of microvilli.[70] Mitochondria and nuclei are affected last. Given such a sick cell, the simplest explanation is that release of EB results from rupture and lysis of the cytoplasmic and inclusion membranes. At the time of C. psittaci release, lysosomal enzymes can be demonstrated free in the host cell cytoplasm together with associated autolytic changes.[71] It is not clear whether autolysis is a cause or a consequence of chlamydial release. Benes[109] describes the explosive release of chlamydial EB from inclusions and suggests that in intact tissue strains of chlamydiae which in vitro require centrifuge-assisted infection may be capable of direct cell to cell transfer between adjacent cells in vivo. More recently it has been postulated that whole chlamydial inclusions could be released from cells by exocytosis without concomitant cell death.[72] Unfortunately, static microscopic observations of dynamic events such as release, based inevitably on different cells can be interpreted in a variety of ways; the requirement is for time-lapse cinemicrography of individual living cells throughout the course of infection. Special opportunities for chlamydial spread arise when the developmental cycle is delayed.

III. DELAYED CHLAMYDIAL DEVELOPMENT

A. In Vitro Aspects

It has long been known that chlamydiae may cause persistent, latent infection in cell culture. Lee and Moulder[73,74] infected mouse McCoy cells with an isolate of C. trachomatis serovar A in the presence of cycloheximide, a reversible inhibitor of host cell-protein synthesis. Cycloheximide was removed, and the infected cells were washed and subcultured weekly. On first infection more than half the cells were productively infected, resulting in chlamydial inclusions. Many of these cells were destroyed. On subculture the fraction of

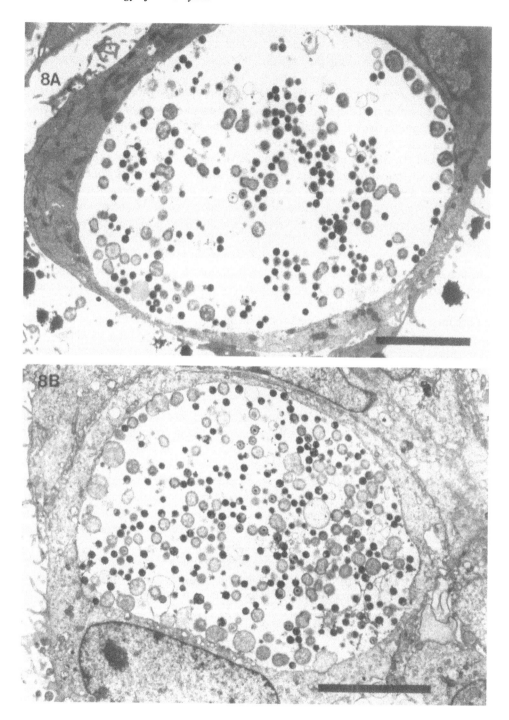

FIGURE 8.   (A) Mature inclusion of *C. trachomatis* showing the common tendency of EB to cluster together. This is expected if multiple EBs are frequently produced from a single RB either as a result of internal reorganization, or intermediate body division or budding. (B) Mature inclusion of *C. trachomatis* L2 42 hr after infection; some $^3/_4$ of the cell volume is occupied by the inclusion. The bars represent 5.0 μm.

inclusion-bearing cells declined to less than 1%, reflecting the greater cloning efficiency of uninfected compared with infected cells.[75] However, after 100 days, the proportion of inclusion-bearing cells rose dramatically, cultures alternating between periods of massive, chlamydial-induced host cell destruction and periods of host-cell proliferation. These reciprocal, fluctuating changes in chlamydial and host cell density were not accompanied by changes in chlamydial invasiveness. Given the fact that in vitro serovar A trachoma agents do not readily infect cells in the absence of centrifugation, the most reasonable explanation was that chlamydial infection was being transferred from cell to cell at cell division. This hypothesis has been confirmed by Richmond who demonstrated by immunofluorescence that chlamydial inclusions can be transferred to progreny cells at cell division.[76] Cell to cell transfer of chlamydial infection can be maintained for long periods of time; cell cultures persistently infected with C. trachomatis have been maintained for 3 years.

Cells persistently infected with C. psittaci[77-79] but not C. trachomatis[73,80] are resistant to superinfection with exogenous C. psittaci. This resistance to superinfection occurred irrespective of whether the cells productively formed inclusions, was related to a failure of the exogenous chlamydiae to attach, and was associated with the altered expression of plasma membrane proteins at the surface of the persistently infected cells.[79] These changes in plasma membrane proteins and resistance to superinfection disappeared when the L-cells were cured of their chronic C. psittaci infection with the antichlamydial drugs chlortetracycline or rifampicin. It was postulated that persistently infected cells were cryptically infected with C. psittaci,[79] indeed in one such clone resistant to superinfection a proportion of the cells eventually gave rise to chlamydial inclusions after 128 days of maintenance culture. The implication is that in such cultures the chlamydial developmental cycle may be delayed at an intermediate stage, with immature chlamydial forms being transferred at cell division to reappear generations later under favorable conditions. Electron microscopic observations on cells persistently infected with C. trachomatis suggest that the cryptic chlamydial structures may be RB or small aberrant forms resembling the mini cell-like forms which are occasionally produced by RB.[81-83]

Transfer of the chlamydial developmental cycle from delayed to productive infection is a multistep, penicillin-sensitive process dependent on host cell density and nutrition. Although chlamydial RB lack a conventional peptidoglycan target for penicillin, this antibiotic nevertheless inhibits RB growth and division, as well as the subsequent differentiation of EB. Thus delays in chlamydial development might ensue from antibiotic treatment,[84] depletion of an essential nutrient below its critical threshold,[55,56] or as a result of interferon action.[85,86] Factors governing the interplay of delayed and productive chlamydial infection might provide important clues to metabolic regulation of the chlamydial development cycle. However, the key question is whether delayed development is a feature of natural chlamydial infection.

B. Clinical Aspects

It is clear both ocular and genital chlamydial infections persist for months in the absence of treatment.[87,88] These infections are often asymptomatic, or, as in trachoma, may be characterized by intermittent periods of activity and chlamydial shedding. Thus symptomatic trachoma amongst American Indian school children living in the South-West U.S. rapidly diminished when these children attended a boarding school with a much higher standard of environmental health than that on their home reservation. Nevertheless, chlamydial antigen detected by immunofluorescence persisted in ocular tissue and did not diminish, even when infected children were treated topically or systemically with antichlamydial drugs.[87] Isolation of C. trachomatis from the genital tracts of female contacts of men with gonococcal urethritis was dependent not only on the presence of chlamydiae in the male partner, but also on actual infection by N. gonorrhoeae in the woman herself. It was postulated that gonococcal infection might reactivate a latent chlamydial infection in some of these women.[88] Thus there is evidence of persisting asymptomatic chlamydial infection at both ocular and genital sites.

Chlamydial activity is considered greatest in host cells which are themselves actively growing and dividing.[76] Thus factors increasing mucosal turnover such as trauma or coincident infection might reactivate delayed chlamydial development. In the absence of chlamydial replication chlamydial antigen is unlikely to persist due to the high turnover rate of conjunctival and genital epithelia. The clinical observations cited might be explained by reactivation of deep-seated chlamydial infection, of low-level asymptomatic infection, or the triggering of productive infection in cells cryptically infected with *Chlamydia*. In the case of trachoma, acute activation of this chronic disease has been attributed to intermittent shedding of chlamydiae which had gained access to deep-seated sites as a result of denudation of conjunctival epithelia or the breakdown of lymphoid follicles. Histopathological studies are essential to distinguish chronic infection due to the cell to cell transfer of chlamydiae in delayed development from clinically inapparent conventional infection. Careful comparison of chlamydial isolation rates (productive infection releasing infectious EB), presence of chlamydial antigen (productive and inapparent infection), and histological studies in patients or animal models with long established infection is needed to define the in vivo role of delayed chlamydial development and cell to cell transfer.

IV. STRUCTURAL AND GENETIC REGULATION OF CHLAMYDIAL DEVELOPMENT

A. Structure and Function in Chlamydial Development

The function of EB is the transfer of chlamydial infection through the hostile extracellular environment. EB are resistant to mechanical disruption or osmotic lysis, despite the absence of the peptidoglycan layer which normally confers structural rigidity to Gram-negative bacteria. How is this achieved?

The cell walls of RB contain much less cysteine than those of the EB. Disulfide cross-linkage of cysteine in EB cell wall proteins might contribute to their structural rigidity.[89,90] This far-seeing concept has essentially been confirmed. EBs of *C. psittaci* contain at least four cysteine-rich proteins. Dominant amongst these is the MOMP described independently by three separate groups in 1981.[9-11] In addition, there are three minor cysteine-rich proteins of relative molecular mass 12, 59, and 62 kdaltons.[91] In *C. trachomatis* two minor proteins of 18 and 32 kdaltons bind to HeLa cell proteins, are more abundant on EB than RB, and may play a role in chlamydial attachment.[12] However, the 32 kdalton protein may be a proteolytic cleavage product of another protein. Extensive proteolysis of whole EB leads to the appearance of the 32 kdalton protein but does not impair infectivity.[12,21] Indeed, proteolytic cleavage by exogenous (host) or endogenous (chlamydial) protease might be required for exposure of an active site as occurs with the F-glycoprotein of paramyxoviruses which mediate virus penetration and cell fusion. Chlamydial attachment might be promoted by proteolytic exposure of critical disulfide residues on the EB surface followed by mixed disulfide formation and exchange with host cell proteins,[12] analogous to the interaction of insulin with its cell surface receptor, or the proposed role of disulfides in the interiorization of diptheria toxin. Covalent interaction of chlamydiae with host cell receptors might influence uptake and prevent endolysosomal fusion.

Once the EB enters the host cell, disulfide cross-linked oligomers of chlamydial protein are reduced to monomeric form. Synthesis of MOMP continues throughout the developmental cycle so that RBs contain virtually as much MOMP as EBs. By contrast, synthesis of the 12, 59, and 62 kdaltons cysteine-rich proteins of *C. psittaci* does not reccommence following reduction until the RB begin to reorganize into EB late in the developmental cycle[91,92] (see

Chapter 3). Elegant work based on reconstitution experiments in liposomes has demonstrated that the outer membrane protein functions as a "porin", with an "exclusion limit" of 850 to 2250 daltons.[93] Opening and closure of the porin is under redox control, the three cysteine residues per molecule of MOMP enabling it to form a trimer of three polypeptide molecules surrounding a pore. On entry into the host cell the EB within its endosome is thought to be exposed to reducing conditions, probably in the form of reduced glutathione.[93] Reduction of the disulfide bonds of MOMP would lead to relaxation of the pore size resulting in increased membrane flexibility and ingress of ATP and other essential nutrients required for development. Later, development of RB might slow as the level of ATP and reducing power (reduced glutathione, NADPH, or both) decreased below a critical threshold. This would result in decreased metabolic activity, the oxidation of free sulfhydryl groups into disulfides, and closure of the outer membrane pores and outer membrane rigidification, all features characteristic of the transition of RB to EB. Electron microscopic studies support this concept of MOMP forming anionic, intramolecular channels in the outer envelope of chlamydial EB.[94,95]

Support for a regulatory role of MOMP and redox potential in chlamydial nutrition and development came from the observation that polyclonal antibodies to MOMP prevent chlamydial development without blocking attachment or uptake.[96] Transformation of RB to EB requires cysteine, presumably to reform the S-S bridges of MOMP and other proteins required for EB stability.[97] However, the most direct evidence comes from the observation that EB exposed to the dithiol reducing agent dithiothreitol undergoes a number of the changes normally associated with their differentiation to RB. These include enhanced rates of glutamate oxidation, reduced infectivity, decreased osmotic stability, and altered staining properties. However, these changes were insufficient to induce the morphological alterations associated with EB differentiation to RB.[98]

Chlamydial development is probably also regulated by cyclic nucleotides. The dibutyryl analogues of cyclic guanosine monophosphate (cGMP) and cyclic adenosine monophosphate (cAMP) increase or depress, respectively, the infectivity of *C. trachomatis* for HeLa cells, suggesting that chlamydial infectivity may be under bidirectional cyclic nucleotide control.[38] Recent work shows that high levels of cAMP added to cells within 12 hr of infection reversibly impairs the maturation of RB to EB. The resulting immature inclusions were noninfectious and the chlamydiae had a protein and antigenic profile characteristic of RB. A cAMP receptor protein was identified which was present in RB but not EB.[99] In other prokaryotes such proteins form a complex with cAMP which binds to DNA at the promoter region of cAMP-dependent operons, stimulating or repressing transcription of certain genes.[100] Thus chlamydial development may also be under cyclic nucleotide control. Cyclical modulation of cyclic nucleotide levels in genital tract epithelia by female sex hormones during the menstrual cycle might directly influence chlamydial replication. In vitro delayed development of *C. trachomatis* in BHK cells can be switched to overt infection by treatment with cGMP.[101]

B. Genetic Regulation of Development

It is clear that oxidation and reduction of disulfide linked proteins cannot be the only regulatory mechanisms controlling chlamydial development. It can be anticipated that precise transcriptional controls must govern the temporal expression of chlamydial genes during the developmental cycle. In EB the DNA genome forms a densely packed nucleoid tightly organized around a histone-like protein. Relaxation of this structure as evident in RB may be necessary for transcription of genes important in chlamydial development.[102] In addition, there may be posttranscriptional control mechanisms, although well-documented examples of these in prokaryotes are uncommon. Unfortunately *Chlamydia* lack the defined mutants, vectors, and transformation systems necessary to investigate the genetic mechanisms (see Chapter 6). In the absence of hard information it is only possible to speculate. An example of a prokaryotic developmental system which has been under genetic investigation for many years is endosporulation in *Bacillus subtilis*.[103] In this Gram-positive organism sporulation is induced by the deprivation of critical nutrients. At least five distinct morphological stages of sporulation are known, each of which is associated with a characteristic biochemical event. A number of different classes of polypeptide can be distinguished on the basis of their time of appearance during sporulation. However, changes in the pattern of pulse-labeled proteins during development do not necessarily reflect changes in gene expression. Thus at least one spore coat polypeptide which appears late in sporulation is known to be derived by proteolytic cleavage from a larger precursor polypeptide synthesized early in development. The direct demonstration of temporally programmed gene expression requires the isolation of cloned probes for individual loci or for specific mRNA. Indeed, for this "simple" prokaryotic development cycle something in the order of 30 loci have been identified, mutation in any of which is capable of effecting sporulation. The approach, therefore, has been to prepare radioactively-labeled RNA as a hybridization probe for identifying and cloning segments of *B. subtilis* DNA that are actively transcribed during sporulation. Of particular interest has been a class of regulatory genes known as the *spoO* loci, which control the initiation stage of sporulation; mutations in any one of these eight regulatory genes block development before the earliest morphological signs of spore formation. A *B. subtilis* gene known as *spo*VG, which is activated at the onset of sporulation has been cloned into *Escherichia coli*. Transcription of this gene is controlled by the products of at least five of the eight known *spoO* loci. In sporulating cells transcription of this gene is initiated from two closely-spaced promoters, P1 and P2. Recognition and binding of *B. subtilis* RNA polymerase to these promoters is controlled by the polymerase-associated sigma factor. In *B. subtilis* at least seven bacterial or associated phage-encoded sigma factors are known which govern the selection of promoters.[104] Thus in this organism RNA polymerase is a heterogeneous enzyme, capable of existing in multiple holo-enzyme forms. This heterogeneity provides *B. subtilis* with a general mechanism for separately controlling transcription of different genes, including developmental genes. Multiple promoters are a common feature of genes whose expression is under complex regulation. Thus the *gal* operon of *E. coli* has two promoters differentially regulated by a cyclic-AMP receptor protein capable of activating the down-stream promoter and inhibiting the up-stream promoter. As the gene products within the operon may each be synthesized more efficiently from one promoter rather than the other, cyclic-AMP levels may have an enhanced influence on levels of protein synthesis. In *B. subtilis* it is likely that competition between two different sigma factors for access to the transcription initiation region of the *spo*VG gene could modulate both transcription of the gene and the translatability of its message. DNA-dependent RNA polymerase activity has been demonstrated in the T'ang strain of *C. trachomatis* and has been shown to be inhibited by rifampicin, indicating the presence of sigma factors.[105-107] From this rudimentary account of the transcription of just one gene concerned with *B. subtilis* sporulation it can be seen that the opportunity for transcriptional controls in the chlamydial developmental cycle could be very complex.

V. CONCLUSIONS

The chlamydial developmental cycle is well understood at a descriptive level, and the last decade has seen the beginnings of a molecular understanding. Perhaps most crucial has been the recent recognition of the structural and regulatory role of disulfide-linked proteins and the identification of cyclic nucleotide dependent regulatory mechanisms. The development cycle is central to an understanding of the biology of *Chlamydia* and can be anticipated to challenge developmental biologists for many years to come.

ACKNOWLEDGMENTS

I thank Mrs. Pam Rickman for typing the manuscript and Miss Sue Cox for printing the electron micrographs.

REFERENCES

1. **Ward, M. E.**, Chlamydial classification, development and structure, *Br. Med. Bull.*, 39, 109, 1983.
2. **Moulder, J. W.**, Intracellular parasitism: life in an extreme environment, *J. Infect. Dis.*, 130, 300, 1974.
3. **Weiss, E.**, Chlamydial evolution, in *Trachoma and Related Diseases*, Nichols, R. L., Eds., Excerpta Medica, Amsterdam, 1971, 3.
4. **McCaul, T. F. and Williams, J. C.**, Developmental cycle of *Coxiella burnetii* structure and morphogenesis of vegetative and sporogenic differentiations, *J. Bacteriol.*, 147, 1063, 1981.
5. **Watt, P. J. and Ward, M. E.**, Adherence of *Neisseria gonorrhoeae* and other *Neisseria* species to mammalian cells, in *Bacterial Adherence*, Beachey, E. H., Ed., Chapman and Hall, New York, 1980, Receptors and Recognition Series B, 6, 253.
6. **Becker, Y.**, The *Chlamydia:* molecular biology of procaryotic obligate parasites of eucaryotes, *Microbiol. Rev.*, 42, 274, 1978.
7. **Kuo, C-c., Wang, S-P., and Grayston, J. T.**, Effect of polycations, polyanions and neuraminidase on the infectivity of trachoma inclusion conjunctivitis and lymphogranuloma verereum organisms in HeLa cells. Sialic acid residues as possible receptors for trachoma inclusion conjunctivitis, *Infect. Immun.*, 8, 74, 1973.
8. **Grayston, J. T. and Wang, S. P.**, New knowledge of chlamydiae and the diseases they cause, *J. Infect. Dis.*, 132, 87, 1975.
9. **Salari, S. H. and Ward, M. E.**, Polypeptide composition of *Chlamydia trachomatis, J. Gen. Microbiol.*, 123, 197, 1981.
10. **Caldwell, H. D., Kromhout, J., and Schachter, J.**, Purification and partial characterization of the major outer membrane protein of *Chlamydia trachomatis, Infect. Immun.*, 31, 1161, 1981.
11. **Hatch, T. P., Vance, D. W., and Al-Hossainey, E.**, Identification of a major envelope protein in *Chlamydia* spp., *J. Bacteriol.*, 146, 426, 1981.
12. **Hackstadt, T.**, Identification and properties of chlamydial polypeptides that bind eucaryotic cell surface components, *J. Bacteriol.*, 165, 13, 1986.
13. **Wyrick, P. B., Sixbey, J. W., Davis, C. H., Rumpp, B., and Walton, L. A.**, Growth of *Chlamydia trachomatis* in human cervical epithelial cell monolayers, in *Chlamydial Infections*, Mardh, P-A., Holmes, K. K., Oriel, J. D., Piot, P., and Schachter, J., Eds., Elsevier, Amsterdam, 1982, 275.
14. **Hutchinson, G. R., Taylor-Robinson, D., and Dourmashkin, R. R.**, Growth and effect of chlamydiae in human and bovine oviduct organ cultures, *Br. J. Vener. Dis.*, 55, 194, 1979.
15. **Bose, S. K., Smith, G. B., and Paul, R. G.**, Influence of lectins, hexose and neuraminidase on the association of purified elementary bodies of *Chlamydia trachomatis* UW-31 with HeLa cells, *Infect. Immun.*, 40, 1060, 1983.
16. **Levy, J. N.**, Wheat germ agglutinin blockage of chlamydial attachment sites. Antagonism by *N*-acetyl-*D*-glucosamine, *Infect. Immun.*, 25, 946, 1979.
17. **Söderlund, G. and Kihlström, E.**, Effect of methylamine and monodansylcadaverine on the susceptibility of McCoy cells to *Chlamydia trachomatis* infection, *Infect. Immun.*, 40, 534, 1983.
18. **Söderlund, G. and Kihlström, E.**, Attachment and internalisation of a *Chlamydia trachomatis* lymphogranuloma verereum strain by McCoy cells: kinetics of infectivity and effect of lectins and carbohydrates, *Infect. Immun.*, 42, 930, 1983.
19. **Byrne, G. I.**, Requirements for ingestion of *Chlamydia psittaci* by mouse fibroblasts (L cells), *Infect. Immun,* 14, 645, 1976.
20. **Byrne, G. I.**, Kinetics of phagocytosis of *Chlamydia psittaci* by mouse fibroblasts (L cells): separation of the attachment and ingestion stages, *Infect. Immun.*, 19, 607, 1978.
21. **Hackstadt, T. and Caldwell, H. D.**, Effect of proteolytic cleavage of surface-exposed proteins on infectivity of *Chlamydia trachomatis, Infect. Immun.*, 48, 546, 1985.
22. **Nano, F. E. and Caldwell, H. D.**, Expression of the chlamydial genus-specific lipopolysaccharide epitope in *Escherichia coli, Science*, 228, 742, 1985.
23. **MacDonald, A. B.**, Antigens of *Chlamydia trachomatis, Rev. Infect. Dis.*, 7, 731, 1985.

24. **Friis, R. R.,** Interaction of L cells and *Chlamydia psittaci:* entry of the parasite and host response to its development, *J. Bacteriol.,* 110, 706, 1972.

25. **Bard, J. A. and Levitt, D.,** Binding, ingestion, and multiplication of *Chlamydia trachomatis* (L2 Serovar) in human leukocyte cell lines, *Infect. Immun.,* 50, 935, 1985.

26. **Ward, M. E. and Murray, A.,** Control mechanisms governing the infectivity of *Chlamydia trachomatis* for HeLa cells: mechanism of endocytosis, *J. Gen. Microbiol.,* 130, 1765, 1984.

27. **Byrne, G. I. and Moulder, J. W.,** Parasite-specific phagocytosis of *Chlamydia psittaci* and *Chlamydia trachomatis* by L and HeLa cells. *Infect. Immun.,* 19, 598, 1978.

28. **Aggeler, J. and Werb, Z.,** Initial events during phagocytosis by macrophages viewed from outside and inside the cell: membrane particle interactions and clathrin, *J. Cell Biol.,* 94, 613, 1982.

29. **Davies, P. J. A., Davies, D. R., Levitzki, A., Maxfield, F. R., Milhaud, P., Willingham, M. C., and Pastan, I. H.,** Transglutaminase is essential in receptor-mediated endocytosis of α_2-macroglobulin and polypeptide hormones, *Nature,* 283, 162, 1980.

30. **Goldstein, J. L., Anderson, R. G. W., and Brown, M. S.,** Coated pits, coated vesicles and receptor-mediated endocytosis, *Nature (London),* 279, 679, 1979.

31. **Pastan, I. and Willingham, M. C.,** Receptor-mediated endocytosis: coated pits, receptosomes and the Golgi, *Trends Biochem. Sci.,* 8, 251, 1983.

32. **Adelstein, R. S.,** Calmodulin and the regulation of actin-myosin interaction in smooth muscle and non muscle cells, *Cell,* 30, 349, 1982.

33. **Salisbury, J. L., Condeelis, J. S., and Satir, P.,** Role of coated vesicles, microfilaments and calmodulin in receptor mediated endocytosis by cultured B lymphoblastoid cells, *J. Cell Biol.,* 87, 132, 1980.

34. **Weltzien, H. U.,** Cytolytic and membrane perturbing properties of lysophosphatidylcholine, *Biochim. Biophys. Acta,* 559, 259, 1979.

35. **Wyrick, P. B., Register, K. B., Moorman, D. R., Hodinka, R. L., Schafer, W., and Spitznagel, J. K.,** The pathogenesis of *Chlamydia trachomatis* infection, in *Chlamydial Infections,* Oriel, D., Ridgway, G., Schachter, J., Taylor-Robinson, D., and Ward, M. E., Eds., Cambridge University Press, Cambridge, 1986, 39.

36. **Kihlström, E. and Söderlund, G.,** Trifluorperazine inhibits the infectivity of *Chlamydia trachomatis* for McCoy cells, *FEMS Microbiol. Letts.,* 20, 119, 1983.

37. **Murray, A. and Ward, M. E.,** Control mechanisms governing the infectivity of *Chlamydia trachomatis* for HeLa cells: the role of calmodulin, *J. Gen. Microbiol.,* 130, 193, 1984.

38. **Ward, M. E. and Salari, S. H.,** Control mechanisms governing the infectivity of *Chlamydia trachomatis* for HeLa cells: modulation by cyclic nucleotides, prostaglandins and calcium, *J. Gen. Microbiol.,* 128, 639, 1982.

39. **Kaul, R. and Wenman, W.,** Chlamydial elementary bodies (EB) possess phosphoprotein phosphatase activity, in *Chlamydial Infections,* Oriel, D., Ridgway, G., Schachter, J., Taylor-Robinson, D., and Ward, M. E., Eds., Cambridge University Press, Cambridge, 1986, 27.

40. **Gregory, W. W., Byrne, G. I., Gardner, M., and Moulder, J. W.,** Cytochalasin B does not inhibit ingestion of *Chlamydia psittaci* by mouse fibroblasts (L cells) and mouse peritoneal macrophages, *Infect. Immun.,* 25, 463, 1979.

41. **Hart, P., Arcy, D., and Young, M. R.,** Manipulations of the phagosomelysosome fusion response in cultured macrophages. Enhancement of fusion by chloroquine and other amines, *Exp. Cell. Res.,* 114, 486, 1978.

42. **Painter, R. G., Whisenand, J., and McIntosh, A. T.,** Effects of cytochlasin B on actin and myosin association with particle binding sites in mouse macrophages: implications with regard to the mechanism of action of cytochalasins, *J. Cell Biol.,* 91, 373, 1981.

43. **Brownridge, E. and Wyrick, P. B.,** Interaction of *Chlamydia psittaci* reticulate bodies with mouse peritoneal macrophages, *Infect. Immun.,* 24, 697, 1979.

44. **Eissenberg, L. G. and Wyrick, P. B.,** Inhibition of phagolysosome fusion is localized to *Chlamydia psittaci*-laden vacuoles, *Infect. Immun.,* 32, 889, 1981.

45. **Jones, T. C. and Hirsch, J. G.,** The interaction between *Toxoplasma gondii* and mammalian cells. II. The absence of lysosomal fusion with phagocytic vacuoles containing living parasites, *J. Exp. Med.,* 136, 1173, 1972.

46. **Kordová, N., Wilt, J. C., and Sadiq, M.,** Lysosomes in L cells infected with *Chlamydia psittaci* 6BC strain, *Can. J. Microbiol.,* 17, 955, 1971.

47. **Eissenberg, L. G., Wyrick, P. B., Davis, C. H., and Rumpp, J. W.,** *Chlamydia psittaci* elementary body envelopes: infestion and inhibition of phagolysosome fusion, *Infect. Immun.,* 40, 741, 1983.

48. **Zeichner, S. L.,** Isolation and characterisation of macrophage phagosomes containing infectious and heat inactivated *Chlamydia psittaci:* two phagosomes with different intracellular behaviours, *Infect. Immun.,* 40, 956, 1983.

49. **Prain, C. J. and Pearce, J. H.,** Endocytosis of chlamydiae by McCoy cells: measurement and effects of centrifugation, *FEMS Microbiol. Letts.,* 26, 233, 1985.

50. **Hackstadt, T. and Williams, J. C.,** Biochemical stratagem for obligate parasitism of eukaryotic cells by *Coxiella burnetii, Proc. Natl. Acad. Sci. USA,* 78, 3240, 1981.

51. **Wisseman, C. L.,** Some biological properties of rickettsiae pathogenic for man, in *Rickettsiae and Rickettsial Disease,* Burgdorfer, W. and Anacker, R. L., Eds., Academic Press, New York, 1981, 293.

52. **Gutter, B. and Becker, Y.,** Synthesis and maturation of ribosomal RNA during the developmental cycle of trachoma agent, a prokaryotic obligate parasite of eukaryotes, *J. Mol. Biol.,* 66, 239, 1972.

53. **Hatch, T. P., Al-Hossainy, E., and Silverman, J. A.,** Adenine nucleotide and lysine transport in *Chlamydia psittaci, J. Bact.,* 150, 662, 1982.

54. **Moulder, J. W.,** Looking at chlamydiae without looking at their hosts, *ASM News,* 50, 353, 1984.

55. **Hatch, T. P.,** Competition between *Chlamydia psittaci* and L cells for host isoleucine pools: a limiting factor in chlamydial multiplication, *Infect. Immun.,* 12, 211, 1975.

56. **Allan, I. and Pearce, J. H.,** Amino acid requirements of strains of *Chlamydia trachomatis* and *C. psittaci* growing in McCoy cells: relationship with clinical syndrome and host origin, *J. Gen. Microbiol.,* 129, 2001, 1983.

57. **Ladany, S. and Sarov, I.,** Recent advances in *Chlamydia trachomatis, Eur. J. Epidemiol.,* 1, 235, 1985.

58. **Garrett, A. J., Harrison, M. J., and Manire, G. P.,** A search for the bacterial mucopeptide component muramic acid in *Chlamydia, J. Gen. Microbiol.,* 80, 315, 1974.

59. **Louis, C., Nicolas, G., Eb, F., Lefebre, J-F., and Orfila, J.,** Modification of the envelope of *Chlamydia psittaci* during its developmental cycle: freeze-fracture study of complementary replicas, *J. Bacteriol.,* 141, 868, 1980.

60. **Richmond, J. R. and Stirling, P.,** Localization of chlamydial group antigen in McCoy cell monolayers infected with *Chlamydia trachomatis* or *Chlamydia psittaci, Infect. Immun.,* 34, 561, 1981.

61. **Pavia, C. S. and Schachter, J.,** Failure to detect cell-mediated cytotoxicity against *Chlamydia trachomatis*-infected cells, *Infect. Immun.,* 39, 1271, 1983.

62. **Wilde, C.,** Paper presented at meeting of American Society of Microbiology, Las Vegas, 1985.

63. **Patterson, S., Oxford, J. S., and Dourmashkin, R. R.,** Studies on the mechanism of influenza virus entry into cells, *J. Gen. Virol.,* 43, 223, 1979.

64. **Fan, V. S. C. and Jenkin, H. M.,** Glycogen metabolism in *Chlamydia*- infected HeLa cells, *J. Bacteriol.,* 104, 608, 1970.

65. **Weigent, D. A. and Jenkin, H. M.,** Contrast of glycogenesis and protein synthesis in Monkey kidney cells and HeLa cells infected with *Chlamydia trachomatis* lymphogranuloma venereum, *Infect. Immun.,* 20, 632, 1978.

66. **Rosenkranz, H. S., Gutter, B., and Becker, Y.,** Studies on the developmental cycle of *Chlamydia trachomatis:* selective inhibition by hydroxyurea, *J. Bacteriol.,* 155, 682, 1973.

67. **Swanson, J., Eschenbach, D. A., Alexander, E. R., and Holmes, K. K.,** Light and electron microscopic study of *Chlamydia trachomatis* infection of the uterine cervix, *J. Infect. Dis.,* 131, 678, 1975.

68. **Gregory, W. W., Gardner, M., Byrne, G. I., and Moulder, J. W.,** Arrays of hemispheric projections on *Chlamydia psittaci* and *Chlamydia trachomatis* observed by scanning electron microscopy, *J. Bacteriol.,* 138, 241, 1979.

69. **Nichols, N., Setzer, P. Y., Pang, F., and Dawson, C. R.,** New view of the surface projections of *Chlamydia trachomatis, J. Bacteriol.,* 164, 344, 1985.

70. **Todd, W. J., Doughri, A. M., and Storz, J.,** Ultrastructural changes in host cellular organelles in the course of the chlamydial developmental cycle, *Zbl. Bakt. Hyg., I. Abt. Orig.,* A236, 359, 1976.

71. **Todd, W. J. and Storz, J.,** Ultrastructural cytochemical evidence for the activation of lysosomes in the cytocidal effect of *Chlamydia psittaci, Infect. Immun.,* 12, 638, 1975.

72. **Todd, W. J. and Caldwell, H. D.,** The interaction of *Chlamydia trachomatis* with host cells: ultrastructural studies of the mechanism of release of a biovar II strain from HeLa 229 cells, *J. Infect. Dis.,* 151, 1037, 1985.

73. **Lee, C. K. and Moulder, J. W.,** Persistent infection of mouse fibroblasts (McCoy cells) with a trachoma strain of *Chlamydia trachomatis, Infect. Immun.,* 32, 822, 1981.

74. **Lee, C. K.,** Interaction between a trachoma strain of *Chlamydia trachomatis* and mouse fibroblasts (McCoy cells) in the absence of centrifugation, *Infect. Immun.,* 31, 584, 1981.

75. **Horoschack, K. and Moulder, J. W.,** Division of single host cells after infection with chlamydiae, *Infect. Immun.,* 19, 281, 1978.

76. **Richmond, S. J.,** Division and transmission of inclusions of *Chlamydia trachomatis* in replicating McCoy cell monolayers, *FEMS Microbiol. Letts.,* 29, 49, 1985.

77. **Moulder, J. W., Levy, N. J., and Schulman, L. P.,** Persistent infection of mouse fibroblasts (L cells) with *Chlamydia psittaci:* evidence for a cryptic chlamydial form, *Infect. Immun.,* 30, 874, 1980.

78. **Moulder, J. W., Levy, N. J., Zeichner, S. L., and Lee, C. K.,** Attachment defect in mouse fibroblasts (L cells) persistently infected with *Chlamydia psittaci, Infect. Immun.,* 34, 285, 1981.

79. **Moulder, J. W., Zeichner, S. L., and Levy, N. J.,** Association between resistance to superinfection and patterns of surface protein labeling in mouse fibroblasts (L cells) persistently infected with *Chlamydia psittaci, Infect. Immun.,* 35, 834, 1982.

80. **Shatkin, A. A., Orlova, O. E., Popov, V. L., Beskina, S. R., Pankratova, V. N., Pogacheva, I. F., Soldatova, S. I., Smirnova, N. S., and Shcherbakova, N. I.,** Persistent chlamydial infection in cell culture, *Vestnik Akad. Med. Sci. USSR,* 3, 51, 1985 (in Russian only).

81. **Kuz'menko, V. P., Kupchinskii, L. G., and Barshtein, Y. A.,** Ultrastructural features of chlamydial reproduction, *Zh. Mikrobiol.,* 40, 64, 1978 (in Russian only).

82. **Lepinay, A., Orfila, J., Anteunis, A., Boutry, J. M., Orme-Roselli, L., and Robineaux, R.,** Etude en microscopie électronique due development et de la morphologie de *Chlamydia psittaci* dans les macrophages de souris, *Ann. Inst. Pasteur,* 119, 222, 1970.

83. **Tanami, Y. and Yamada, Y.,** Miniature cell formation in *Chlamydia psittaci, J. Bacteriol.,* 114, 408, 1973.

84. **Kramer, M. J. and Gordon, F. B.,** Ultra structural analysis of the effects of penicillin and chlortetracycline on the development of a genital tract *Chlamydia, Infect. Immun.,* 3, 333, 1971.

85. **Byrne, G. I. and Krueger, D. A.,** Lymphokine-mediated inhibition of *Chlamydia* replication in mouse fibroblasts is neutralized by antigamma infection immunoglobulin, *Infect. Immun.,* 42, 1152, 1983.

86. **Shemer, Y. and Sarov, I.,** Inhibition of growth of *Chlamydia trachomatis* by human gamma interferon, *Infect. Immun.,* 48, 592, 1985.

87. **Hanna, L., Dawson, C. R., Briones, O., Thygeson, P., and Jawetz, E.,** Latency in human infections with TRIC agents, *J. Immunol.,* 101, 43, 1968.

88. **Oriel, J. D. and Ridgway, G. L.,** Studies of the epidemiology of chlamydial infection of the human genital tract, in *Chlamydial Infection,* Mardh, P-A., Holmes, K. K., Oriel, J. D., Piot, P., and Schachter, J., Eds., Elsevier, Amsterdam, 1982, 425.

89. **Tamura, A. and Manire, G. P.,** Preparation and chemical composition of developmental reticulate forms of meningopneumonitis organisms, *J. Bacteriol.,* 94, 1184, 1967.

90. **Tamura, A., Matsumoto, A., and Higashi, N.,** Purification and chemical composition of reticulate bodies of the meningopneumonitis organism, *J. Bacteriol.,* 93, 2003, 1967.

91. **Hatch, T. P., Vance, D. W., and Al-Hosainey, E.,** Identification of a major envelope protein in *Chlamydia* spp., *J. Bacteriol.,* 146, 426, 1981.

92. **Newhall, W. J. V. and Jones, R. B.,** Disulfide linked oligomers of the major outer membrane protein of chlamydiae, *J. Bacteriol.,* 154, 998, 1983.

93. **Bavoil, P., Ohlin, A., and Schachter, J.,** Role of disulphide bonding in outer membrane structure and permeability in *Chlamydia trachomatis, Infect. Immun.,* 44, 479, 1984.

94. **Chang, J. J., Leonard, K., Arad, T., Pitt, T., Zhang, Y. X., and Zhang, L. H.,** Structural studies of the outer envelope of *Chlamydia trachomatis* by electron microscopy, *J. Mol. Biol.,* 161, 579, 1982.

95. **Schiefer, H. G., Krauss, H., and Schummer, U.,** Anionic sites on *Chlamydia* membranes, *FEMS Microbiol. Letts.,* 15, 41, 1982.

96. **Caldwell, H. D. and Perry, L. J.,** Neutralization of *Chlamydia trachomatis* infectivity with antibodies to the major outer membrane protein, *Infect. Immun.,* 38, 745, 1982.

97. **Stirling, P., Allan, I., and Pearce, J. H.,** Influence of cysteine on transformation of chlamydiae from reproductive to infective body forms, *FEMS Microbiol. Letts.,* 19, 133, 1983.

98. **Hackstadt, T., Todd, W. J., and Caldwell, H. D.,** Disulfide-mediated interactions of the chlamydial major outer membrane protein: role in the differentiation of chlamydiae?, *J. Bacteriol.,* 161, 25, 1985.

99. **Kaul, R. and Wenman, W.,** Cyclic AMP inhibits developmental regulation of *Chlamydia trachomatis, J. Bacteriol.,* 168, 722, 1986.

100. **Botsford, J. L. and Dressler, M.,** The cAMP receptor protein and regulation of cyclic AMP synthesis in *Escherichia coli, Mol. Gen. Genet.,* 165, 47, 1978.

101. **MacDonald, A. B., Than, K. A., and Stuart, E. S.,** Persistent infection of BHK cells with *Chlamydia trachomatis* can be switched to an overt infection by treatment with cyclic nucleotides, in *Chlamydial Infections,* Oriel, J. D. et al., Eds., Cambridge University Press, Cambridge, 1986, 67.

102. **Wagar, E. A. and Stevens, R. S.,** Evidence for a Histone-like nucleoprotein in Chlamydia in *Chlamydial Infections,* Oriel, J. D. et al., Eds., Cambridge University Press, Cambridge, 1986, 51.

103. **Losick, R.,** Sporulation genes and their regulation, in *The Molecular Biology of the Bacilli,* Dubnau, D. A., Ed., Academic Press, New York, 1982, 179.

104. **Johnson, W. C., Moran, C. P., Jr. and Losick, R.,** Two RNA polymerase sigma factors from *Bacillus subtilis* discriminate between overlapping promoters for a developmentally regulated gene, *Nature (London),* 302, 800, 1983.

105. **Becker, Y., Asher, Y., Himmel, N., and Zakay-Rones, Z.,** Rifampicin inhibition of trachoma agent *in vivo, Nature (London),* 224, 33, 1969.

106. **Keshishyan, H., Hanna, L., and Jawetz, E.,** Emergence of rifampicin-resistance in *Chlamydia gtrachomatis, Nature (London),* 244, 173, 1973.

107. **Sarov, I. and Becker, Y.,** Deoxyribonucleic acid-dependence ribonucleic acid polymerase activity in purified trachoma elementary bodies: effect of sodium chloride on ribonucleic acid transcription, *J. Bacteriol.,* 107, 593, 1971.

108. **Ward, M. E.,** unpublished data.

109. **Benes, S.,** Intercellular propagation of a non-lymphogranuloma venereum strain of *Chlamydia trachomatis* in McCoy cells, in *Chlamydial Infections,* Oriel, J. D. et al., Eds., Cambridge University Press, Cambridge, 1986, 59.

Chapter 5

METABOLISM OF CHLAMYDIA

Thomas P. Hatch

TABLE OF CONTENTS

I. INTRODUCTION

A. Procaryotic Nature of Chlamydiae

Chlamydia trachomatis and *Chlamydia psittaci* are obligate intracellular parasitic bacteria. In nature, they multiply within a membrane-bound vacuole (inclusion) in the cytoplasm of eucaryotic host cells; in the laboratory, they may be propagated in chicken embryos and a variety of continuous cell lines, but not on artificial medium. The metabolic capabilities of chlamydiae are limited; most noticeably, they appear to lack the ability to generate net ATP and are speculated to be "energy parasites." The exact nature of the host-supplied factors which permit chlamydiae to grow is not known, and probably will remain unknown until chlamydiae are propagated on axenic media. Host-cell dependence does not extend to macromolecular synthesis; chlamydiae possess all of the cell machinery for procaryotic DNA, RNA, and protein synthesis, although they probably are dependent on their hosts for many precursors such as nucleotides and amino acids.

B. Problems in Identifying Metabolic Activities of Chlamydiae

Definitive studies on the metabolism of chlamydiae are extraordinarily difficult. On the one hand, it can always be argued that detection of an activity in chlamydial preparations is the result of host contamination, unless the activity is absent in host cells or can be demonstrated to be different from the corresponding activity in host cells. On the other hand, the failure to detect an activity may be attributed to the use of improper assay conditions or the incorrect life cycle form. Early studies are particularly subject to these criticisms. Purification schemes were restricted to differential centrifugation and digestion with proteases to reduce host-cell material, and powerful analytical techniques, such as the *in situ* assays of enzymes separated on nondenaturing gels by electrophoresis, were not available. Moreover, most of these studies were carried out with the metabolically inactive elementary body (EB) form rather than with the vegetative reticulate body (RB) form of chlamydiae.

II. IN VIVO METABOLISM

A. Energy Metabolism

Moulder[1] speculated that chlamydiae are energy parasites incapable of generating net ATP. All of the evidence which supports this hypothesis is indirect since chlamydiae have not been propagated outside host cells. Most of the early supporting evidence was of a negative nature: failure to detect enzymatic reactions or an electron transport chain in extracts or whole chlamydiae which could lead to the net generation of ATP.

Ormsbee and Weiss[2] and Weiss et al.[3] reported that suspensions of *C. psittaci* (meningopneumonitis) and *C. trachomatis* (TE55, an LGV strain) could convert glucose to CO_2 and pyruvate; however, these studies were all conducted with partially purified EBs and with ATP in the incubation mixture. It was later demonstrated that ATP was required for CO_2 production from glucose but not from glucose-6-phosphate.[4] Moulder et al.[5] detected three enzymes of the pentose phosphate pathway, glucose-6-phosphate dehydrogenase, 6-phosphogluconate dehydrogenase, and phosphoglucose isomerase, in extracts of *C. psittaci* (meningopneumonitis), but failed to detect hexokinase, 1,6-biphosphofructokinase, and pyruvate kinase activities. Using a more sensitive radioisotopic assay, Vender and Moulder[6] detected hexokinase activity in EBs; however, this activity was associated with the EB membrane and could be eliminated by treatment of intact EBs with Pronase or trypsin. The authors suggested that the enzyme was of host origin. A complete understanding of the catabolic activities of chlamydiae is still lacking; however, chlamydiae appear to be capable of converting glucose-6-phosphate (but not glucose) to pyruvate and pentose by the Embden-Meyerhoff (or Entner-Doudoroff) and pentose phosphate pathways, respectively.[3-6] Pyruvate may be oxidatively decarboxylated to acetate, but does not enter the citric acid cycle.[7] Glutamate is converted to succinate by both species of chlamydiae, but is not further metabolized.[7] NADH-cytochrome C reductase activity has been detected in *C. psittaci*.[8,9] The physiological significance of this activity is uncertain in that flavoproteins and cytochromes have not been detected in chlamydiae.[10] In summary, chlamydiae possess fragments of the glycolytic, pentose phosphate, and citric acid pathways, but appear to lack enzymes required for the net generation of ATP.

The unambiguous demonstration that chlamydiae cannot generate useful levels of ATP through catabolic reactions is difficult because few procaryotic- or eucaryotic-specific inhibitors of energy metabolism are available. Moulder[11] and Gill and Stewart[12] presented evidence that the enhanced rate of aerobic glycolysis when L cells are infected with *C. psittaci* represents a response to infection rather than independent parasite catabolic activity. Gill and Stewart[13] also found that antimycin, an inhibitor of electron transport in eucaryotes, reduced the yield of *C. psittaci* grown in L cells. Similarly, Becker and Asher[14] found that EBs of *C. trachomatis* failed to develop in host cells treated with a concentration of ethidium bromide which is believed to preferentially inhibit transcription of mitochondrial DNA, and hence possibly alter host nucleotide levels.

The requirement of an exogenous source of ATP and other nucleotides for macromolecular synthesis and the discovery of a specific ATP transport system in chlamydiae represent more positive supporting evidence for the energy parasite concept.[15-21] These host-free studies will be discussed in more detail below. It must be emphasized, however, that another obligately intracellular parasite, *Rickettsia prowazekii*, can also transport ATP, yet can generate ATP from glutamate by oxidative phosphorylation.[22] It is possible that intracellular *R. prowazekii* organisms rely on their host for ATP while extracellular organisms catabolize glutamate in order to generate an energized membrane which is required for entry of the parasite into host cells.[23] A similar requirement for extracellular generation of ATP would be unnecessary for chlamydiae which gain entry to host cells by an endocytic mechanism that is not dependent on parasite metabolic activity.[24]

Hatch[25] demonstrated that the 6BC strain of *C. psittaci* multiplying within L cells incorporates exogenous uridine and adenine labels into RNA at rates consistent with the parasite, drawing exclusively on the host's ribonucleoside triphosphate pools. Similar in vivo evidence that chlamydiae may draw on host nucleotides was provided by Ceballos and Hatch[26] who found that chlamydiae could incorporate exogenous guanine into their nucleic acids if grown in hypoxanthine guanine phosphotransferase-containing HeLa cells, but not if grown in transferase-deficient host cells. It was concluded that intracellular chlamydiae were capable of using guanine only if the guanine were first elevated to the nucleotide level by the host cell.

B. Synthesis of Glycogen and Lipids

Although chlamydiae appear to rely on their hosts for a wide variety of nutrients, including many precursors of macromolecules which other bacteria synthesize for themselves, chlamydiae are likely as capable as other microorganisms of synthesizing complex lipids, polysaccharides, proteins, and nucleic acids. The lipid and phospholipid content of chlamydiae is discussed in Chapter 3: Macromolecular and Antigenic Composition of Chlamydiae. The pathways, including the nature of the precursors obtained from host cells, by which chlamydiae synthesize lipids are not known. Experiments with isolated chlamydiae suggest that relatively simple precursors, pyruvate, glutamate, asparate, and isoleucine, may be used for lipid biosynthesis.[16,27] On the other hand, Fan and Jenkin[28] showed that an LGV strain of *C. trachomatis* can divert phosphotitic acid synthesis in the host cell for the formation of phosphatidyl ethanolamine. Although chlamydiae cannot use glucose, they appear capable of transporting glucose-6-phosphate and may use this host-derived substrate as a starting point for synthesis of carbohydrates and lipids.[16,27] Because chlamydiae possess unusual transport mechanisms, it is possible, but not yet demonstrated, that they glean nucleoside diphosphate hexoses, such as UDP- and ADP-glucose, directly from the host cell. Jenkin and Fan[29] showed that deposition of glycogen in *C. trachomatis* inclusion vacuoles is mediated by a chlamydia-specific glycogen synthetase which uses ADP-glucose as substrate.

C. Amino Acid Metabolism

Chlamydiae may draw on host soluble pools for most amino acids. Many investigators have examined the effects of amino acid starvation on the growth of chlamydiae in tissue culture cells; their results indicate that specific amino acid requirements are highly strain-dependent (for example, see Bader and Morgan,[30] Ossowski et al.,[31] Karayiannis and Hobson,[32] and Allan and Pearce[33]). Omission of cysteine was found to have a curious effect on the development of a serotype E strain of *C. trachomatis*: RB multiplication was not noticeably affected, but reorganization of RBs to EBs was severely retarded.[34,35] EBs possess several cysteine-rich proteins in their outer membranes which are not present in rapidly dividing RBs.[36] Thus, the cysteine requirement of RBs may be met by the turnover of host proteins under starvation conditions, while a greater demand for cysteine for outer membrane synthesis during RB to EB reorganization may require additional cysteine in the growth medium.[35]

Consistent with the earlier findings of Bader and Morgan,[30] Hatch[37] found that the 6BC strain of *C. psittaci* remained dormant in L cells incubating in isoleucine-deficient medium. It was found that the concentration of isoleucine required to activate minimal growth of *C. psittaci* also minimally stimulated uninfected cells to divide. The addition of cycloheximide to infected cells incubating in deficient medium also activated the latent chlamydial infection because it stimulated the incorporation of host protein-derived isoleucine into chlamydial protein. It was suggested that the chlamydial parasite and the L-cell host compete for the isoleucine in the soluble pool of the host cell and that the parasite is capable of sequestering isoleucine for its own biosynthetic needs only when the concentration of isoleucine in the host pool rises above the level required to maintain the host in the stationary state. Cycloheximide is commonly used in the culturing of *C. trachomatis* from clinical specimens.[38] Improved growth in the presence of cycloheximide almost certainly is related to the sparing effect that this inhibitor has on the availability of amino acids to chlamydiae. Whether or not host-parasite competition for nutrients plays an important role in natural latent infections is not known. However, the recent observations of Byrne et al.[39] suggest that one mechanism of host defense against chlamydiae may be to induce tryptophan deficiency in infected cells. These workers found that gamma interferon, produced in response to chlamydial infection, induces tryptophan dioxygenase activity in human fibroblasts. The failure of *C. psittaci* to grow in gamma interferon-treated cells was shown to be related to this activity since the addition of high concentrations of tryptophan to the incubation medium reversed the static effect of interferon.

The ability of chlamydiae to grow in cells deprived of certain essential amino acids suggests that chlamydiae may be capable of synthesizing at least some amino acids; however, the role of turnover of host proteins in supplying amino acids cannot be ruled out in these types of studies. The capability of chlamydiae to synthesize two amino acids has been directly demonstrated, however. Moulder et al.[40] detected diaminopimelate decarboxylase in the meningopneumonitis strain of *C. psittaci*, an observation which suggests the occurrence of a typical bacterial pathway for lysine biosynthesis in this microorganism. Treuhaft and Moulder[41] found that L cells infected with the mouse pneumonitis strain of *C. trachomatis* or the meningopneumonitis strain of *C. psittaci* incorporated [14C]glutamate and [14C]ornithine into the arginine fraction of infected L-cell protein. Similar incorporation was not detected in uninfected cells, and it was concluded that the two chlamydial strains tested were capable of arginine biosynthesis. The caveat that individual chlamydial strains may possess different biosynthetic abilities is manifested by the observations of Ossowski et al.[31] who noted that *C. psittaci* (6BC) was capable of growth in arginine-free medium while the TE55 strain of *C. trachomatis* required arginine for growth.

D. Cofactor and Vitamin Metabolism

Just as chlamydiae may rely on their hosts for most of their amino acids, they probably also sequester vitamins and cofactors from their hosts' soluble pools. Weiss[4,7] demonstrated that the metabolism of pyruvate and glucose-6-phosphate by whole *C. psittaci* (meningo-pneumonitis) and *C. trachomatis* (TE55) organisms required, or was stimulated by, NAD and NADP suggesting that chlamydiae can transport host-derived pyrimidine nucleotides. In addition, Weiss[7] found that alpha lipoic acid stimulated the decarboxylation of pyruvate by intact chlamydiae. A number of chlamydial strains, including all known strains of *C. trachomatis* and the 6BC strain of *C. psittaci*, are susceptible to sulfonamides and, consequently, must be capable of synthesizing folic acid. The production of folic acid by the mouse pneumonitis strain of *C. trachomatis* was directly demonstrated by Colon.[42,43] The synthesis or transport of other cofactors has not been demonstrated.

E. Synthesis of RNA, DNA, and Protein

The first indication that chlamydiae contain both RNA and DNA was provided by the cytochemical studies of Starr et al.[44] and Collier.[45] Purification of EBs of *C. psittaci* (men-ingopneumonitis agent) on sucrose gradients permitted Tamura and Higashi[46] to chemically detect chlamydia-specific nucleic acids. Subsequent purification of RBs revealed that the ratio of RNA to DNA in RBs was about 5:1 while that in EBs was 1:1.[47,48] This difference in ratios reflects the relative deficiency of ribosomes in EBs. Sucrose velocity gradient centrifugation analysis revealed that chlamydiae possess 21, 16, 4, and 5S RNA species.[49,50] More accurate size analysis by polyacrylamide electrophoresis indicates that the large rRNA species is 23S, rather than 21S, as is the case in other procaryotes.[51,52] The genes encoding 23S and 16S rRNA in both *C. trachomatis* and *C. psittaci* have been cloned, and the nucleotide sequence of 16S RNA of both species has been determined.[53] Although the sequences are fundamentally similar to other Gram-negative bacteria, computer analysis revealed that chlamydiae represent a distinct phylogenetic group which is only distantly related to other eubacteria. In contrast to DNA hybridization studies which demonstrated less than 10% homology between *C. trachomatis* and *C. psittaci*,[54] sequence analysis of 16S RNA revealed greater than 95% homology and strongly confirms the inclusion of the two species within a single genus. Both the 23S and 16S rRNA genes of *C. psittaci* and *C. trachomatis* reside within a 10 kb *Bam*H1 restriction fragment.[53] It is likely, therefore, that they comprise a single operon which, as in other bacteria, is processed to individual stable RNA species. The pulse-chase experiments of Gutter and Becker[51] and Gutter et al.[52] support the supposition that chlamydial ribosomal RNAs are the products of posttranscriptional cleavage events.

Various methods have been used to determine the genome size of chlamydiae. The most accurate determinations revealed a genome size of approximately 660 mdaltons and G + C content of between 41 and 44% for both species.[55,56] A 7 kb cryptic plasmid has been detected in most strains investigated thus far.[57-62]

The lack of dependence of chlamydiae on their hosts' macromolecular synthetic apparatus was first demonstrated by the effects of metabolic inhibitors on chlamydial growth. Cycloheximide and emetine specifically inhibit eucaryotic protein synthesis, the cessation of which ultimately also results in inhibition of RNA and DNA synthesis.[51,52,63] In the presence of moderate concentrations of these compounds, chlamydiae continue to multiply and synthesize protein, RNA, and DNA. The failure of chlamydiae to grow in the presence of high concentrations of cycloheximide and emetine probably is related to the toxic effects of these substances on the host cell on which chlamydiae are dependent for a continuous supply of metabolites such as nucleoside triphosphates. In contrast, procaryotic-specific inhibitors, including chloramphenicol, tetracycline, and rifampicin, quickly inhibit chlamydial growth and synthesis of macromolecules.[51,52,63,64] The lack of a need for continuous host cell transcription was dramatically demonstrated by the ability of chlamydiae to grow in enucleated host cells.[37,65]

The relationship between host and parasite macromolecular synthesis is most clearly revealed by studies conducted with chlamydiae which have been isolated from host material. Tamura[15] demonstrated that RBs of *C. psittaci* incorporated radiolabeled GTP into RNA in the presence of ATP, UTP, and CTP. Curiously, GTP was incorporated into poly G in the absence of the other nucleotides. Sarov and Becker[17] reported the synthesis of ribosomal and 4S RNA by mercaptoethanol-treated EBs of *C. trachomatis*, and my own laboratory has detected protein, DNA, and RNA synthesis in host-free preparations of both species of chlamydiae.[10-21] These studies will be discussed in detail below. An important feature of all of these investigations is that chlamydiae are capable of synthesizing macromolecules in a host-independent system, but only when supplied appropriate precursors including ATP.

Early studies indicated that chlamydiae inhibited host protein, RNA, and DNA synthesis within 10 to 20 hr postinfection.[46,66-69] It now appears that cessation of host macromolecular synthesis is not an obligatory step in the chlamydial life cycle. Rather, cessation of host synthesis is probably nonspecific and related to the general cytopathic effects of the parasite competing for the host's resources. At low multiplicities of infection, host cells continue their normal metabolic activities until the burden of supporting an intracellular parasite with limited metabolic capabilities becomes intolerable. Three studies show that cells infected with a single infectious particle are actually capable of dividing.[70-72] At high multiplicities of infection, cessation of host macromolecular synthesis is more rapid, perhaps reflecting a more immediate crisis resulting from the host and parasite competing for limited resources. At very high multiplicities (10 for *C. trachomatis*; 100 for *C. psittaci*), infected cells die within 10 hr as a result of a cytotoxin associated with the cell walls of EBs.[73-78]

The length of the chlamydial developmental cycle varies, depending upon the cultural conditions and strain in question. As a general rule, *C. psittaci* strains and LGV strains of *C. trachomatis* possess a 48 hr cycle, while trachoma biotypes of *C. trachomatis* complete a cycle in about 72 hr. EBs reorganize to RBs during the first 4 to 10 hr of the cycle. Nothing is known about the intracellular environmental signals which trigger this reorganiztion. It is clearly a period of active protein and RNA synthesis because the peptide and RNA composition of RBs varies considerably from that of EBs; however, the exact synthetic events which take place early in the cycle have not been analyzed because of the difficulty of detecting the activities of a few chlamydial organisms within the background of an infected host cell. Once EBs have reorganized to RBs, RBs divide logarithmically with a generation time as short as 2 hr.[68] Not surprisingly, peak levels of chlamydial DNA, RNA, and protein synthesis have been detected during the logarithmic phase which extends to 20 to 30 hr postinfection.[49,52,68] EBs may be detected within host cells as early as 18 hr postinfection, although substantial numbers of infectious EBs usually are not detected before 24 to 30 hr.[79-81] Host cells infected with moderate doses (multiplicities of infection of 1 to 5) of *C. psittaci* and LGV strains lyse between 30 and 48 hr postinfection. Cells infected with trachoma biovars tend to resist lysis and appear to persist in a nonviable but intact state for several days after the chlamydial cycle is complete. The temporal sequence of synthesis of only a few chlamydial proteins has been documented. The major outer membrane protein, while synthesized throughout the cycle, is most actively produced during the logarithmic phase.[82] Three cysteine-rich outer membrane proteins have been shown to be synthesized during RB to EB reorganization between 20 and 48 hr postinfection by *C. psittaci*.[82] These proteins are believed to contribute to the osmotic stability of EBs by forming an extensive network of interpeptide disulfide-bond cross-links with themselves and other outer membrane proteins.[36,83,84] Stokes[85] detected a chlamydial protease which may play a role in host-cell lysis late in the developmental cycle of *C. psittaci*. Glycogen is deposited within the inclusion vacuole of *C. trachomatis* strains late in the cycle, some time after 24 hr postinfection.[86-88] Peak levels of rRNA synthesis have been detected during the logarithmic phase;[49,51,52,66,67] however, it is clear that ribosomal proteins and RNAs must also be synthesized during EB to RB reorganization.

III. HOST-FREE ACTIVITIES OF CHLAMYDIAE

A. Nucleotide Transport

Various studies indirectly indicate that chlamydiae possess a mechanism for the transport of ATP. Weiss and co-workers[3-4,16,27] found that an exogenous source of ATP either stimulated or was required by isolated chlamydiae for carbohydrate catabolism and the incorporation of low molecular weight substrates into lipids, and Tamura[15] and Sarov and Becker[17] noted that host-free synthesis of RNA required the addition of ATP. Hatch et al.[18] directly demonstrated that host-free RBs of *C. psittaci* (6BC) could transport ATP and ADP by an ATP-ADP exchange mechanism similar to that found in mitochondria[89] and *R. prowazekii*.[22] The Km of transport for both ATP and ADP was approximately 5 μM, and the calculated V_{max} for both was about 1 nmol of nucleotide transported per min per mg of protein. ADP was a competitive inhibitor of ATP transport, but other nucleotides were without effect. No transport activity could be detected in host-free EBs. When RBs were incubated in phosphate buffer, transported ATP was rapidly hydrolyzed to ADP by a magnesium-dependent chlamydial ATPase. Breakdown of ATP was associated with the generation of an energized chlamydial membrane which was demonstrated to be required for the concentration of an amino acid, lysine, against the electrochemical gradient. It was suggested that chlamydiae behave in the reverse manner of mitochondria. For example, mitochondria import ADP from the host cytoplasm and export ATP, whereas chlamydiae import ATP and export ADP; and mitochondria use a proton motive force (obtained through catabolic reactions and electron transport) to drive the net generation of ATP from ADP, whereas chlamydiae hydrolyze ATP to ADP to generate a proton motive force which can be used to garner other host-supplied compounds. An ATP-ADP translocation mechanism with identical kinetic characteritics has been found in an LGV strain of *C. trachomatis*.[97]

Host-free RBs of *C. psittaci* (6BC) and *C. trachomatis* (L2) also possess a GTP-GDP translocation system which is independent of the ATP-ADP translocase.[98] Interestingly, the GTP-GDP translocase is noncompetitively inhibited by ATP. The inhibition is partial, and GTP is reduced by only 50% even at high concentrations of ATP. It is likely that chlamydiae possess a complicated network of nucleotide transport mechanisms which permit the parasite to draw on alternative host-supplied energy-rich compounds. For example, when host ATP pools are high, the parasite may use GTP mainly as a substrate for RNA synthesis and thus may not require maximal transport of GTP. When host ATP pools are low, chlamydiae may use GTP to regenerate ATP from ADP via a nucleotide diphosphokinase reaction as well as for RNA synthesis. Under this condition, GTP transport would be maximal.

B. RNA Synthesis

Tamura[15] was the first to demonstrate that RBs of chlamydiae (meningopneumonitis strain of *C. psittaci*) were capable of synthesizing macromolecules outside of host cells. Radiolabeled GTP and UTP were found to be incorporated into acid-insoluble material by two different mechanisms. GTP and UTP were incorporated into homopolymers (poly G and poly U) in the absence of orthophosphate and other nucleoside triphosphates. The reaction was not inhibited by actinomycin D and presumably was mediated by polynucleotide phosphorylase. The significance of this reaction is unknown, but we have made similar observations with RBs of *C. psittaci* (6BC). When all four ribonucleoside triphosphates were present and the incubation buffer contained orthophosphate, radiolabled GTP and UTP were incorporated into RNA by an actinomycin D-sensitive mechanism mediated by a DNA-dependent chlamydial RNA polymerase. RNA synthesis was first detected in host-free *C. trachomatis* organisms by Sarov and Becker.[17] These investigators found that radiolabled UTP was incorporated into stable RNA species by EBs treated with mercaptoethanol. Incorporation was dependent on the presence of all four ribonucleoside triphosphates and was sensitive to actinomycin D and rifampin. The function of mercaptoethanol may have been to increase the permeability of EBs. Bavoil et al.[90] have presented evidence that complexes of the major outer membrane protein of chlamydiae may have porin activity which is activated by the reduction of interpeptide disulfide bonds.

Workers in my laboratory have confirmed the results of Tamura,[15] detecting RNA synthesis in host-free RBs of *C. psittaci* (6BC) and *C. trachomatis* (L2), but have not been able to detect RNA synthesis in EBs of either species. Our failure to detect any host-free activities (ATP transport, and RNA, DNA, and protein synthesis) in EBs (in the presence or absence of reducing agents) may be related to our method of purifying EBs. We treat EBs with 0.5% Nonidet P-40, a nonionic detergent, to lyse any contaminating RBs and osmotically sensitive intermediate forms which may contribute to host-free activities. The components of our RB host-free RNA synthesis system include RBs, all four ribonucleoside trisphosphates, an ATP regenerating system, magnesium and potassium ions, orthophosphate (to inhibit polynucleotide phosphorylase), and isosmotic buffer at pH 7. RNA synthesized in the host-free system hybridizes to genomic DNA, cryptic plasmid DNA, and cloned major outer membrane protein DNA. Synthesis is inhibited by rifampicin and actinomycin D.

C. DNA Synthesis

Hatch et al.[21] reported that host-free RBs of *C. psittaci* (6BC) and *C. trachomatis* (L2) incorporate radiolabeled deoxyribonucleoside triphosphates (dNTPs) into DNA over a 2 hr time period. Southern blot analysis revealed that DNA synthesized in the host-free system hybridized to plasmid DNA, *Eco*RI restriction fragments of chlamydial genomic DNA, but not to host cell (HeLa) DNA. In contrast to host-free RNA and protein synthesis, host-free DNA synthesis was inhibited by ATP and this inhibition was found to be reversed by proton ionophores. The mechanism by which ATP inhibits DNA synthesis has not been determined, but it may be at the level of entry of dNTPs into RBs. All four dNTPs, including dTTP, were required for optimal synthesis. Several investigators have reported that thymidine either is not incorporated or is poorly incorporated into the DNA of chlamydiae, although thymidine is readily incorporated into DNA of infected host cells.[91-95] It has been shown that chlamydiae lack thymidine kinase activity, and it has been suggested that chlamydiae synthesize dTTP from dUMP via thymidylate synthetase.[92,95] These observations suggest that the host dTTP pool is not readily accessible to chlamydiae and are inconsistent with our finding that host-free chlamydiae can incorporate exogenous dTTP into DNA. It is possible that intracellular chlamydiae synthesize their own substrates for DNA synthesis, and that use of exogenous dNTPs by host-free RBs is accomplished by nonspecific diffusion of the nucleotides across the chlamydial cytoplasmic membrane. The supposition that RBs transport dNTPs poorly is supported by the observation that DNA synthesis by lysates of RBs is approximately 5-fold higher than by intact RBs.[21] Tanaka[96] previously observed DNA polymerase in cell-free extracts of EBs of *C. psittaci* (meningopneumonitis).

D. Protein Synthesis

Host-free RBs of *C. psittaci* and *C. trachomatis* have been shown to incorporate amino acids into protein over a 2 to 4 hr period.[19,20] Incorporation was totally dependent on an exogenous source of ATP and an ATP-regenerating system. Although the inclusion of UTP, CTP, and GTP in the host-free incubation mixture did not significantly stimulate or prolong incorporation, host-free protein synthesis was inhibited by rifampicin. It is possible that endogenous levels of ribonucleotides other than ATP are sufficient to generate the mRNA needed in the host-free system. The sodium dodecyl sulfate polyacrylamide gel profile of proteins synthesized by RBs isolated at 21 hr postinfection was shown to be similar to the profile of proteins synthesized by intracellular RBs at 21 hr postinfection.[20] The reason for the cessation of host-free RB protein synthesis after 2 to 4 hr is not known, but is an inherent function of the host-free RBs; addition of fresh RBs to a 2 hr incubation mixture restored synthesis. It is possible that the RBs dedifferentiate to an EB-like state, exhaust some endogenous factor required for synthesis, or simply are damaged after a few hours of incubation at 37°C in a host-free environment.

IV. PROSPECTS OF ACHIEVING HOST-FREE GROWTH OF CHLAMYDIAE

Development of a manageable method of growing chlamydiae on an artificial culture medium would serve several useful functions. It might permit definitive diagnosis of clinical infections, although culture techniques are unlikely to supercede the increasingly accurate rapid diagnostic methods. Large-scale axenic growth of chlamydiae certainly would facilitate basic studies. Nutritional requirements could be defined and the problem of distinguishing between chlamydial components and host contamination would be obviated. The generation of mutants would be facilitated if chlamydiae could be propagated free of host cells. An operation as rudimentary as biological cloning is a Herculean task at present and almost certainly would be simplified in a cell-free system. Mutations might be obtained which block intracellular growth but do not interfere with host-free multiplication. In this way, chlamydial virulence factors could be defined.

The prospects of achieving long-term synthesis and growth outside of a host cell are dim. The demonstration of only short-term macromolecular synthesis in host-free chlamydiae has been particularly disappointing. Chlamydiae may be the ultimate auxotrophs, and it is impossible to predict the entire constellation of nutrients required by chlamydiae. Requirements may include unusual lipid and carbohydrate precursors as well as nucleotides, vitamins, and amino acids. Another problem is controlling the developmental cycle. It may be necessary to establish at least three separate conditions for host-free metabolism of chlamydiae: one in which EBs reorganize to RBs, a second in which RBs multiply, and a third in which RBs reorganize to EBs. Finally, chlamydiae undergo their developmental cycle within a membrane-bound vacuole. Nothing is known about the environment of this vacuole; consequently, there is no rational approach to reproducing this environment in vivo. It is even conceivable that the vacuole itself is required for growth.

REFERENCES

1. **Moulder, J. W.**, *The Biochemistry of Intracellular Parasitism*, The University of Chicago Press, Chicago, Ill., 1962, 122.
2. **Ormsbee, R. A. and Weiss, E.**, Trachoma agent: glucose utilization by purified suspensions, *Science*, 2, 1077, 1963.
3. **Weiss, E., Myers, W. F., Dressler, H. R., and Chun-Hoon, H.**, Glucose metabolism by agents of the psittacosis-trachoma group, *Virology*, 22, 551, 1964.
4. **Weiss, E.**, Adenosine triphosphate and other requirements for the utilization of glucose by agents of the psittacosis-trachoma group, *J. Bacteriol.*, 90, 243, 1965.
5. **Moulder, J. W., Grisso, D. L., and Brubaker, R. R.**, Enzymes of glucose catabolism in a member of the psittacosis group, *J. Bacteriol.*, 89, 810, 1965.

6. **Vender, J. and Moulder, J. W.,** Initial step in catabolism of glucose by the meningopneumonitis agent, *J. Bacteriol.,* 94, 1967.

7. **Weiss, E.,** Transaminase activity and other enzymatic reactions involving pyruvate and glutamate in *Chlamydia* (psittacosis-trachoma group), *J. Bacteriol.,* 93, 177, 1967.

8. **Allen, E. G. and Bovarnick, M. R.,** Enzymatic activity associated with meningopneumonitis, *Ann. N.Y. Acad. Sci.,* 98, 229, 1962.

9. **Tamura, A. and Manire, G. P.,** Cytochrome C reductase activity of meningopneumonitis organisms at different stages of development, *Proc. Soc. Exp. Biol. Med.,* 129, 390, 1968.

10. **Weiss, E. and Kiesow, L. A.,** Incomplete citric acid cycle in agents of the psittacosis-trachoma group, *Bacteriol. Proc.,* 85, 1966.

11. **Moulder, J. W.,** Glucose metabolism in L cells before and after infection with *Chlamydia psittaci., J. Bacteriol.,* 104, 1189, 1970.

12. **Gill, S. D. and Stewart, R. B.,** Glucose requirements of L cells infected with *Chlamydia psittaci, Can. J. Microbiol.,* 16, 997, 1970.

13. **Gill, S. D. and Stewart, R. B.,** Effect of inhibitors on the production of *Chlamydia psittaci* by infected L cells, *Can. J. Microbiol.,* 16, 1079, 1970.

14. **Becker, Y. and Asher, Y.,** Obligate parasitism of trachoma agent: lack of trachoma development in ethidium bromide-treated cells, *Antimicrob. Agents Chemother.,* 1, 171, 1972.

15. **Tamura, A.,** Studies on RNA synthetic enzymes associated with meningopneumonitis agent, *Annu. Rep. Inst. Virus Res., Kyoto Univ.,* 10, 26, 1967.

16. **Weiss, E., and Wilson, N. N.,** Role of exogenous adenosine triphosphate in catabolic and synthetic activities of *Chlamydia psittaci, J. Bacteriol.,* 97, 719, 1969.

17. **Sarov, I. and Becker, Y.,** Deoxyribonucleic acid-dependent ribonucleic acid polymerase activity in purified trachoma elementary bodies: effect of sodium chloride on ribonucleic acid transcription, *J. Bacteriol.,* 107, 593, 1971.

18. **Hatch, T. P., Al-Hossainy, E., and Silverman, J. A.,** Adenine nucleotide and lysine transport in *Chlamydia psittaci, J. Bacteriol.,* 150, 662, 1982.

19. **Hatch, T. P.,** Host-free activities of chlamydia, in *Chlamydia Infections,* Mardh, P.-A., Holmes, K. K., Oriel, J. D., Piot, P., and Schachter, J., Eds., Elsevier, New York, 1982, 25.

20. **Hatch, T. P., Miceli, M. M., and Silverman, J. A.,** Synthesis of protein in host-free reticulate bodies of *Chlamydia psittaci* and *Chlamydia trachomatis, J. Bacteriol.,* 162, 938, 1985.

21. **Hatch, T. P., Plaunt, M., and Sublett, J.,** DNA synthesis by host-free chlamydia, in *Chlamydial Infections,* Oriel, J. D., Ridgway, G., Schachter, J., Taylor-Robinson, D., and Ward, M., Eds., Cambridge University Press, London, 1986, 47.

22. **Winkler, H. H.,** Rickettsial permeability. An ADP-ATP transport system, *J. Biol. Chem.,* 251, 389, 1976.

23. **Winkler, H. H.,** Inhibitory and restorative effects of adenine nucleotides on rickettsial adsorption and hemolysis, *Infect. Immun.,* 9, 119, 1974.

24. **Byrne, G.,** Requirements for ingestion of *Chlamydia psittaci* by mouse fibroblasts (L cells), *Infect. Immun.,* 14, 645, 1976.

25. **Hatch, T. P.,** Utilization of L-cell nucleoside triphosphates by *Chlamydia psittaci* for ribonucleic acid synthesis, *J. Bacteriol.,* 122, 393, 1975.

26. **Ceballos, M. M. and Hatch, T. P.,** Use of Hela cell guanine nucleotides by *Chlamydia psittaci, Infect. Immun.,* 25, 98, 1979.

27. **Gaugler, R. W., Neptune, E. M., Adams, G. M., Sallee, T. L., Weiss, E., and Wilson, N. N.,** Lipid synthesis by isolated *Chlamydia psittaci, J. Bacteriol.,* 100, 823, 1969.

28. **Fan, V. S. C. and Jenkin, H. M.,** Biosynthesis of phospholipids and neutral lipids of monkey kidney cells (LLC-MK-2) infected with *Chlamydia trachomatis* strain lymphogranuloma venereum, *Proc. Soc. Exp. Biol. Med.,* 148, 351, 1975.

29. **Jenkin, H. M. and Fan, V. S. C.,** Contrast of glycogenesis of *Chlamydia trachomatis* and *C. psittaci* strains in HeLa cells, in *Trachoma and Related Disorders Caused by Chlamydial Agents,* Nichols, R. L., Ed., Excerpta, Medica Int. Congr. Ser. No. 223, New York, 1971, 52.

30. **Bader, J. P. and Morgan, H. R.,** Latent viral infection of cells in tissue culture. VI. Role of amino acids, glutamine, and glucose in psittacosis virus propagation in L cells, *J. Exp. Med.,* 106, 617, 1958.

31. **Ossowski, L., Becker, Y., and Bernkopf, H.,** Amino acid requirements of trachoma strains and other agents of the PLT group in cell cultures, *Isr. J. Med. Sci.,* 1, 186, 1965.

32. **Karayiannis, P. and Hobson, D.,** Amino acid requirements of a *Chlamydia trachomatis* genital strain in McCoy cell cultures, *J. Clin. Microbiol.,* 13, 427, 1981.

33. **Allan, I. and Pearce, J. H.,** Amino acid requirements of strains of *Chlamydia trachomatis* and *C. psittaci* growing in McCoy cells: relationship with clinical syndrome and host origin, *J. Gen. Microbiol.,* 129, 2001, 1983.

34. **Sterling, P., Allan, I., and Pearce, J. H.,** Interference with transformation of chlamydiae from reproductive to infective body forms by deprivation of cysteine, *FEMS Micrbiol. Lett.,* 19, 133, 1983.

35. **Allan, I., Hatch, T. P., and Pearce, J. H.,** Influence of cysteine deprevation on chlamydial differentiation from reproductive to infective life-cycle forms, *J. Gen. Micribiol.,* 131, 3171, 1985.

36. **Hatch, T. P., Allan, I., and Pearce, J. H.,** Structural and polypeptide differences between envelopes of infective and reproductive life cycle forms of *Chlamydia* spp., *J. Bacteriol.,* 157, 13, 1984.

37. **Hatch, T. P.,** Competition between *Chlamydia psittaci* and L cells for host isoleucine pools: a limiting factor in chlamydial multiplication, *Infect. Immun.,* 12, 211, 1975.

38. **Ripa, K. T. and Mardh, P.-A.,** Cultivation of *Chlamydia trachomatis* in cycloheximide-treated McCoy cells, *J. Clin. Microbiol.,* 6, 328, 1977.

39. **Byrne, G. I., Lehmann, L. K., and Landry, G. J.,** Induction of tryptophan catabolism is the mechanism for gamma-interferon-mediated inhibition of intracellular *Chlamydia psittaci* replication in T24 cells, *Infect. Immun.,* 53, 347, 1986.

40. **Moulder, J. W., Novosel, D. L., and Tribby, I. C.,** Diaminopimelic acid decarboxylase of the agent of meningopneumponitis, *J. Bacteriol.,* 85, 701, 1963.

41. **Treuhaft, M. W. and Moulder, J. W.,** Biosynthesis of arginine in L cells infected with chlamydiae, *J. Bacteriol.,* 96, 2004, 1968.

42. **Colon, J. I.,** Enzymes for formation of citrovorum factor in members of the psittacosis group of microorganisms, *J. Bacteriol.,* 79, 741, 1960.

43. **Colon, J. I.,** The role of folic acid in the metabolism of members of the psittacosis group of microorganisms, *Ann. N.Y., Acad. Sci.,* 98, 234, 1962.

44. **Starr, T. J., Pollard, M., Tanami, Y., and Moore, R. W.,** Cytochemical studies with psittacosis virus by fluorescence microscopy, *Texas Rep. Biol. Med.,* 18, 501, 1960.

45. **Collier, L. H.,** Growth characteristics of inclusion blenorrhea virus in cell cultures, *Ann. N. Y. Acad. Sci.,* 98, 42, 1962.

46. **Tamura, A. and Higashi, N.,** Purification and chemical composition of meningopneumonitis virus, *Virology,* 20, 596, 1963.

47. **Tamura, A., Matsumoto, A., and Higashi, N.,** Purification and chemical composition of reticulate bodies of the meningopneumonitis organisms, *J. Bacteriol.,* 93, 2003, 1967.

48. **Manire, G. P. and Tamura, A.,** Preparation and chemical composition of the cell walls of mature infectious dense forms of meningopneumonitis organisms, *J. Bacteriol.,* 94, 1178, 1967.

49. **Tamura, A. and Iwanaga, M.,** RNA synthesis in cells infected with meningopneumonitis agent, *J. Mol. Biol.,* 11, 97, 1965.

50. **Sarov, I. and Becker, Y.,** RNA in the elementary bodies of trachoma agent, *Nature (London),* 217, 849, 1968.

51. **Gutter, B. and Becker, Y.,** Synthesis and maturation of ribosomal RNA during the developmental cycle of trachoma agent, a prokaryotic obligate parasite of eukaryotocytes, *J. Mol. Biol.,* 66, 239, 1972.

52. **Gutter, B. Asher, Y., Cohen, Y., and Becker, Y.,** Studies on the developmental cycle of *Chlamydia trachomatis*: isolation and characterization of the initial bodies, *J. Bacteriol.,* 115, 691, 1973.

53. **Weisberg, W. G., Hatch, T. P., and Woese, C. R.,** Eubacterial origin of chlamydiae, *J. Bacteriol.,* 167, 570, 1986.

54. **Kingsbury, D. T. and Weiss, E.,** Lack of deoxyribonucleic acid homology between species of the genus *Chlamydia, J. Bacteriol.,* 96, 1421, 1968.

55. **Sarov, I. and Becker, Y.,** Trachoma agent DNA, *J. Mol. Biol.,* 42, 581, 1969.

56. **Higashi, N.,** Studies on chlamydia and togaviruses, *Annu. Rep. Inst. Virus Res. Kyoto Univ.,* 18, 3, 1975.

57. **Lovett, M., Kuo, C.-C., Holmes, K. K., and Falkow, S.,** Plasmids of the genus *Chlamydia,* in *Current Chemotherapy and Infectious Disease,* Vol. 2. Nelson, J. D. and Grassi, C., Eds., American Society for Microbiology, Washington, D. C., 1980, 1250.

58. **Peterson, E. M. and de la Maza, L. M.,** Characterization of *Chlamydia* DNA by restriction endonuclease cleavage, *Infect. Immun.,* 41, 604, 1983.

59. **Hyypia, T., Larsen, S. H., Stahlberg, T., and Terho, P.,** Analysis and detection of chlamydial DNA, *J. Gen. Microbiol.,* 130, 3159, 1984.

60. **Joseph, T., Nano, F. E., Garon, C. F., and Caldwell, H. D.,** Molecular characterization of *Chlamydia trachomatis* and *Chlamydia psittaci* plasmids, *Infect. Immun.,* 51, 699, 1986.

61. **Palmer, L. and Falkow, S.,** A common plasmid of *Chlamydia trachomatis, Plasmid,* 16, 52, 1986.

62. **Clarke, I. N. and Hatt, C.,** *In vitro* transcription/translation analysis of cloned plasmid DNA from *Chlamydia trachomatis* serovar L1, in *Chlamydial Infections,* Oriel, J. D., Ridgway, G., Schachter, J., Taylor-Robinson, D., and Ward, M., Eds., Cambridge University Press, London, 1986, 85.

63. **Alexander, J. J.,** Separation of protein synthesis in meningopneumonitis agent from that in L cells by differential susceptibility to cycloheximide, *J. Bacteriol.,* 95, 327, 1968.

64. **Tribby, I. I. E., Fri., R. R., and Moulder, J. W.,** Effect of chloramphenicol, rifampicin, and naladixic acid on *Chlamydia psittaci* growing in L cells, *J. Infect. Dis.,* 127, 155, 1973.

65. **Crocker, T. T. and Eastwood, J. M.,** Subcellular cultivation of a virus: growth of ornithosis virus in nonnucleate cytoplasm, *Virology,* 19, 23, 1963.

66. **Schechter, E. M., Tribby, I. I. E., and Moulder, J. W.,** Nucleic acid metabolism in L cells infected with a member of the psittacosis group, *Science,* 145, 819, 1964.

67. **Schechter, E. M.,** Synthesis of nucleic acid and protein in L cells infected with the agent of meningopneumonitis, *J. Bacteriol.,* 91, 2069, 1966.

68. **Alexander, J. J.,** Effect of infection with the meningopneumonitis agent on deoxyribonucleic acid and protein synthesis by its L-cell host, *J. Bacteriol.,* 97, 327, 1969.

69. **Tribby, I. I. E. and Moulder, J. W.,** Inhibition of deoxyribonucleic acid synthesis in synchronized populations of L cells infected with *Chlamydia psittaci, Infect. Immun.,* 3, 363, 1971.

70. **Bose, S. K. and Liebhaber, H.,** Deoxyribonucleic acid synthesis, cell cycle progression, and division of *Chlamydia*-infected HeLa 229 cells, *Infect. Immun.,* 24, 953, 1979.

71. **Horoschak, K. D. and Moulder, J. W.,** Division of single host cells after infection with chlamydiae, *Infect. Immun.,* 19, 281, 1978.

72. **Richmond, S. J.,** Division and transmission of inclusions of *Chlamydia trachomatis* in replicating McCoy cell monolayers, *FEMS Microbiol. Lett.,* 29, 49, 1985.

73. **Taverne, J., Blyth, W. A., and Ballard, R. C.,** Interactions of TRIC agents with macrophages: effects on lysosomal enzymes of the cell, *J. Hyg.,* 72, 297, 1974.

74. **Moulder, J. W., Hatch, T. P., Byrne, G. I., and Kellogg, K. R.,** Immediate toxicity of high multiplicities of *Chlamydia psittaci* for mouse fibroblasts (L cells), *Infect. Immun.,* 14, 277, 1976.

75. **Kellogg, K. R., Horoschak, K. D., and Moulder, J. W.,** Toxicity of low and moderate multiplicities of *Chlamydia psittaci* for mouse fibroblasts (L cells), *Infect. Immun.,* 18, 531, 1977.

76. **Kuo, C.-C.,** Immediate cytotoxicity of *Chlamydia trachomatis* for mouse peritoneal macrophages, *Infect. Immun.,* 20, 613, 1978.

77. **Wyrick, P. B., Brownridge, E. A., and Ivins, B. E.,** Interaction of *Chlamydia psittaci* with mouse peritoneal macrophages, *Infect. Immun.,* 19, 1061, 1978.

78. **Wyrick, P. B. and Davis, C. H.,** Elementary body envelopes from *Chlamydia psittaci* can induce immediate cytotoxicity in resident mouse macrophages and L-cells, *Infect. Immun.,* 45, 297, 1984.

79. **Bernkopf, H. and Mashlah, P.,** The growth cycle of a trachoma agent in FL cell cultures, *J. Immunol.,* 88, 62, 1962.

80. **Higashi, N., Tamura, A., and Iwanaga, M.,** Developmental cycle and reproductive mechanism of the meningopneumonitis virus in strain L cells, *Ann. N.Y. Acad. Sci.,* 98, 100, 1962.

81. **Friis, R. R.,** Interaction of L cells and *Chlamydia psittaci:* entry of the parasite and host response to its development, *J. Bacteriol.,* 110, 706, 1972.

82. **Hatch, T. P., Miceli, M. M., and Sublett, J. E.,** Synthesis of disulfide-bonded outer membrane proteins during the developmental cycle of *Chlamydia psittaci* and *Chlamydia trachomatis, J. Bacteriol.,* 165, 379, 1986.

83. **Newhall, W. J. V. and Jones, R. B.,** Disulfide-linked oligomers of the major outer membrane protein of chlamydiae, *J. Bacteriol.,* 154, 998, 1983.

84. **Hackstadt, T., Todd, W. J., and Caldwell, H. D.,** Disulfide-mediated interaction of the chlamydial major outer membrane protein: role in the differentiation of chlamydiae? *J. Bacteriol.,* 161, 25, 1985.

85. **Stokes, G. V.,** Proteinase produced by *Chlamydia psittaci* in L cells, *J. Bacteriol.,* 118, 616, 1974.

86. **Bernkopf, H., Mashlah, P., and Becker, Y.,** Correlation between morphological and biochemical changes and the appearance of infectivity in FL cell cultures infected with trachoma agent, *Ann. N.Y. Acad. Sci.,* 98, 62, 1962.

87. **Fan, V. S. C. and Jenkin, H. M.,** Glycogen metabolism in chlamydia-infected HeLa cells, *J. Bacteriol.,* 104, 608, 1970.

88. **Garrett, A. J. and Harrison, M. J.,** The development of a TRIC agent *(Chlamydia trachomatis)* and its associated polysaccharide in suspended cell cultures, *J. Gen. Microbiol.,* 78, 297, 1973.

89. **Duee, E. D. and Vignais, P. V.,** Exchange between extra- and intra-mitochondrial adenine nucleotides, *Biochim. Biophys. Acta,* 107, 1814, 1965.

90. **Bavoil, P., Ohlin, A., and Schachter, J.,** Role of disulfide bonding in outer membrane structure and permeability in *Chlamydia trachomatis, Infect. Immun.,* 44, 479, 1984.

91. **Crocker, T. T., Pelc, S. R., Nielson, B. I., Eastwood, J. M., and J. Banks,** Population dynamics and deoxyribonucleic acid synthesis in HeLa cells infected with ornithosis agent, *J. Infect. Dis.,* 115, 105, 1965.

92. **Lin, H.-S.,** Inhibition of thymidine kinase activity and deoxyribonucleic acid synthesis in L cells infected with meningopneumonitis agent, *J. Bacteriol.,* 96, 2054, 1968.

93. **Starr, T. J. and Sharon, N.,** Autoradiography with the agents of psittacosis and trachomas, *Proc. Soc. Exp. Biol. Med.,* 113, 912, 1963.

94. **Tribby, I. I. E. and Moulder, J. W.,** Availability of bases and nucleosides as precursors of nucleic acids in L cells and in the agent of meningopneumonitis, *J. Bacteriol.,* 91, 2362, 1966.

95. **Hatch, T. P.,** Utilization of exogenous thymidine by *Chlamydia psittaci* growing in thymidine kinase-containing and thymidine kinase-deficient L cells, *J. Bacteriol.,* 125, 706, 1976.

96. **Tanaka, A.,** Detection of DNA polymerase activity in *Chlamydia psittaci, Jpn J. Exp. Med.,* 46, 181, 1976.
97. **Hatch, T. P.,** unpublished data.
98. **Hatch, T. P.,** unpublished data.

Chapter 6

CHLAMYDIAL GENETICS

Richard S. Stephens

TABLE OF CONTENTS

I. INTRODUCTION

Genetic studies investigate heredity, variation, and the biochemical and molecular mechanisms involved in the regulation and expression of phenotype. Chlamydial genetics by this definition is in its infancy. There are two important reasons which account for the paucity of information regarding classic genetic study of chlamydial organisms. First, these organisms are obligate intracellular bacteria, consequently, the isolation of meaningful mutants has been difficult against the background of host cell activities. Second, and most importantly, no method of DNA transformation has been developed for chlamydiae, preventing the study of the action of specific genes that could be removed, manipulated, and reintroduced. Nevertheless, the body of knowledge concerning the DNA and RNA of these organisms has been growing steadily, and recently recombinant DNA techniques have been employed to study specific structural genes. These new directions promise to accelerate our understanding of chlamydial genetics in the years to come.

The potential applications of molecular genetic approaches for chlamydial studies are numerous. At the outset, the expression of unlimited quantities of chlamydial components in heterologous hosts such as *Escherichia coli*, provides a method of obtaining sufficient quanitities of these components for molecular research. Heretofore this was nearly impossible given the limited and obligate growth of chlamydiae within its host cell. The cloning of structural genes and their DNA sequence analysis can readily provide the primary amino acid sequence of chlamydial proteins. Sequence information can be used to develop testable hypotheses concerning the potential structure and function of these molecules.

Recombinant technology also has application in developing diagnostic probes. While immunodiagnostic methods have become established for chlamydial diagnosis, these procedures currently do not provide diagnostic systems more sensitive than existing culture methods. The need for diagnostic systems that are capable of detecting low levels of chlamydial infection in low-prevalence and asymptomatic populations is described in Chapter 8. The use of gene probes and new developments for amplification of specific signals in DNA based diagnostic systems promises rapid and sensitive detection of chlamydiae. Furthermore, gene probes could easily be manipulated to detect specific virulence determinants among chlamydial strains.

The use of recombinant DNA technology will also play an ever increasing role in identifying components that are essential for pathogenesis and for induction of important immune pathways. The ability to express defined antigenic determinants that are recognized by B-cells, and those that are recognized by T-cells, can be investigated in animal model systems to help elucidate the complexities of host immune responses to chlamydial infections. Detailed structural knowledge of those determinants necessary for the induction of immunity to chlamydial agents permits recombinant expression of these determinants in other hosts. This provides large quantities of predefined immunogen separate from chlamydial immunopathogenic antigens. The type-specificity of protective immunity that has been observed in primate[1] and human[2] trials suggests that a multivalent vaccine will be necessary. Molecular biological approaches provide a means for construction of multiple determinants in one polypeptide. A mosaic construction that can induce immunity to multiple serovars is both theoretically feasible and highly desirable.

Certainly one of the most unique biological attributes of chlamydiae is their biphasic growth cycle. Very simply, this cycle is represented by two morphologically distinct forms of the organism; the extracellular and metabolically inactive elementary body (EB), and the intracellular and metabolically active reticulate body (RB). Current data for the membrane differences and the latest views concerning biochemical mediation of these developmental forms is presented in detail in Chapters 4 and 5. The penultimate target for these mediators must be at the genetic level wherein mechanisms are triggered which alter fundamental gentic processes. Regulation of transcription of specific early and late genes is necessary for the systematic differentiation of these developmental forms. Another genetic regulatory mechanism is at the level of chromosomal organization. Chlamydial EB's display an ordered chromosome structure. Such an ordered structure is incapable of participating in general transcription or replication processes, hence metabolic inactivity can be efficiently maintained in the EB. Early in the developmental transition from EB to RB, the secondary structure of the chromosome is relaxed which permits generalized transcription and replication processes to begin.

The importance of a better understanding of chlamydial genetic regulation is two-fold. First, cell-free growth of chlamydiae would have a profound impact upon isolation and growth of these organisms and greatly simplify all chlamydial research. For the development of a cell-free growth system, one must first initiate and maintain those events that regulate transcriptional activity. Second, there is provocative in vivo and in vitro evidence, for both *Chlamydia trachomatis* and *C. psittaci*, that demonstrates a proclivity for latent and persistent infection.[3-6] An understanding of the genetic regulatory mechanisms of chalmydiae will provide essential information concerning the molecular mechanisms of chlamydial latency and "cryptic" developmental forms.[6]

In recent years our knowledge of the basic genetic and molecular biology of chlamydiae has increased dramatically. These advances provide the opportunity to expand research in chlamydial genetics. In this chapter investigations will be reviewed concerning the DNA and RNA of chlamydiae, the components that have been cloned, and the genetic approaches employed to investigate the mechanisms involved with regulation of the chlamydial developmental cycle.

II. DNA CONTENT

The DNA content of chlamydial EBs and RBs is 3.4% of the total dry weight.[7,8] Both *C. psittaci* and *C. trachomatis* DNA consist of a closed circular double stranded chromosome[9] and plasmid.[10] The molar percentages of guanine plus cytosine have been determined by evaluations of bouyant densities and thermal elution or denaturation temperatures. Although different techniques give different absolute values, the concensus observation is that *C. psittaci* strains generally display lower percentages with a mean of 41% and a range of 39 to 43%, while *C. trachomatis* strains have a mean of 44% and a range of 41 to 45%[11-14] Thus, among representative strains within the genus no distinct groupings are found.

A. Genome

The size of the chlamydia chromosome for both species is very similar and, next to the mycoplasmas, is one of the smallest prokaryote genomes. As a point of reference, the chlamydial genome is about one-fourth the size of the genome of *E. coli*. The molecular weight of the chlamydial chromsome was initially evaluated using gentle DNA release methods and sucrose gradient centrifugation. Using these methods, Tamura[15] reported the molecular weight of a *C. psitaci* strain to be 9.5×10^8 and Sarov and Becker[9] estimated the molecular weight of a *C. trachomatis* strain to be $7.5 \times 10.^8$ Because the DNA molecule is circular it has unique sedimentatin properties, thus Sarov and Becker modified their estimates and recalculated a molecular weight of 6.6×10^8 for *C. trachomatis* DNA. Sarov and Becker[9] also measured the contour length of the *C. trachomatis* chromosome (342.5 μm) and Higashi[16] measured the length of a *C. psittaci* chromosome (345.5 μm). These measurements represent an estimated molecular weight of 6.6×10^8 which corresponds to 1.1×10^6 nucleotide pairs. Kingsbury[17] measured the rate of reassociation of denatured DNA and reported somewhat lower molecular weights of 3.3×10^8 for *C. trachomatis* and 4.7×10^8 for *C. psittaci* DNA. It can be concluded that the chlamydial chromosome has the capacity to encode 400 to 600 proteins, assuming nonoverlapping open reading frames and no bidirectional transcription.

Genetic relatedness among organisms can be evaluated by measuring the extent and rate of heteroduplex formation between single stranded DNAs after reannealing of heterologous DNA. This measurement provides a rough evaluation of relatedness which can be refined by measuring the thermal stabilities of hybrid duplexes. Because DNA duplex stability is dependent upon salt concentrations and temperature, varying one of these parameters, such as measuring the temperature at which these duplexes separate (T_m), indicates the number of mismatched base pairs. Studies by Gerloff et al.[13,18] demonstrated similar amounts of heteroduplex formation among *C. psittaci* strains. Kingsbury and Weiss[12] evaluated both the amount of heteroduplex formation and the thermal stabilities of these DNA hybrids between two strains of *C. psittaci*, three strains of *C. trachomatis*, and several strains of *Neisseria* species. Their findings indicated virtually complete annealing between the *C. psittaci* strains (Mn and 6BC) and complete annealing among the *C. trachomatis* strains. Each of these were also thermostable. When *C. psittaci* DNA was tested with DNA from each of the *C. trachomatis* strains only 7 to 10% of the DNA annealed and, significantly, these duplexes were thermolabile. This level of homology was the same as the amount observed between *Chlamydia* and *Neisseria*.[12] Consequently, the genetic relatedness between the two chlamydial species is incredibly low given the morphological, biological, biochemical, and antigenic attributes they both share.

Weiss et al.[14] extended these studies using more strains, including the mouse pneumonitis (MoPn) strain which is taxonomically placed within the *C. trachomatis* species. Comparisons between *C. psittaci* strains and MoPn demonstrated less than 10% homology as did human *C. trachomatis* strains. However, the relationship between MoPn and human *C. trachomatis* strains was not close. At moderate stringency (i.e., low temperatures) 62% of the DNA from MoPn bound human *C. trachomatis* DNA, and at higher stringency only 29% binding was detected. None of the binding was thermostable indicating that despite moderate capacity for binding, the DNA that did form heteroduplexes had many base pair mismatches.

The conclusions that have been drawn from DNA homology studies include;

1. *C. psittaci* and *C. trachomatis*, including MoPn, are not related and these species evolved from different ancestors or diverged long ago.[12]
2. *C. psittaci* strains display greater differences among themselves than *C. trachomatis* strains.
3. There is very limited relatedness between human *C. trachomatis* strains and the murine MoPn strain. If there exists a common ancestral relationship between these strains, most of the genome has not been conserved.[14]

It must be emphasized that study of total genomic homology is only one measurement of molecular relatedness and divergence. As information becomes available concerning the sequence variation and relationships of specific genes, we can expect to account for the metabolic and phenotypic similarities and differences among these organisms (Chapter 1).

Incorporation of labeled precursors into chlamydial DNA has shown that DNA synthesis increases between 12 to 40 hours after infection and then declines.[19,20] Hydroxyurea is an inhibitor of DNA synthesis. Addition of hydroxyurea at different times during the growth cycle demonstrated that the development and multiplication of RBs require DNA synthesis and there is inhibition of EB formation.[21] Thus, DNA synthesis occurs in RBs, is coincident with RB development, and is coupled to the formation of infectious EB (Figure 1). In vitro studies have demonstrated that chlamydiae have a DNA-dependent DNA polymerase for replication and repair of the genome.[22] The genetics of replication, however, have not been studied. It is of considerable interest that attempts to transform *E. coli* with the chlamydial

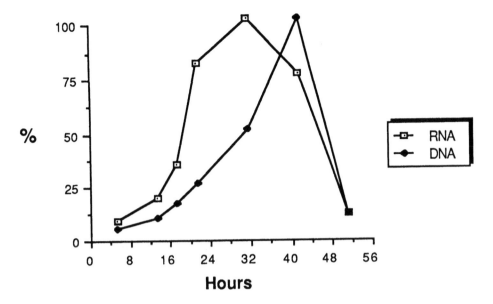

FIGURE 1. DNA and RNA synthesis during developmental cycle of chlamydiae. This is a schematic representation based upon cytochemical, incorporation, and hybridization data.[19-21,33,37-39,43,65,69]

plasmid have not been successful.[23] As with other organisms, this suggests that chlamydial replication origins are not recognized by heterologous hosts and are unique to these organisms.

B. Plasmid

The observation and isolation of plasmids from *Chlamydia* was first reported by Lovett et al.[10] DNA isolated from *C. trachomatis* and *C. psittaci* EBs displayed a satellite band upon cesium chloride gradient centrifugation. When this extrachromosomal DNA was analyzed by agarose gel electrophoresis, the DNA from *C. trachomatis* and *C. psittaci* displayed a 4.4 Mdalton plasmid.[10] In addition, the plasmid from *C. trachomatis* strains also existed as higher molecular weight multimers. Endonuclease restriction of these plasmids from *C. trachomatis* serovars B and C showed identical patterns, with plasmid from serovar L_2 also showing nearly identical restriction fragments. In contrast, plasmid from *C. psittaci* (Mn) displayed a very different restriction profile compared to the *C. trachomatis* plasmid.[10]

Hyypia et al.[24] cloned the plasmid from serovar L_2 into pBR327. They estimated the size of the plasmid to be 6.7 kilobase pairs (kb) in length (ie., ~4.4 Mdaltons) and provided a restriction endonuclease map for a *C. trachomatis* plasmid. Homologs to each of the 15 serovars were detected with labeled plasmid used as a hybridization probe. They also confirmed the existence of plasmid multimers in *C. trachomatis* DNA preparations from EBs when the plasmid probe detected three bands, excluding the chromosome band, in southern blots of total genomic DNA. Joseph et al.[25] also cloned the plasmid from serovar L_2 and additionally cloned the plasmid from a *C. psittaci* strain (Mn). Southern analysis confirmed the significant lack of DNA homology between the *C. trachomatis* and *C. psittaci* plasmids. The possible exception was detectable, but unanalyzed, hybridization between two *Cla* 1 fragments. Comparison of the polyacrylamide gel electrophoresis patterns of plasmid encoded polypeptides, using both in vitro and in vivo systems, revealed common products of 48,000 and 38,000 m.w. for the *C. trachomatis* plasmid, while the *C. psittaci* plasmid produced products of 58,000, 38,000, and 22,000 m.w. Notably none of these proteins were immunoreactive with immune rabbit sera. Recently, Clarke and Hatt[26] cloned the *C. trachomatis* plasmid from serovar L_1. They isolated a 7.2 kb plasmid that produced polypeptides of 46,000, 37,000, 29,000, and 15,000 m.w. in an in vitro transcription/translation system.

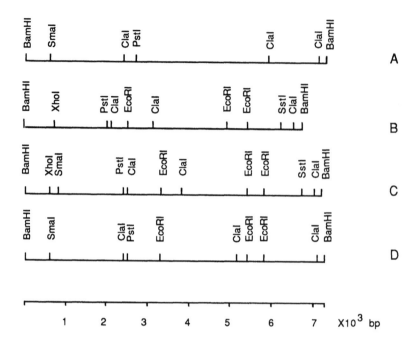

FIGURE 2.   Restriction endonuclease restriction maps of the *C. trachomatis* plasmid (A) from reference 25; (B) from reference 24; (C) from reference 26; (D) from reference 23.

These results compare favorable to those of Joseph et al.[25] who also observed lower molecular weight proteins, although only in their in vitro transcription/translation system.

Palmer and Falkow[23] comprehensively evaluated the *C. trachomatis* plasmid by cloning the plasmid from eight serovars and comparing the restriction endonuclease profiles, and transcription (mRNA) and translation products. Comparison of cloned plasmids from serovars B, C, D, E, F, G, K, L_1, and L_2 each had a 7.4 kb plasmid, which when cleaved with nine restriction enzymes revealed identical restriction patterns on all plasmids except for the lack of a single *Hind* III site in plasmid DNA for serovars D, G, K, and L_2. As in earlier studies they also observed dimer forms (14.8 kb) of the plasmid by gel electrophoresis and they confirmed these observations by electron microscopy. Furthermore, they estimated that the 7.4 kb plasmid was present in ten copies per EB. It is not known what the plasmid copy number is for RBs or if in vitro culture history alters the copy number. These investigators have detected homologous nucleotide sequences in each *C. trachomatis* serovar, except for the murine MoPn strain, and they have found these sequences in all of over 200 human isolates. Using both in vivo and in vitro systems they observed the production of 8 to 9 protein products, ranging in molecular weight from 11,000 to 48,000. Palmer and Falkow[23] also examined the synthesis of plasmid specific mRNA during the chlamydial developmental cycle. Transcripts of ~480, 680, 1300, 3500, and 5000 nucleotides were detected by Northern hybridization. The maximum size of translation products from the smaller transcripts are 17,000, 24,000, and 47,000 m.w., and the larger transcripts probably represent polycistronic mRNA that encode for more than one protein.

In general, the plasmid data generated from different laboratories are similar. Four published restriction maps are presented in Figure 2. The size of the plasmids obtained by restriction endonuclease mapping range from 6.7 to 7.4 kb. Palmer and Falkow[23] measured the contour length of the plasmid by electron microscopy and found it compatible with the 7.4 kb size. Restriction endonuclease studies by Peterson and de la Maza[27] demonstrated that the plasmids from serovars L_1, L_2, and L_3 were identical in size, and that only the L_3 plasmid showed different endonuclease restriction fragments. These findings are supported by Palmer and Falkow[23] and demonstrate that these plasmids are highly conserved. The restriction maps represented in Figure 2 show that there is substantial agreement for all sites except for the placement of *Cla* 1 sites. Unfortunately, none of the investigations evaluated whether or not the *Bam* HI site used for cloning disrupted an open reading frame in that portion of the plasmid.

The functions of these plasmids remain cryptic.[10] The data demonstrates that the *C. trachomatis* plasmid has been strongly conserved in both the trachoma and LGV biovars, thus the products of these plasmid genes must play an essential role in the virulence or biology of these organisms. Significantly, the presence of a similar size plasmid in *C. psittaci* that also produces analogous polypeptides is consistent with the conclusion that the chlamydial plasmids mediate similar convergent functions for the two species. The findings of Palmer and Falkow[23] that plasmid specific RNA is coordinately transcribed during the developmental cycle (see below) suggest important roles of plasmid proteins for the growth and differentiation of the organism.

The fact that these genes reside exclusively on extrachromosomal DNA implies that plasmid copy number during the developmental cycle could be important for amplification of needed gene products, or alternatively, that there is some reason that these genes need to be available for mobilization. Clearly more work is needed in this important area. Evaluation of plasmid differences among strains, especially *C. psittaci* strains which represent a more diverse group, will help define those genes that are indispensable. An appreciation of the role of plasmid encoded products and copy number throughout the developmental cycle may shed light upon the function of these gene products.

Careful mapping of open reading frames and regulatory sequence boundaries may be necessary to effectively use these plasmids as vectors (see below). It will also be important to establish compatibility of various plasmids with different chlamydial strains. Of additional interest is the compatibility or exclusion of *C. trachomatis* plasmids in murine (MoPn) and *C. psittaci* strains and vice versa.

C. Restriction Endonuclease Analysis

Serotyping *C. trachomatis* isolates continues to be a powerful approach for understanding the pathogenesis and natural history of human chlamydial infections. The serotyping determinants detected by the micro-IF test, as well as broadly reacting determinants, reside on the major outer membrane protein (MOMP) (see Chapter 3), thus serological categorization is restricted to the polymorphisms in one portion of one gene. In contrast, restriction endonuclease analysis of the entire genome provides a broad molecular view of the similarities and differences between isolates over many gene loci. This approach defines a different set of criteria for evaluation of the relationships among isolates because restriction endonuclease analysis of genomic DNA reflects biological, phenotypic, and antigenic characteristics inclusively. Restriction analysis of *C. trachomatis* DNA was initiated by Peterson and de la Maza.[27] For these studies they used serovars L_1, L_2, L_3, B, C, and D and a *C. psittaci* strain. Consistent with the results of genomic DNA hybridization studies, they could readily differentiate the *C. trachomatis* strains from *C. psittaci*. Serovar L_1 and B were more closely related than serovars L_1 and L_2. Surprisingly, serovar D appeared to be the most divergent among the *C. trachomatis* serovars tested. Recently, these observations have been extended to include all 15 *C. trachomatis* serovars.[28] Using different restriction enzymes each serovar could be differentiated. In addition, strains within serovars could be differentiated. The ability to identify and differentiate specific isolates of *C. trachomatis* offers a new source of data for understanding the epidemiology and pathogenesis of eye and genital tract diseases.

In contrast to the well established serological differentiation of *C. trachomatis* strains, the relationship among the very diverse *C. psittaci* species are less understood. McClenaghan et al.[29] approached this issue by comparing the restriction endonuclease profiles of *C. psittaci* strains representing ovine abortion, ovine arthritis, avian, and Cal 10 (Mn) isolates. They readily differentiated each of the four groups with *Eco* R1 digests; whereas, multiple isolates from ovine abortion and ovine arthritis displayed identical patterns within groups. The data for the two avian isolates suggests more diversity within this group. The profiles obtained with the Cal10 strain showed many similarities to the avian isolates, thereby supporting the conclusion that this strain originated from an avian source. The heterogeneity of the *C. psittaci* species has been appreciated by evaluation of other parameters; however, it seems clear that restriction endonuclease analysis provides a powerful approach toward developing the epidemiology and taxonomy for this species. Such a task is formidable, nevertheless, this approach may be a better alternative to using a serological approach because it should provide more information within a short period of time.

Restriction endonuclease methods could also be used for chromosomal mapping of cloned genes. Although this has yet to be thoroughly investigated, the use of restriction enzymes that cleave DNA infrequently generate a limited number of chromosomal DNA fragments. A restriction endonuclease enzyme that has an eight base pair recogition sequence would be expected to produce 15 to 30 fragments from a complete digest of the chlamydial chromosome.[30] Using pulse-field electrophoretic techniques, large fragments can accurately be resolved.[31] These fragments may then be probed by Southern hybridization to localize specific genes. This would provide an alternative to standard mapping techniques which require defined mutants and systems of transduction or conjugation. Mapping of chromosomal genes provides information concerning the physical organization of the genome, and nonselectable genes can be manipulated if they are linked to known selectable markers.

III. RNA CONTENT

The quantitative RNA content varies greatly between developmental forms of chlamydiae. The metabolically active RB contains at least 3 to 4-fold more RNA than the metabolically inactive EB.[8,32] Relatively little information has been presented since Becker[33] and Storz and Spears[34] comprehensively reviewed the state of the art in 1978. Early studies concentrated upon characterization of various ribosomal RNA species (rRNA) which firmly placed *Chlamydia* as a prokaryote. While the RNA content for ribosomal species is much higher in RBs than EBs, it can also be assumed that this is the case for messenger RNA (mRNA), although this has not been directly established. The DNA cloning and characterization of specific chlamydial genes (see below) now provide probes for detection and quantitation of chlamydial mRNA, thus only recently has any information been forthcoming for mRNA species in chalmydiae.

A. rRNA

The first analysis of chlamydial RNA evaluated the sedimentation coefficients of predominant RNA species. Tamura and Iwanga[35] identified 21S, 16S, and 4S RNA fractions in *C. psittaci*. Interestingly, the 21S and 16S RNA were more abundant in RBs, whereas the 4S predominanted in EBs. Studies by Sarov and Becker[36] found similar RNA species for *C. trachomatis*, and they also isolated 52S and 30S ribonucleoproteins. The 23S (21S) RNA was associated with 52S particles and the 16S RNA was associated with 30S particles. Using polyacrylamide gel electrophoresis, Gutter et al.[37,38] analyzed the synthesis and maturation of rRNA species. These studies demonstrated that 23S and 16S RNA were the mature RNA species which were derived from 32S and 16S RNA were the mature RNA species which were derived from 32S and 17.5S precursors. Thus, chlamydiae have ribosomes, rRNA, and tRNA species that resemble all other bacteria. Recently two groups have cloned and sequenced chalmydial 16S rRNA.[39,40] These studies have shown that chlamydiae are a unique eubacterial group and that there is a phylogenetic relationship between the *C. trachomatis* and *C. psittaci* species (see Chapter 1).

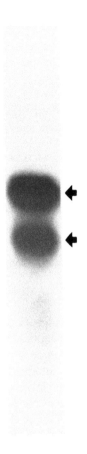

FIGURE 3. Northern hybridization of serovar L_2 RNA probed with DNA from *omp*1L2. Two *omp*1 gene RNA transcripts are shown by arrows. (From Stephens, R. S., Mullenbach, G., Sanchez-Pescador, R., and Agabian, N., *J. Bacteriol.*, 168, 1277, 1986. With permission.)

B. mRNA

The recent availability of specific chlamydial DNA clones finally makes the analysis of mRNA in chlamydiae possible. Northern analysis consists of size separation of all RNA by gel electrophoresis, and specific messages are visualized by hybridization with labeled DNA probes. Stephens et al.[41] detected *omp1* specific mRNA species that encode the MOMP by Northern hybridization using *omp1* DNA gene probes (Figure 3). This gene is constitutively expressed throughout the chlamydial growth cycle. Interestingly, two mRNA species are detected. Detailed analysis of these transcripts shows that the *omp1* gene has two promoters.[42] Palmer and Falkow[23] performed Northern analysis using the entire chlamydial plasmid as a probe. Unlike the *omp1* gene, plasmid genes are temporally regulated and different plasmid genes are transcribed only at specific times during the developmental cycle (see below).

The detection, analysis, and characterization of specific mRNA promises to provide considerable information regarding the genetics of chlamydial growth and differentiation. Identification of the genes, and gene products, that are temporally regulated should also lead to a better understanding of the unique biology of these organisms.

IV. MOLECULAR CLONING

The obligate intracellular habitat of chlamydiae and their limited growth in these systems has frustrated many molecular research investigations as well as attempts at genetic studies.

Consequently, an attractive approach to these problems is the application of recent advances in molecular cloning techniques. Two basic strategies have been employed for cloning chlamydial genes. The first is the use of plasmid or bacteriophage vectors for cloning relatively large DNA fragments in *E. coli* and detecting recombinants that express chlamydial antigens using immune sera. Success depends upon the ability of the transcription and translation machinery of the host organism to express chlamydial genes. The first examples of this approach were reported by Wenman and Lovett[43] and Stephens et al.[44] Since these reports, several chlamydial components have been cloned and expressed in *E. coli*; however, several prominent proteins, especially the MOMP, were not readily detected using this type of methodology.

An alternative strategy is to use vector systems which provide host promoters and other regulatory factors, such as the bacteriophage lambda gt11 system[45] or the pUC family plasmids.[46] These cloning systems rely on the fusion of a chlamydial gene to an *E. coli* cistron. This results in controlled expression of the hybrid gene which, once induced, usually produces large quantities of product. The product is a chimeric protein that consists of an *E. coli* protein, such as betagalactosidase, fused to a chlamydial polypeptide. Using this strategy, Stephens et al.[47] cloned the *C. trachomatis* MOMP gene as well as other chlamydial proteins.

Application of molecular cloning techniques to contemproary problems in chlamydial research provides virtually unlimited quantities of predefined protein for immunological, biological, and structure-function studies, free of other chlamydial and mammalian host cell components. In addition, DNA sequence information of gene and its flanking sequences can address issues of the genetic mechanisms involved with temporal expression and the molecular basis of diversity of specific chlamydial genes.

A. MOMP

While the MOMP gene was not the first chlamydial gene to be cloned, it has certainly been the focus of research efforts because of its interesting antigenic, structural, and functional properties (see Chapter 3). Allan et al.[48] identified a recombinant clone from a lambda 1059 library made from *C. trachomatis* serovar L$_1$ DNA. This recombinant clone in *E. coli* expressed a doublet of proteins of 40,000 and 41,000 m.w. on polyacrylamide gels. These proteins each showed three V8 protease fragments of similar molecular weight in one-dimensional polyacrylamide gels, and these fragments comigrated with analogous peptides obtained from purified MOMP. The doublet character of these cloned polypeptides and the doublet character of MOMP is intriguing (Chapter 3). In their initial studies using the lambda 1059 vector, Stephens et al.[44], were unable to detect recombinants expressing sufficient quantities of chlamydial MOMP from serovar L$_2$ for immunological detection. However, using the lambda gt11 fusion vector they selected a recombinant that expresses chlamydial MOMP antigens which were identified with MOMP-specific monoclonal antibodies.[47] This recombinant fusion protein reacted strongly with *C. trachomatis* species-specific and sub-species-specific monoclonal antibodies,[49] and when the chlamydial DNA insert was excised from the vector it hybridized to homologous nucleotide sequences in each of the 15 *C. trachomatis* serovars. Interestingly, a weak reaction was observed with a *C. psittaci* strain (Mn) which suggests that sequence homology may be maintained within the members of the *C. psittaci* MOMP family as well. DNA sequencing of this insert revealed a ~340 bp open reading frame, thus this clone consists of the carboxyl terminal third of MOMP fused to betagalactosidase. Southern analysis of genomic DNA obtained from serovars L$_2$, B, and C displayed single *Bam* HI fragments of similar molecular weight when probed with the insert DNA from the lambda gt11 recombinant.[47] This demonstrates that the MOMP gene is chromosomal and occupies a similar genomic context for each of the serovars.

Stephens et al.[41] used this lambda gt11 insert DNA to identify genomic clones of *Bam*

HI fragments from serovar L_2. The cloned gene was mapped with restriction endonuclease enzymes and sequenced.[41] The structural gene for the MOMP from serovar L_2 consists of a 1182 bp open reading frame that encodes 394 amino acids. This open reading frame is preceded by regulatory-like sequences and is punctuated by three stop codons and a transcription termination sequence. The gene was designated CTR *omp1*L2 to indicate that this is a *C. trachomatis* outer membrane protein gene for the principal protein for serovar L_2. The amino terminus of the L_2 MOMP has been determined by Edman degradation.[50] The amino-terminal amino acid of the mature protein aligns with amino acid residue 23 of the translated amino acid sequence.[41] Thus, MOMP has a typical signal or leader sequence to mediate vectorial transport to its surface location. The primary amino acid sequence and amino acid composition of MOMP displays characteristics compatible with other porin proteins and is consistent with those characteristics reported by Bavoil et al.[51] Significantly, computer assisted comparisons of amino acid or DNA sequences between MOMP and other porin proteins do not reveal outstanding homologies.[41] Also unique with respect to other porins are the number of cysteine residues per molecule. Porins from other organisms have one or fewer cysteine residues, whereas the L_2 MOMP sequence contains nine cysteine residues which perform unique and essential roles for chlamydial porin regulation.[41] Nano et al.[50] derived two plasmid clones, one was selected because it showed homology to a concensus oligonucleotide probe derived from the amino-terminal amino acid sequence of purified MOMP, and the other was selected because it expressed a polypeptide that reacts with a MOMP-specific monoclonal antibody. These plasmids, however, do not represent the structural gene for MOMP, because they do not match the *omp1*L2 gene described by Stephens et al.,[41] and additional studies by Nano have shown that these plasmid clones do not hybridize to *omp1*L2 but may represent vestigial gene components for MOMP.[52] It is difficult to evaluate the MOMP clone described by Allan et al.[48] because they did not provide specific immunological or restriction data that could be used for comparisons.

The MOMP genes for serovars B and C have recently been cloned and sequenced by Stephens et al.[53] Comparisons of these three genes (*omp1*L2, *omp1*B, and *omp1*C) are of interest because serovars L_2 and B are closely related antigenically, yet represent different biovars, while serovar C is the most antigenically divergent. Thus the *omp1*C gene product represents the maximum potential differences likely to be encountered within the structural constraints of MOMP. The calculated molecular weights of MOMP from translated DNA sequences for serovars L_2 and B were the same (40,282) and correlated with estimates obtained by polyacrylamide gel electrophoresis. The *omp1*C gene encodes three additional amino acids with a calculated molecular weight of 40,607. This appears to be insufficient to account for the higher molecular weight observed for serovar C MOMP by polyacrylamide gel electrophoresis (i.e., about 1,000 m.w. higher than L_2 values). This discrepancy needs to be resolved; it may simply be caused by unusual sieving properties of these proteins in electrophoretic systems, or other factors could be considered such as carbohydrate or lipid moieties associated with MOMP.

Comparisons of *omp1* genes provide the opportunity to address the molecular basis of the antigenic diversity among *C. trachomatis* serovars as well as to consolidate structural and functional concepts proposed for this protein. Comparisons of these three genes define four sequence variable domains which are separated by highly conserved segments (Figure 4).[53] The even distribution and alternation of variable and conserved domains is remarkable and may be related to the structural symmetry observed for this protein (see Chapter 2). As expected from the antigenic relationships of these serovars, the number of amino acid changes is greatest for the serovar C sequence. Of the four sequence variable domains, calculations of secondary structure predicts that variable segment 2 has the highest degree of turn and loop potential and variable segment 2 also displays the greatest number of amino acid differences. Thus, this segment may represent a predominant type-specific antigenic and immunogenic domain.[53,42]

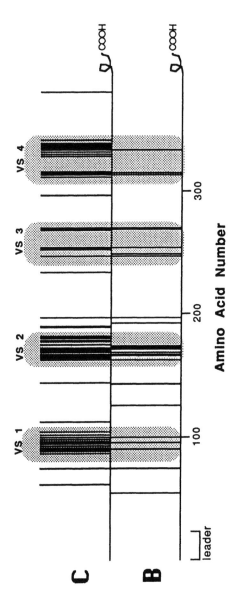

FIGURE 4.   Comparison of amino acid differences translated from *omp*1L2, *omp*1B, and *omp*1C. Each line represents an amino acid difference for either serovar B or C MOMP sequence compared to the sequence for serovar L$_2$. (From Stephens, R. S., Sanchez-Pescador, R., Wagar, E., Inouye, C., and Urdea, M., *J. Bacteriol.*, 169, 3879, 1987. With permission.)

The genetic basis of variation among the *omp*1 structural genes is founded in clustered nucleotide substitutions for closely related serovars such as L_2 and B, and insertions or deletions for distantly related serovars (i.e., serovar C).[53] The molecular mechanisms that account for the development of *omp*1 gene diversity appear to be the result of accumulated point mutations in closely related serovars; however, recombination may play a role in the development of the dramatic nucleotide differences observed for *omp*1C. It is apparent that a few additional *omp*1 gene sequences from other serovars will be needed for a more comprehensive understanding of these genetic mechanisms.

The nucleotide sequence differences of these structural genes are in marked contrast to the differences observed for the 5′ flanking sequences.[53] The changes within the structural genes suggest that serovars B and C are each equidistant from serovar L_2, and that *omp*1C is very distant from *omp*1B. In contrast, the flanking sequences demonstrate a much closer relationship among the trachoma biovar strains (B and C) that differ as a group from the LGV biovar. These observations have important implications for the hypothesis that the gene diversity of this family has been immunologically driven. This data supports the conclusion that the structural *omp*1 genes are diverging much more rapidly than noncoding and nonregulatory flanking regions. Like other parasitic systems that rely upon antigenic variation for survival in their mammalian hosts, the trachoma biovar may ensure the survival of its "species" because of the enrichment over time of a large repertoire of variant antigenic strains. This is consistent with the lack of invasiveness of these organisms which are frequently exposed to host mucosal surface immune defenses prior to infecting a new host cell. Unlike the trachoma biovar, the LGV biovar is much more invasive and is capable of cell to cell transmission. (i.e., as evidenced by plaque formation), thereby reducing exposure to host immune defenses, and perhaps consequently only three LGV serovars have evolved over a similar evolutionary period.

B. LPS

A most significant finding was the recombinant expression of the genus-specific epitope of chlamydial lipopolysaccharide in *E. coli*. Nano and Caldwell[54] identified a recombinant plasmid in a gene bank of serovar L_2 DNA that displayed an additional LPS band to the parent host strain, and this band reacted with a monoclonal antibody specific to the chlamydial LPS. By immunoblot this recombinant LPS species migrated with increased electrophoretic mobility compared to native chlamydial LPS. Furthermore, they demonstrated that this chlamydial LPS antigen was exposed on the surface of the host *E. coli*. A reasonable conclusion for these observations is that the 6.5 kb chamydial DNA insert expresses a gene(s) that encode a glycosyl transferase which adds a chlamydial carbohydrate moiety onto *E. coli* LPS.

Because chlamydial LPS is a prominent surface antigen and has been implicated as a major component in pathogenesis (Chapter 11), this recombinant will be very useful for studies of the role of chlamydial LPS in unique biological mechanisms and its role in immunopathogenesis.

C. 16S rRNA

A strong technique for molecular taxonomy is the comparison of slowly evolving genes such as the 16S ribosomal RNA genes. Palmer et al.[39] cloned the 16S rRNA gene from *C. trachomatis* and Weisburg et al.[40] cloned this gene from a *C. psittaci* strain (6BC). Similar cloning strategies were employed by both groups; 16S rRNA was isotopically labeled and used to probe gene libraries for clones containing homologous sequences. Palmer et al.[39] used these clones for hybridization studies and found that the 16S rRNA genes are present in two copies on the chromosome.[55] These hybridization studies also demonstrated related rRNA gene sequences to *C. psittaci* and, surprisingly, some relationship was found to *Legionella pneumophila*, *Vibrio cholerae*, and *V. vulnificus*.[39,55]

Weisburg et al.[40] sequenced the 16S rRNA gene for *C. psittaci* and compared this sequence to over 400 other eubacterial 16S rRNA sequences. Like the hybridization studies, sequence comparisons found no strong relationships between *Chlamydia* and other organisms; however, a remote relationship may have been found to planctomyces. Interestingly, planctomyces lack peptidoglycan although this is the only phenotypic characteristic in common with chlamydiae. Additional hybridization, sequence, and higher-ordered structure comparisons are necessary to firmly establish the relationship between chlamydial species as well as the relationship of chlamydiae to other prokaryotes. The importance of these evaluations is that they may identify relationships to other organisms which could facilitate genetic and biochemical studies of chlamydiae. Equally significant is that if no relationships are found, then the study of *Chlamydia* as a unique bacterial system is, in its own right, most worthy of study.

D. 74K, 12K, and 19K Proteins

The chlamydial 74,000 m.w. (74K) antigen expressed in *E. coli* was first described by Stephens et al.[44] and has also been studied by Kaul and Wenman[56] and Palmer and Falkow.[55] The function of this protein is unknown, but it is immunogenic for humans and rabbits although the ontogeny of the responses to this antigen has not been studied. This antigen appears to be well represented in gene libraries that contain large DNA inserts; thus, unlike most other chlamydial genes or gene products, this protein displays unusually efficient expression and/or is very stable in *E. coli*. Of considerable interest is the observation that the 74K gene is transcribed only early during the developmental cycle of chlamydiae.[55] Thus this gene represents one of those components that is temporally regulated and, in this case, necessary for early biological events in the developmental cycle.

The significance and the role the 60K and 12K cysteine rich proteins in chlamydial outermembrane structure for the EB developmental form is described in detail in Chapter 3. Bavoil et al.[57] have cloned the gene for the 12K protein which will ultimately provide important information concerning the genetic mechanisms of regulation and structural properties of this protein. A clone containing this gene synthesizes both a 15K and 12K proteins in in vitro translation/transcription systems. This observation suggests that both structural genes are genetically linked and, in fact, they may be linked in a single regulatory operon.[57]

Another protein that is always represented in chlamydial genomic bacteriophage libraries is the 19K antigen described by Wenman and Lovett.[43] This protein was also present in libraries evaluated by Stephens[42] and Palmer and Falkow.[55] Unfortunately nothing is known concerning the function of this antigen.

V. DEVELOPMENTAL CYCLE REGULATION

The developmental cycle of chlamydiae is one of the major attributes that separates these organisms from other bacteria. This cycle consists of an extracellular phase and an intracellular growth phase, each with distinct morphological forms of the organism. The EB is the extracellular form and has a rigid outermembrane, a compacted nucleoid, and is metabolically inactive. The intracellular form is the RB, which has a plastic and fragile outermembrane, a dispersed nucleoid, and is metabolically active. Following entry into a host cell the EB begins a process of differentiation into the RB form. The RB, after 10 to 12 rounds of replication, begins another process of differentiation into the EB form. It should be noted that while the EB and RB developmental forms represent stages of mature differentiation, forms at intermediate stages can also be identified. There is growing evidence that biochemical changes trigger differentiation; nevertheless, the processes of differentiation into EB and RB developmental forms must be mediated by genetic events. A complete description of the developmental cycle is covered in Chapter 4, and the biochemistry associated with potential triggers is covered in Chapter 5. In this section, I will concentrate on the potential genetic mechanisms that underlie the regulation and expression of the developmental cycle.

The process of differentiation is a fundamental biological property of chlamydiae. Although the knowledge of these processes is very limited, the available information suggests that, despite their evolutionary distance, both *C. trachomatis* and *C. psittaci* utilize analogous systems. Following entry into a host cell two ultrastructural changes of the infectious EB form have been documented in detail (see Chapter 2). The first change observed is an alteration in the organization and structure of the EB nucleoid. The second ultrastructional change is the reorganization of the EB outer membrane structure. Coincident with these ultrastructural changes is metabolic activity. In recent years, considerable efforts have focused upon outer membrane changes and these are reviewed in detail in Chapter 3.

The genetic regulatory events involved with the differentiation of chlamydiae include the sequential transcription of specific genes which in turn modulate, and are modulated by, the structural organization of the chromosome. The significance of investigating these regulatory mechanisms include: (1) understanding the molecular basis of latency and persistence and their related "cryptic" developmental forms, (2) the ability to develop axenic growth systems, and (3) fundamentally unique processes may identify specific targets for chemotherapeutic intervention.

A. Chromosome Structure

The chromosomes of eukaryotes are organized in higher ordered structures by histones and other proteins. Chromosomal organization permits efficient packaging of large DNA molecules, facilitates regulatory functions, and protects DNA from damage. Prokaryotes organize their chromosomes in nucleoid structures, presumably for similar purposes. Several histone-like proteins have been identified in *E. coli*, and these interact with chromosomal DNA and alter its structure.[58] Unlike eukaryotes, prokaryotes lack a nuclear membrane, hence the term "nucleoid" has been applied to these structures. The bacterial nucleoid consists of a circular, double-stranded DNA chromosome which is supercoiled, highly folded, and interacts with some membrane associated proteins.[58]

The electron dense nucleoid of chlamydial EBs has been a morphological hallmark in electron microscopic studies of chlamydiae. Costerton et al.[59] specifically addressed the ultrastructure of chlamydial nucleoids. They used a variety of fixation techniques and compared these morphological observations to nucleoids of other bacteria. Generally, bacterial nucleoids consist of a meshwork of fibers located in central areas of the cell and form small, condensed foci. In resting forms, such as a bacterial endospore, the nucleoid is slightly condensed yet individual fibers can readily be seen. The nucleoid of the metabolically active RB is a dispersed fibrillar structure, very similar to nucleoids of other vegetative bacterial cells (Figure 5). In contrast, the nucleoid of the metabolically inactive EB is a highly compacted, electron dense mass which lacks fibrillar character (Figure 5). It is eccentrically located within the cell and is localized from a ribosome dense area. Costerton et al.[59] concluded from these comparative studies that the EB nucleoid is so morphologically distinct that "they have no parallel in other bacteria". Using ultrastructural cytochemical approaches, Popov et al.[60] demonstrated that nucleic acids were precisely located in the respective RB and EB nucleoids. It can be concluded that the unique EB nucleoid provides a structural organization to the EB chromosome which differs significantly from that of other bacteria, and these higher ordered structures provide genomic integrity to the quiescent developmental form of chlamydiae.

Wagar and Stephens[61] have found evidence for histone-like nucleoproteins in chlamydiae. DNA-associated proteins were detected following micrococcal nuclease digestion and gel electrophoresis. Using this approach two predominant bands were identified — a ~15,000

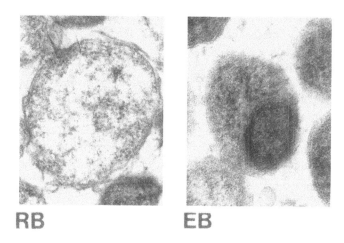

FIGURE 5.   Electron micrograph of an RB and EB. Compare the dispersed RB nucleoid to the compact EB nucleoid. (From Costerton, J. W., Poffenroth, L., Wilt, J. C., and Kordova, N., *Can. J. Bot.*, 22, 16, 1976. With permission.)

m.w. band represented in DNA obtained from either RBs or EBs, and a ~25,000 m.w. band specific for DNA obtained from EBs.[61] Recently, Wagar and Stephens[62] have identified other EB-specific DNA-binding proteins using protein transfer techniques of chlamydial lysates. This technique permits the detection of several chlamydia specific DNA-binding proteins. Significantly, major species can be identified that are only represented in lysates of *C. trachomatis* EB developmental forms. *C. trachomatis* chromosomes isolated by sucrose fractionation displayed a prediminant ~17,000 m.w. EB-specific protein complex (Figure 6). *C. psittaci* (Mn) lysates have analogous DNA-binding proteins, although with different electrophoretic mobilities (Figure 6). *C. psittaci* (Mn) lysates have analogous DNA-binding proteins, although with different electrophoretic mobilities (Figure 6). While our studies have concentrated upon the identification of EB-specific chromosomal proteins, it remains to be demonstrated that these proteins interact with chlamydial DNA to promote compaction, regulate gene expression, or serve a structural role for the chlamydial nucleoid. Nevertheless, these findings are provocative and we propose that these protein-DNA interactions provide the molecular basis for an essential step in mediating developmental differentiation, including pathways for latent or persistent developmental forms.

Despite evidence of limited metabolic activity, attempts at inducing cell free growth of chlamydial EBs have not succeeded in altering the nucleoid structure of these organisms.[63] Knowledge of the molecular mechanisms involved with EB-specific nucleoproteins should identify means to initiate nucleoid structure differentiation. This is seen as a requisite step for initiating axenic growth of chlamydiae. Additionally, it is a reasonable expectation that these protein-DNA interactions may be unique from other eubacterial and eukaryotic regulatory processes, thus they could be targets for new chemotherapeutic approaches for control of chlamydial infections.

B. Transcription

Prokaryote mRNA has a half-life on the order of minutes, thus it is reasonable to assume that little, if any, mRNA is present in chlamydial EBs. At the genetic level, once an EB enters a host cell the initiation of growth must begin with the synthesis of mRNA transcripts. These assumptions are supported by the recent demonstration by Hatch et al.[64] that reduction of MOMP in *C. psittaci* EBs to its monomeric form occurs within 1 hr of entry, and that this step and subsequent differentiation is inhibited by chloramphenicol. Four basic ap-

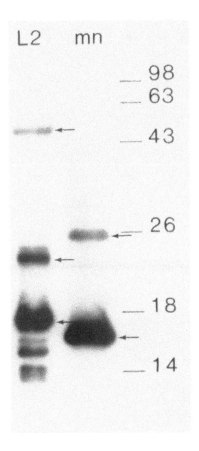

FIGURE 6. DNA-binding proteins of *C. trachomatis* serovar L₂ and *C. psittaci* (Mn). Protein lysates were electrophoretically separated in polyacrylamide gels, transferred to nitrocellulose, and probed with ^{32}P-labeled *C. trachomatis* DNA. Arrows indicate DNA-binding proteins unique to the EB developmental form.

proaches have been used to study RNA synthesis in chlamydiae: (1) cytochemical studies, (2) measurements of incorporation of labeled precursors, (3) effects of inhibiting transcription initiation with rifampicin, and (4) Northern hybridization techniques.

Cytochemical studies have shown that RNA can be detected as early as 3 hr following infection of host cells.[65-67] Similarly, studies using labeled precursors have demonstrated early initial synthesis of RNA that steadily increases to a maximum at approximately 20 hr and then declines (Figure 1).[68] Interestingly, the profile for incorporation of labeled precursors into DNA is later in the cycle and coincides with the conversion of RBs to EBs (Figure 1). More detailed information concerning the timing of RNA synthesis was obtained by studying the effects of rifampicin on chlamydiae.[69-71] Rifampicin selectively inhibits RNA polymerases of prokaryotes. When rifampicin was added to *C. trachomatis* cultures at the time of infection it completely blocked the differentiation of EBs. Sarov and Becker[71] studied the effect of rifampicin on chlamydiae in vitro and documented that there was a 15 min lag before polymerase activities were completely inhibited. They concluded that the chlamydial DNA-dependent RNA polymerase molecules may be attached to EB DNA in an initiated form and these are only inhibited after completion of a round of RNA synthesis. While these conclusions should be cautiously evaluated in light of the possible contamination of purified EB preparations with RB or intermediate forms,[64] it is apparent that a chlamydia-specific DNA-dependent RNA-polymerase is activated within minutes after infection.

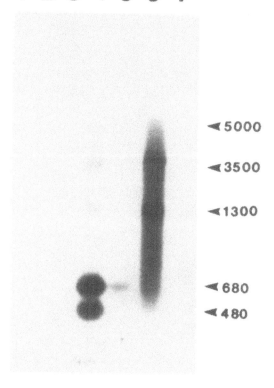

FIGURE 7.   Northern hybridization of serovar L_2 RNA probed with the chlamydial plasmid. Lanes: (1) uninfected HeLa cells; (2) 5 min postinfection; (3) 12 hr postinfection; (4) 24 hr postinfection; (5) 36 hr postinfection; (6) 48 hr postinfection; and (7) 60 hrs postinfection. (From Palmer, L. and Falkow, S., *Plasmid*, 16, 52, 1986. With permission.)

The cloning of chlamydial genes provides the opportunity to investigate the synthesis of specific mRNA species. Northern analysis of the mRNA for the gene that encodes the MOMP has shown that these transcripts can be detected as early as 4 hr following infection of serovar L_2 in L929 cells, and that this gene is transcribed throughout the growth period.[42] Another chromosomal gene, one that encodes a 74,000 m.w. protein, has been used to probe mRNA and this gene is transcribed early during infection.[23] Because message is present only early during the developmental cycle this protein may be involved with differentiation. Palmer and Falkow[23] have used the entire chlamydial plasmid to identify specific transcripts during the developmental cycle (Figure 7). These results provide the first clues toward understanding the "cryptic" functions of the chlamydial plasmid. Two predominant plasmid genes are transcribed at approximately 24 hr following infection; a time of maximum RB multiplication and just preceeding the initiation of differentiation pathways into EBs. Other plasmid genes are not detected until approximately 48 hr, a point in the developmental cycle where maximum numbers of organisms are differentiating into EBs. These results suggest that the highly conserved chlamydial plasmids encode functions that are essential for the unique developmental cycle of chlamydiae.

Cloned chlamydial genes have also been used to detect specific mRNA transcripts from gene loci that represent transcriptional units (i.e., operons). The mRNA transcript for the CTR *omp*1 gene is approximately 1400 nucleotides.[41] Upstream of the structural gene are promoter and ribosome binding site domains. Downstream of the structural gene is an 11 nucleotide dyad immediately followed by a run of poly-T nucleotides, which together represent a rho-independent terminator. The size of the *omp*1 transcript is consistent with the size of the structural gene plus the flanking regulatory sequences, thus this is a monocistronic gene.[41] Evidence for polycistronic gene systems has been presented by Palmer and Falkow[23] and Bavoil et al.[57] The *C. trachomatis* plasmid transcripts detected by Northern hybridization included transcripts of 5000 and 3500 nucleotides, yet the largest plasmid encoded proteins are approximately 48,000 m.w. Since a 48,000 m.w. protein could readily be encoded by transcripts of less than 2000 nucleotides, then these large transcripts most likely represent polycistronic mRNAs. If the protein products for each transcript are linked by one regulatory unit, then they probably have cooperative biochemical functions. It is of interest that the 3500 nucleotide molecule is predominantly transcribed only late in the developmental cycle. The characterization of its *cis*-acting regulatory elements and protein products remains an important issue for chlamydial plasmid genetics. Bavoil et al.[57] described a DNA clone that produces both the 15,000 and 12,000 m.w. cysteine rich proteins. Because these are encoded on a small DNA fragment, they concluded that these two genes are probably genetically linked. This is compatible with the facts that these two proteins are only expressed late in the developmental cycle and that they may participate in the disulfide bonded structure of the outer membranes of chlamydiae.[57] The regulatory and structural elements of other chromosomal genes, especially those that are temporally regulated (i.e., 60K outer membrane protein), need to be investigated to begin to understand the genetics of developmental cycle regulation.

Evidence is growing that the chlamydial DNA-dependent RNA polymerase interacts with promoter recognition sequences which are quite different from those of other eubacteria. Studies by Palmer and Falkow[55] confirm the observations of others[47,50] that chlamydial genes are often not well expressed in *E. coli* hosts. The differences in expression products between in vivo and in vitro transcription/translation systems may reflect the failure of *E. coli* DNA-dependent RNA polymerase to efficiently recognize chlamydial promoter sequences. Although such observed differences could be caused by high rates of catabolic turnover of foreign proteins in *E. coli*, preliminary evaluation of promoter regions for chlamydial *omp*1 genes suggest that these regions do not show high correlation to *E. coli* promoter concensus sequences.[42] Furthermore, when CTR *omp*1 promoters are used to direct synthesis of *E. coli* genes, high levels of expression are not observed.[42] Attempts to detect genetic complementation of *E. coli* mutants using cloned chlamydial DNA have also been unsuccessful.[55] However, the cloning and expression of some chlamydial genes, in particular, the chlamydial glycosyl transferase,[54] the 74K antigen,[44,55,56] and the 19K antigen[43] have been very successful. Consequently, genetic complementation approaches may prove useful for investigations of some chlamydial metabolic pathways. Together these findings suggest that chlamydial and *E. coli* RNA polymerases differ in their gene class recognition of promoter structures. This is of interest because holenzyme specificities could play a role in the sequential and temporal transcription and expression of chlamydial genes, as has been shown for the temporal regulation of sporulation pathways for *Bacillus subtilis*.[72]

Northern analysis of transcriptional units are beginning to provide information concerning the organization of chlamydial genes, as well as their temporal sequence of expression. The identification and characterization of operons and operator DNA will provide important information concerning the genetics of chlamydial regulation of transcription. In addition, coordinately expressed protein products from polycistronic operons should lead to an understanding of the putative functional interrelationships of these proteins. mRNA chromosomal mapping will ultimately be important in developing linkage relationships that also may identify significant functional interactions. Identification of specific mRNA species that are sequentially and temporally regulated are providing important new directions for genetic research into fundamental biological problems of chlamydiae.

VI. FUTURE DEVELOPMENT OF CHLAMYDIAL GENETICS

To date, chlamydial genetics are rudimentary. However, the recent application of molecular genetic techniques to chlamydial research is beginning to provide the opportunity to develop specific reagents to examine the macromolecular and intramolecular organization of the chlamydial genome, insights into the regulation of specific genes, and the genetics and molecular structure of outer membrane components. This has all been accomplished without a genetic system. The development of a system for genetic exchange and complementation in chlamydiae depends upon the construction and use of appropriate vectors to introduce and exchange genetic information, as well as the identification and selection of defined mutants. An appreciation of the evolutionary and molecular distance of chlamydiae from other eubacteria, suggest that transformation systems must rely upon the use of chlamydial regulatory and replication sequences to ensure success. With these tools essential questions may be addressed regarding the genetics of unique elements of chlamydial biology, pathogenesis, and virulence.

A. Vectors

Two potential vectors are currently available which might be developed to introduce DNA into chlamydiae: (1) a chlamydial bacteriophage and (2) the chlamydial plasmid. Richmond et al.[73] have provided electron microscopy data that a *C. psittaci* strain has a bacteriophage. Passage of this phage to a different *C. psittaci* strain has been described by Bevan and Labram.[74] The introduction of DNA using bacteriophage systems is the most rapid and efficient approach for most bacterial systems. Unfortunately, nothing is known about the biology or genetic structure of the chlamydial phage. Consequently much more information is needed to use this approach for transfection of chlamydiae. In contrast, considerable information is being gathered concerning the chlamydial plasmids.

Mapping of essential plasmid open reading frames and regulatory and replication sequences will identify potential cloning sites in which genes may be introduced without disruption of essential functions. Actually, because all chlamydiae contain this plasmid, most functions would be complemented, thus the origin of replication and any transfer and compatibility functions are probably the only essential components that must be maintained. It seems unlikely that chlamydial EBs would be competent hosts for the uptake of naked DNA, and RBs are not infectious, thus plasmid DNA would have to be introduced either at the time of infection where it may be endocytized along with EBs, or infected cells could be directly transformed using techniques developed for mammalian cell transformation. Chlamydial genes that contained flanking regulatory sequences could be cloned directly into the plasmid. For example, cloning a trachoma biovar *omp*1 gene into a serovar L$_2$ plasmid and transforming serovar L$_2$ organisms, transformants could be identified and isolated using monoclonal antibodies to detect mosaic presentation of both MOMP species. Foreign genes may likely require chlamydial promoter sequences. The use of *omp*1 gene promoters would be good candiates for use in developing a useful plasmid vector construction because they are expressed early in the growth cycle and, for chlamydiae, are strong promoters.[42] For positive selection a plasmid containing a chloramphenicol resistance gene, with *omp*1 promoters and a multiple cloning site, is a good candidate vector construction. While these or other constructions can be envisioned, it is clear that techniques for transforming chlamydiae are feasible and will be an important area of research.

B. Mutants

The identification and characterization of specific chlamydial mutants would facilitate studies concerning biochemical pathways, genetic regulatory systems, recombination systems, and virulence determinants. The obligate intracellular growth of chlamydiae has made the identification of phenotypic and biochemical markers difficult, and this has generally frustrated efforts to isolate defined mutants. Nevertheless, chlamydial mutants have been generated. Drug resistant strains have been selected for *C. trachomatis*[75] and *C. psittaci*.[76] The general approach is to make repeated passages of organisms with increasing concentrations of drug. Drug resistant has been demonstrated for sulfonamides,[76] penicillin,[77] and chlortetracycline.[75,77] Although genetic studies with these resistant strains have not been done, it is assumed that drug resistance arises by selection of spontaneous mutants, presumably by mechanisms similar to those defined for other bacteria. Multiple resistance has also been obtained. Gordon et al[78] used mixed cultures of a sulfadiazine-resistant strain and a chlortetracycline-resistant strain and readily obtained cloned organisms that were doubly resistant. Evaluations of the growth curves for each strain and the rapidity of selection of double mutants favors the proposition that these double mutants arose from transfer and recombination of genetic information.[78] Molecular genetic studies have yet to be evaluated for these strains; nevertheless, the implications of these findings are that chlamydiae can, and probably do, share genetic information. Pertinent to this issue is whether different chlamydial strains or species can superinfect the same host cell and/or share the same intracellular endocytic vacuole, since this would facilitate the exchange of genetic information. This can now be investigated with the availability of serovar-specific monoclonal antibodies capable of detecting and localizing specific chlamydial strains.

Temperature sensitive (*ts*) mutants are particularly useful for genetic studies. Recently, *ts* mutants for an ovine abortion strain of *C. psittaci* have been selected.[79] For these studies, Rodolakis and Bernard[79] mutagenized infected cell cultures with nitrosoguanidine. Following growth at the permissive temperature (35°C) organisms were harvested, replicate plated, and replicate plates were grown at 35°C and 39.5°C. Plaque purified *ts* mutants were found that displayed lowered titers, lowered plating efficiency, and unusual plaque and inclusion morphology when grown at the restrictive temperature. Although the molecular basis for these mutants have not been evaluated they have been effectively used to vaccinate ewes.[80]

An uninvestiged research area of interest is the identification and role of mobile genetic elements (i.e., transposons and insertion sequence (IS) elements) in chlamydial genetics. Transposons and related elements are defined genetic entities that translocate to different, and often specific, chromosomal or extrachromosomal sites. This results in DNA rearrangements, altered expression of adjacent genes, and some transposons carry antibiotic resistance genes. Evaluation of characteristic repetitive DNA segments in chlamydial genomic DNA could lead to the identification of transposons and IS elements. Significantly, these can be used to study chlamydial chromosomal organization and used to generate chlamydial mutants. Because mobile genetic elements universally provide an important source of variability, it will also be of interest to study these functions as they relate to the generation of chlamydial diversity.

As the knowledge increases regarding specific molecular components, selective amino acid requirements, and biochemical pathways of chlamydiae, this information adds to the repertoire of phenotypic markers that could be used for genetic studies of chlamydiae. It can certainly be anticipated that with the growing use of molecular genetic tools in chlamydial research, and the likelihood of developing effective genetic transformation systems, that research into the genetic mechanisms and organization of chlamydiae will have a profound impact upon our knowledge of the biological and pathogenic properties of this unique class of intracellular parasites.

REFERENCES

1. **Taylor, H. R.,** Ocular models of chlamydial infections, *Rev. Infect. Dis.,* 7, 737, 1985.
2. **Grayston, J. T., Wang, S. P., Yeh, L. J., and Kuo, C. C.,** Importance of reinfection in the pathogenesis of trachoma, *Rev. Infect. Dis.,* 7, 717, 1985.
3. **Hanna, L., Dawson, C. R., Briones, O., Thygeson, P., and Jawetz, E.,** Latency in human infections with TRIC agents, *J. Immunol.,* 101, 43, 1968.
4. **Meyer, K. F. and Eddie, B.,** Latent psittacosis infections in mice, *Proc. Soc. Exp. Biol. Med.,* 30, 483, 1933.
5. **Lee, C. K. and Moulder, J. W.,** Presistent infection of mouse fibroblasts (McCoy cells) with a trachoma strain of *Chlamydia trachomatis, Infect. Immun.,* 32, 822, 1981.
6. **Moulder, J. W., Levy, N. J., and Schulman, L. P.,** Persistent infection of mouse fibroblasts (L cells) with *Chlamydia psittaci:* evidence for a cryptic chlamydial form, *Infect. Immun.,* 30, 874, 1980.
7. **Jenkin, H. M.,** Preparation and properties of cell walls of the agent of meningopneumonitis, *J. Bacteriol.,* 80, 639, 1960.
8. **Moulder, J. W.,** Structure and chemical composition of isolated particles, *Ann. N.Y. Acad. Sci.,* 90, 92, 1962.
9. **Sarov, I. and Becker, Y.,** Trachoma agent DNA, *J. Mol. Biol.,* 42, 581, 1969.
10. **Lovett, M., Kuo, C. C., Holmes, K., and Falkow, S.,** Plasmids of the genus *Chlamydia,* in *Current Chemotherapy and Infectious Disease,* Vol. 2, Nelson, J. D. and Grassi, C., Eds., American Society for Microbiology, Washington, D.C., 1250, 1980.
11. **Moulder, J. W.,** The relation of the psittacosis group *(Chlamydia)* to bacteria and viruses, *Annu. Rev. Microbiol.,* 20, 107, 1966.
12. **Kingsbury, D. T. and Weiss, E.,** Lack of deoxyribonucleic acid homology between species of the genus *Chlamydia, J. Bacteriol.,* 96, 1421, 1968.
13. **Gerloff, R. K., Ritter, D. B., and Watson, W. O.,** Studies on thermal denaturation of DNA from various chlamydiae, *J. Infect. Dis.,* 121, 65, 1970.
14. **Weiss, E., Schramek, S., Wilson, N. N., and Newman, L. W.,** Deoxyribonucleic acid heterogeneity between human and murine strains of *Chalmydia trachomatis, Infect. Immun.,* 2, 24, 1970.
15. **Tamura, A.,** Biochemical studies on the mechanism of infection of meningopneumonitis agent, *Annu. Rep. Inst. Virus Res. Kyoto Univ.,* 7, 1, 1964.
16. **Higashi, N.,** Studies on chlamydia and togaviruses, *Annu. Rep. Inst. Virus Res. Kyoto Univ.,* 18, 3, 1975.
17. **Kingsbury, D. T.,** Estimate of the genome size of various microorganisms, *J. Bacteriol.,* 98, 1400, 1969.
18. **Gerloff, R. K., Ritter, D. B., and Watson, R. O.,** DNA homology between the meningopneumonitis agent and related microorganisms, *J. Infect. Dis.,* 116, 65, 1966.
19. **Schechter, E. M.,** Synthesis of nucleic acid and protein in L cells infected with the agent of meningopneumonitis, *J. Bacteriol.,* 91, 2069, 1966.
20. **Becker, Y., Loker, H., Sarov, I., Asher, Y., Gutter, B., and Zakay-Rones, Z.,** Studies on the molecular biology of trachoma agent, in *Trachoma and Related Disorders Caused by Chlamydial Agents,* Nichols, R. L., Ed, Excepta Medica, New York, 1971, 13.
21. **Rosenkranz, H. S., Gutter, B., and Becker, Y.,** Studies on the developmental cycle of *Chlamydia trachomatis:* selective inhibition of hydroxyurea, *J. Bacteriol.,* 115, 682, 1973.
22. **Tanaka, A.,** Detection of DNA polymerase activity in *Chlamydia psittaci, Jpn. J. Exp. Med.,* 46, 181, 1976.
23. **Palmer, L. and Falkow, S.,** A common Plasmid of *Chlamydia trachomatis, Plasmid,* 16, 52, 1986.
24. **Hyypia, T., Larsen, S. H., Stahlberg, T., and Terho, P.,** Analysis and detection of chlamydial DNA, *J. Gen. Microbiol.,* 130, 3159, 1984.
25. **Joseph, T., Nano, F. E., Garon, C. F., and Caldwell, H.,** Molecular characterization of *Chlamydia trachomatis* and *Chlamydia psittaci* plasmids, *Infect. Immun.,* 51, 699, 1986.
26. **Clarke, I. N. and Hatt, C.,** In vitro transcription/translation analysis of cloned plasmid DNA from *Chlamydia trachomatis* serovar L1, in *Chlamydial Infections,* Oriel, D., Ridgway, G., Schachter, J., Taylor-Robinson, D., and Ward, M., Eds., Cambridge University Press, London, 1986, 85.
27. **Peterson, E. M. and de la Maza, L. M.,** Characterization of *Chlamydia* DNA by restriction endonuclease cleavage, *Infect. Immun.,* 41, 604, 1983.
28. **Peterson, E. M., and de la Maza, L. M.,** Restriction endonuclease analysis of chlamydia DNA, in *Chlamydial Infections,* Oriel, D., Ridgway, G., Schachter, J., Taylor-Robinson, D., and Ward, M., Eds., Cambridge University Press, London, 1986, 81.
29. **McClenagan, M., Herring, A. J., and Aitken, I. D.,** Comparison of *Chlamydia psittaci* isolates by DNA restriction endonuclease analysis, *Infect. Immun.,* 45, 384, 1984.
30. **McClelland, M., Kessler, L. G., and Bittner, M.,** Site-specific cleavage of DNA at 8- and 10-base pair sequences, *Proc. Natl. Acad. Sci. U.S.A.,* 81, 983, 1984.

31. **Schwartz, D. C. and Cantor, C. R.,** Separation of yeast chromosome-sized DNAs by pulsed field gradient gel electrophoresis, *Cell,* 37, 67, 1984.
32. **Tamura, A. and Manire, G. P.,** Preparation and chemical composition of the cell membranes of developmental reticulate forms of meningopneumonitis organisms, *J. Bacteriol.,* 94, 1184, 1967.
33. **Becker, Y.,** The chlamydiae: molecular biology of procaryotic obligate parasites of eucaryocytes, *Microbiol. Rev.,* 42, 274, 1978.
34. **Storz, J. and Spears, P.,** Chlamydiales: properties, cycle of development and effect on eukaryotic host cells, in *Microbiology and Immunology,* Springer-Verlag, New York, 1978, 167.
35. **Tamura, A. and Iwanaga, M.,** RNA synthesis in cells infected with meningopneumonitis agent, *J. Mol. Biol.,* 11, 97, 1965.
36. **Sarov, I. and Becker, Y.,** RNA in the elementary bodies of trachoma agent, *Nature (London),* 217, 849, 1968.
37. **Gutter, B., Asher, Y., Cohen, Y., and Becker, Y.,** Studies on the developmental cycle of *Chlamydia trachomatis:* isolation and characterization of the initial bodies, *J. Bacteriol.,* 115, 691, 1973.
38. **Gutter, B. and Becker, Y.,** Synthesis and maturation of ribosomal RNA during the developmental cycle of trachoma agent, a prokaryotic obligate parasite of eukaryocytes, *J. Mol. Biol.,* 66, 239, 1972.
39. **Palmer, L., Klevan, K., and Falkow, S.,** 16S ribosomal RNA genes of *Chlamydia trachomatis,* in *Chlamydial Infections,* Oriel, D., Ridgway, G., Schachter, J., Taylor-Robinson, D., and Ward, M., Eds., Cambridge University Press, London, 1986, 89.
40. **Weisburg, W. G., Hatch, T. P., and Woese, C. R.,** Eubacterial origin of chlamydiae, *J. Bacteriol.,* 167, 570, 1986.
41. **Stephens, R. S., Mullenbach, G., Sanchez-Pescador, R., and Agabian, N.,** Sequence analysis of the major outer membrane protein gene from *Chlamydia trachomatis* serovar L2, *J. Bacteriol.,* 168, 1277, 1986.
42. **Stephens, R. S., Edman, U., and Wagar, E.,** Developmental regulation of tandem promoters for the major outer membrane protein gene of *Chlamydia trachomatis, J. Bacteriol.,* in press.
43. **Wenman, W. M. and Lovett, M. A.,** Expression in *E. coli* of *Chlamydia trachomatis* antigen recognized during human infection, *Nature (London),* 269, 68, 1982.
44. **Stephens, R. S., Kuo, C. C., and Agabian, N.,** Expression of a 74,000 m.w. *Chlamydia trachomatis* protein in *E. coli,* Annu. Meet. Am. Soc. Microbiol., Washington, D.C., Abstr. B29, 35, 1982.
45. **Young, R. A. and Davis, R. W.,** Efficient isolation of genes by using antibody probes, *Proc. Natl. Acad. Sci. U.S.A.,* 80, 1194, 1983.
46. **Vieira, J. and Messing, J.,** The pUC plasmids, and M13 mp7-derived system for insertion mutagenesis and sequencing with synthetic universal primers, *Gene,* 19, 259, 1982.
47. **Stephens, R. S., Kuo, C. C., Newport, G., and Agabian, N.,** Molecular cloning and expression of *Chlamydia trachomatis* major outer membrane protein antigens in *Escherichia coli, Infect. Immun.,* 47, 713, 1985.
48. **Allan, I., Cunningham, T. M., and Lovett, M. A.,** Molecular cloning of the major outer membrane protein of *Chlamydia trachomatis, Infect. Immun.,* 45, 637, 1984.
49. **Stephens, R. S., Tam, M. R., Kuo, C. C., and Nowinski, R. C.,** Monoclonal antibodies to *Chlamydia trachomatis:* antibody specificites and antigen characterization, *J. Immunol.,* 128, 1083, 1982.
50. **Nano, F. E., Barstad, P. A., Mayer, L. W., Coligan, J. E., and Caldwell, H. D.,** Partial amino acid sequence and molecular cloning of the encoding gene for the major outer membrane protein of *Chlamydia trachomatis, Infect. Immun.,* 48, 372, 1985.
51. **Bavoil, P., Ohlin, A., and Schachter, J.,** Role of disulfide bonding in outer membrane structure and permeability in *Chlamydia trachomatis, Infect. Immun.,* 44, 479, 1984.
52. **Nano, F. E.,** personal communication, 1986.
53. **Stephens, R. S., Sanchez-Pescador, R., Wagar, E., Inouye, C., and Urdea, M.,** Diversity of the major outer membrane proteins of *Chlamydia trachomatis, J. Bacteriol.,* 169, 3879, 1987.
54. **Nano, F. E. and Caldwell, H. D.,** Expression of the chlamydial genus-specific lipopolysaccharide epitope in *Escherichia coli, Science,* 228, 742, 1985.
55. **Palmer, L. and Falkow, S.,** Characterization of cloned genes from *Chlamydia trachomatis,* in *Microbiology 1986,* American Society for Microbiology, Washington, D. C., 1986, 91.
56. **Kaul, R. and Wenman, W. M.,** Cloning and expression in *Escherichia coli* of a species-specific *Chlamydia trachomatis* outer membrane antigen, *FEMS Microbiol. Lett.,* 27, 7, 1985.
57. **Bavoil, P., Palmer, L., Gump, D., and Falkow, S.,** Characterization and cloning of cysteine-rich proteins from *Chlamydia trachomatis:* a preliminary report, in *Chlamydial Infections,* Oriel, D., Ridgway, G., Schachter, J., Taylor-Robinson, D., and Ward, M., Eds., Cambridge University Press, London, 1986, 97.
58. **Pettijohn, D. E.,** Structure and properties of the bacterial nucleoid, *Cell,* 30, 667, 1982.
59. **Costerton, J. W., Poffenroth, L., Wilt, J. C., and Kordova, N.,** Ultrastructural studies of the nucleoids of the pleomorphic forms of *Chlamydia psittaci* 6BC: a comparison with bacteria, *Can. J. Microbiol.,* 22, 16, 1976.

60. **Popov, V., Eb, F., Lefebvre, J. F., Orfila, J., and Viron, A.,** Morphological and cytochemical study of *Chlamydia* with EDTA regressive technique and Gautier staining in ultrathin frozen section of infected cell cultures: a comparison with embedded material, *Ann. Microbiol. (Inst. Past.),* 129 B, 313, 1978.
61. **Wagar, E. A. and Stephens, R. S.,** Evidence for a histone-like nucleoprotein in chlamydia, in *Chlamydial Infections,* Oriel, D., Ridgway, G., Schachter, J., Taylor-Robinson, D., and Ward, M., Eds., Cambridge University Press, London, 1986, 51.
62. **Wagar, E. A. and Stephens, R. S.,** unpublished data, 1986.
63. **Hackstadt, T., Todd, W. J., and Caldwell, H. D.,** Disulfide-mediated interactions of the chlamydial major outer membrane protein: role in the differentiation of chlamydiae?, *J. Bacteriol.,* 161, 25, 1985.
64. **Hatch, T. P., Miceli, M., and Sublett, J. E.,** Synthesis of disulfide-bonded outer membrane proteins during the developmental cycle of *Chlamydia psittaci* and *Chlamydia trachomatis, J. Bacteriol.,* 165, 379, 1985.
65. **Pollard, M., Starr, T. J., Moore, R. W., and Tanami, Y.,** Cytochemical changes in human amnion cells infected with psittacosis virus, *Nature (London),* 188, 770, 1960.
66. **Bernkopf, H., Mashiah, P., and Becker, Y.,** Correlation between morphological and biochemical changes and the appearance of infectivity in FL cell cultures infected with trachoma agent, *Ann. N.Y. Acad. Sci.,* 98, 62, 1962.
67. **Pollard, M. and Tanami, Y.,** Cytochemistry of trachoma virus replication in tissue cultures, *Ann. N.Y. Acad. Sci.,* 98, 50, 1962.
68. **Higashi, N., Tamura, A., Iwanaga, M.,** Developmental cycle and reproductive mechanism of the meningopneumonitis virus in L cells, *Ann. N.Y. Acad. Sci.,* 98, 100, 1962.
69. **Becker, Y. and Zakay-Rones, Z.,** Rifampicin—a new antitrachoma drug, *Nature (London),* 222, 851, 1969.
70. **Sarov, I. and Becker, Y.,** DNA-dependent RNA polymerase in trachoma elementary bodies, in *Trachoma and Related Disorders Caused by Chlamydial Agents,* Nichols, R. L., Ed., Excerpta Medica, New York, 1971, 27.
71. **Sarov, I. and Becker, Y.,** Deoxyribonucleic acid-dependent ribonucleic acid polymerase activity in purified trachoma elementary bodies: effect of sodium chloride on ribonucleic acid transcription, *J. Bacteriol.,* 107, 593, 1971.
72. **Trempy, J. E., Morrison-Plummer, J., and Haldenwang, W. G.,** Synthesis of sigma-29, an RNA polymerase specificity determinant, is a developmentally regulated event in *Bacillus subtilis, J. Bacteriol.,* 161, 340, 1985.
73. **Richmond, S. J., Stirling, P., and Ashley, C. R.,** Virus infecting the reticulate bodies of an avian strain of *Chlamydia psittaci, FEMS Microbiol. Lett.,* 14, 31, 1982.
74. **Bevan, B. J. and Labram, J.,** Laboratory transfer of a virus between isolates of *Chlamydia psittaci, Vet. Rec.,* 112, 280, 1983.
75. **Moulder, J. W., Novosel, D. L., and Tribby, I.,** Changes in mouse pneumonitis agent associated with development of resistance to chlortetracycline, *J. Bacteriol.,* 89, 17, 1965.
76. **Golub, O. J.,** Acquired resistance of psittacosis virus to sulfandiazine and effects of chemical antagonists on sulfonamide activity, *J. Lab. Clin. Med.,* 33, 1241, 1948.
77. **Gordon, F. B., Andrew, V. W., and Wagner, J. C.,** Development of resistance to penicillin and to chlortetracycline in psittacosis virus, *Virology,* 4, 156, 1957.
78. **Gordon, F. B., Mamay, H. K., and Trimmer, R. W.,** Studies with drug-resistant strains of psittacosis virus. II. Derivation of strains with dual drug resistance from mixed culture of singly resistant strains, *Virology,* 11, 486, 1960.
79. **Rodolakis, A.,** In vitro and in vivo properties of chemically induced temperature-sensitive mutants of *Chlamydia psittaci* var. ovis: screening in a murine model, *Infect. Immun.,* 42, 525, 1983.
80. **Rodolakis, A. and Bernard, F.,** Vaccination with temperature-sensitive mutant of *Chlamydia psittaci* against enzootic abortion of ewes, *Vet. Rec.,* 114, 193, 1984.

Chapter 7

HOST CELL RELATIONSHIPS

Gerald I. Byrne

TABLE OF CONTENTS

"The invasion of the . . . cell represents an incursion of one genetic system into the precincts of another; the invaded cell capitulates and becomes virtually a new biologic unit complete with its own natural history . . . "

from Helliconia Spring
by Brian W. Aldiss

I. INTRODUCTION

Successful pathogens proceed by a succession of highly specialized adaptations that culminate in a unique relationship with their hosts. Interactions that have evolved between intracellular pathogens and susceptible host cells are sophisticated, with the attendant compromises and concessions that characterize any long-term relationship between individuals. The successful pathogen may cause changes that at first almost go unnoticed. They find a foothold, they become established, and they begin anew the vital processes that lead to the perpetutation of their own existence. This survival strategy works for any organism well suited to its particular niche and for chlamydiae the process is a well orchestrated movement that seems masterfully accomplished. The relationship that chlamydiae have established with their host cells is the story of an intimate association that has evolved between two very different life forms. But, as is true for many intimate relationships, one member often is hurt because the other takes advantage of the situation. Chlamydiae are superb advantage takers, and without external intervention they almost always get their way. The host cell seems powerless to resist their advances and time after time the chlamydiae make contact and penetrate into the very substance of the host. For a time, once inside the host cell, chlamydia appear almost conciliatory. They share nutrients, they do little damage, and even may go unnoticed. But eventually, after the chlamydiae have gotten all they need from the host cell, they appear to turn ruthless. They use the host cell, then change form and destroy it, leaving behind debris as a reminder that initially benign encounters can become malignant.

The purpose of this chapter is to describe how chlamydiae establish this intimate association, how they maintain themselves once they gain entry into the host, and how all of this contributes to the remarkable success of these ubiquitious intracellular pathogens.

II. THE PROCESS OF INTRACELLULAR PARASITISM

A. Problems to be Solved by the Parasite

It is easy to describe the process of intracellular parasitism as a study in ecology. Moulder[1] has used this analogy in his descriptions of faculative and obligate intracellular pathogens. In ecological terms the parasite is thought of as having evolved specialized adaptations to exploit a particular niche, the inside of eukaryotic host cell, for the purposes of acquiring essential nutrients and perpetuating its own gene pool. To do this effectively intracellular pathogens must first make contact with susceptible host cells (attachment stage), then gain entry into the host cell (uptake stage), compete with the host cell for what may be a limiting supply of nutrients (replicative stage), get out of the host cell (exit stage), and finally, get to a new susceptible host or at the very least to a new susceptible host cell (transmission stage).

The particular mechanisms that intracellular pathogens have evolved to efficiently carry out of these stages of development are nearly as varied as are the numbers of different intracellular pathogens. There is no single theme that accurately describes the mechanisms which result in the successful completion of the life processes for intracellular pathogens in general. Indeed, even with the genus *Chlamydia* a single description of the stages is not possible because differences between the several chlamydial biovars exist. Often the differences may appear subtle or trivial but may in fact actually result in very different organismal manifestations that in effect define the various disease syndromes associated with the chlamydiae. For example, *C. psittaci* and the lymphogranuloma venereum (LGV) biovar of *C. trachomatis* do not exhibit a strictly preferred host cell type. Both can cause productive infections within a variety of cell types, including mononuclear phagocytes, and this property no doubt contributes to the systemic spread of these organisms within an infected individual. Other *C. trachomatis* biovars exhibit tropism for mucosal epithelium; therefore, diseases caused by the oculogenital *C. trachomatis* biovars tend to remain localized, or if spread occurs, it is by direct extension along mucosal epithelium rather than by the lymphatic or hematogenous routes. Throughout this chapter when clear distinctions between chlamydial biovars have been described involving interactions between host cells and chlamydiae, they will be detailed, otherwise the reader may assume that all chlamydiae, as far as we know, behave identically with respect to a particular function.

B. The Host Cell as an Environment and a Competing Element

Moulder[2] has developed a theory that considers the intracellular location for chlamydial growth as a hostile rather than a nutrient rich environment. This is principally because although the concentration of essential nutrients in cytoplasmic pools may be plentiful, their availability may be limiting. In a large part this is due to the fact that the host cell is not merely a passive supplier of nutrients essential for chlamydial growth, but rather an active competitor for the available metabolites that are required by both the chlamydiae and the host cell. Thus, there is competition for essential nutrients, and as we shall see, sometimes the host cell has the upper hand and sometimes the chlamydiae do. It is difficult to determine which member of the relationship will win out in any case, but it is clear that the competitive balance in effect nearly always depends upon the particular nutrient. Restriction of essential nutrients may have broad sweeping ramifications with the respect to the interaction between chlamydiae and their host cells and this competition may be a major factor contributing to chlamydial persistence, a consistent feature of virtually all chlamydial infections.

III. MAKING CONTACT WITH THE HOST CELL AND GAINING ENTRY TO THE INTRACELLULAR HABITAT

The chlamydial developmental cycle has been described in a previous chapter (see Chapter 4) and the structural components present on chlamydial elementary bodies that mediate attachment to host cells also has been discussed in detail (see Chapter 3). But the relationship between chlamydiae and their host cells would not be complete without making some mention of the early events that initiate this process. The entire concept of intracellular parasitism centers around the attachment and entry process, and intracellular pathogens have evolved a variety of mechanisms to accomplish this feat. Some time ago it was proposed that chlamydia specify their internalization by virtue of heat-labile adhesins that also function to initiate binding to susceptible host cells.[3,4] Identification of the chlamydial adhesin had proved elusive, but recent evidence implicates two surface proteins (relative molecular mass 31 to 32 kdaltons, 17 to 18 kdaltons) as adhesins for *C. trachomatis*[5,6] and a single 17 to 19 kdalton surface adhesin for *C. psittaci*.[5] Identification and structural analysis of these adhesins has lagged behind other progress in our understanding of the molecular ultrastructural of chlamydiae, probably because the adhesins are not plentiful on intact elementary bodies and their identification was not made until binding-in-gel assays were developed to demonstrate their specific association with host cell surface glycoproteins.[5,6]

Despite the fact that similar adhesins have been identified on the surface of elementary bodies for both *C. psittaci* and *C. trachomatis*, the host cell ligand and the actual endocytic process may be distinct for each chlamydial species. *C. psittaci* enters host cells by a mechanism independent of cytoskeletal elements[7] and is localized to areas of the cytoplasmic membrane that appear to ultrastructurally resemble coated pits.[8] In contrast, *C. trachomatis* uptake requires functional host cell cytoskeletal elements[9] and is not associated with coated pits.[5,9] Endocytosis of *C. psittaci* (note coated pits) and *C. trachomatis* (note microvillus involvement) are compared in Figures 1 and 2.

IV. THE INTRACELLULAR HABITAT

A. Early Events

Once entry of chlamydiae has been effected, two remarkable events occur. Fusion of the chlamydiae-containing endocytic vesicle with lysosomes is inhibited and chlamydiae undergo a metamorphosis that changes the relatively stable, metabolically feeble infectious form (elementary body; EB) to a structurally fragile cell (reticulate body; RB) that is metabolically active and can divide by binary fission. The entire process of differentiation, growth, and division takes place within the membrane-bound vesicle (see Chapter 4 for details), and for every ingested EB, a microcolony of replicating chlamydial RBs results.

Chlamydiae are not unique in their growth being restricted to within vesicles. In fact, most obligate intracellular pathogens do this. A notable exception is the rickettsiae,[10] which grow free in the cytoplasm or, for some species, even in the nucleus. But of the intracellular pathogens that grow within cytoplasmic vesicles, only three inhibit lysosome fusion.

For plasmodium, avoiding lysosomes is not a concern since mammalian erythrocytes are devoid of these intracellular organelles. *Coxiella*,[11] *Leishmania*,[12] and some species of *Mycobacteria*[13] do not inhibit lysosome fusion and even thrive in the acidic conditions that are present subsequent to the fusion event. But certain *Mycoplasma*,[14] *Toxoplasma*,[15] and *Chlamydia*[16] actually remain completely aloof from direct contact with the innards of the host cell. Lysosome fusion does not occur and these pathogens never reach the host cell cytoplasm. They reside within the host cell yet remain apart from the intimacy of the host cell itself. How they manage to do this is just not known. Numerous hypotheses have attempted to explain inhibition of lysosome fusion. Perhaps a metabolite is released by the parasite to prevent fusion or perhaps the pathogens bind host cell membrane ligands to prevent lateral mobility of proteins that may be a prerequisite for fusion. It may be that the segment of the plasma membrane that is initially internalized around the parasites is somehow incapable of fusing with lysosomes. Solving the secret of fusion inhibition is a difficult experimental riddle and no clear cut answer to this question has been definitively found.

An essential point related to the early interactions between chlamydiae and the host cell is that prior to the differentiation of the metabolically inert EB to the metabolically active RB, an event that is not complete until several hours after uptake, any controlling influences exerted by the chlamydiae must be mediated by their ultrastructural composition rather than by their metabolic activity. Thus far, no enzymatic activity has been associated with una-dulterated EB, but if chlamydial enzymes do act to modify the host during the uptake process, then one would predict that these enzymes also function as integral structural components of the chlamydial outer envelope.

B. Intracellular Growth and Development: Modifying the Host Cell

Once chlamydiae establish intracellular residence within vesicles of the eucaryotic host cell, the process of differentiation begins. This differentiation step is the central feature of the chlamydial developmental cycle, and the details have been discussed in Chapter 4. A key event in this process appears to be the reduction of disulfide bonds that extensively crosslink the major outer membrane protein (MOMP) of chlamydial EB.[17] Loss of disulfide-linked MOMP multimers may trigger the differentiation of EB to RB,[18] and also effect changes that result in acquisition of porin activity by RB.[19] Differentiation of EB to RB also

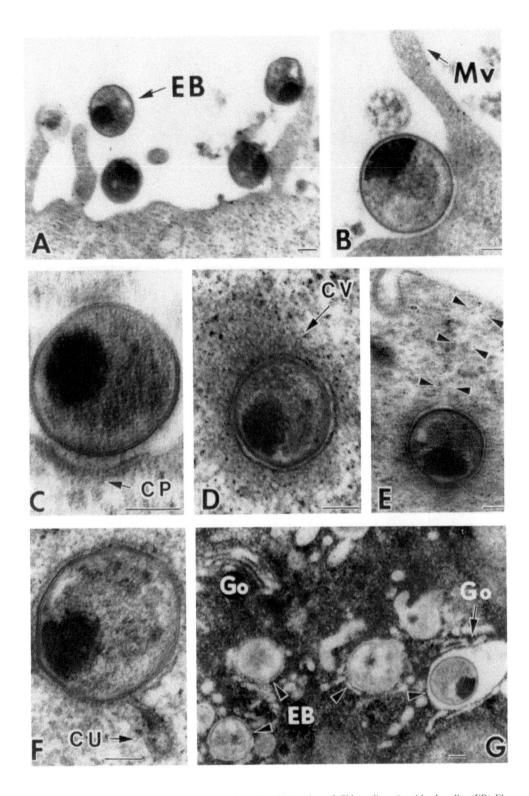

FIGURE 1. Electron micrograph showing attachment and ingestion of *Chlamydia psittaci* by L cells. (EB) Elementary body; (Mv) microvilli; (Cp) coated pit; (Cv) coated vesicle; (Cu) curl; (Go) golgi complex. Bar represents 0.1 μm. (From Hodinka, R. L. and Wyrick, P. B., *Microbiology*, American Society for Microbiology, Washington, D.C., 1986. With permission.)

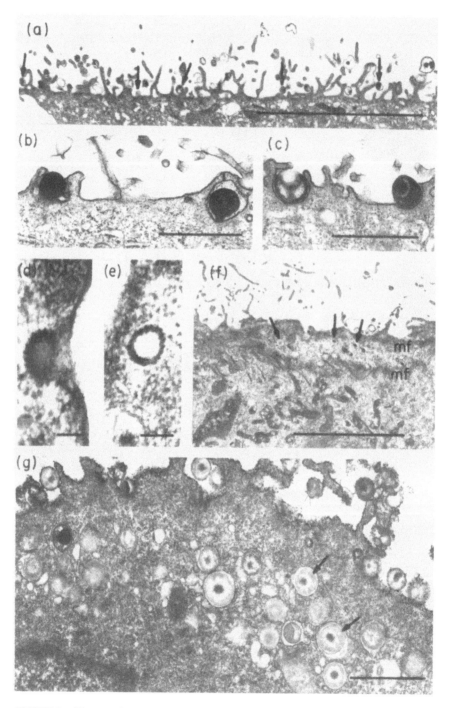

FIGURE 2.   Electron micrograph showing attachment and ingestion of *C. trachomatis* by Hela cells. (a) Attached *Chlamydia* 1 hr after infection at 8°, bar = 5 μm; (b) ingestion 5 min after warming to 36°C, bar = 1 μm; (c) tight fitting endocytic cup, bar = 1 μm; (d and e) clathrin coated vesicles did not contain chlamydiae, bar = 0.1 μm; (f) numerous ingested chlamydiae after incubation at 30°C, bar = 5 μm; and (g) surface labelled thorium hydroxide was not associated with chlamydiae vesicles indicating true endocytic location, bar = 1 μm. (From Ward, M. E. and Murray, A. J., *J. Gen. Microbiol.*, 130, 1765, 1984. With permission.)

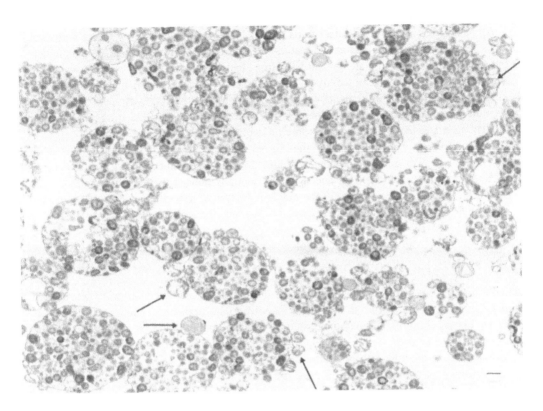

FIGURE 3. Electron micrograph showing isolated chlamydial inclusions with mitochondria (arrows) closely associated with the inclusion membrane. Bar = 1 μm. (From Matsumoto, A., *J. Bacteriol.*, 145, 609, 1981. With permission.)

results in increases of outer membrane fluidity,[20] a feature that is consistent with a loss of structural rigidity, loss of infectivity, and an increase in a membrane permeability, each of which no doubt contributes to the increase of metabolic activity of intracellular RB.

Changes in the morphologic features of the chlamydial cell are not the only alterations that occur as RB begin their replicative phase. Host cell architecture becomes modified as well. For example, the inclusion membrane increases in size to accommodate the expanding RB microcolony. The origin of new inclusion membrane at present is not known. It is probably host derived, but since the increase in membrane occurs in the presence of cycloheximide, an inhibitor of de novo host cell protein synthesis, it must be derived from preformed host cell membranes.[21] There is circumstantial evidence that an inclusion membrane may be derived from michondrial membranes. Matsumoto[22] has provided ultrastructural evidence that mitochondria are closely associated with inclusions in preparations of disrupted infected mouse fibroblasts (Figure 3). If mitochondrial membrane is used to form the inclusion membrane then this may help explain how ATP is transported and made available as an energy source for the developing population of RB.

There is also evidence to indicate that chlamydial lipopolysaccharide (LPS) becomes incorporated into the host cell plasma membrane during the replicative stage of development. Richmond and Sterling[23] proposed that chlamydial LPS is secreted in the form of membraneous blebs, and Wilde et al.[24] found that changes in host cell membrane fluidity were compatible with the hypothesis that chlamydial LPS accumulates as an integral host plasma membrane component during the course of chlamydial development. Functional alterations that accompany these host cell changes have not been described. But these observations indicate that changes in the host cell, other than the actual presence of intracellular chlamydiae, do occur as a result of the interaction between chlamydiae and their host cells.

C. Intracellular Growth and Development: Competing with the Host Cell for Nutrients

All pathogens must exhibit fitness traits that enable them to effectively compete with their hosts for essential metabolites and cofactors. The presence of iron binding proteins (siderophores) that exhibit a greater affinity for iron than host iron binding proteins is a well studied example of how extracellular pathogens compete for essential metabolities.[25] For intracellular pathogens, effective competition with the host cell for essential metabolites is absolutely critical to the success of the parasite, since in most cases (including chlamydiae), the pathogen and the host utilize the same intracellular soluble metabolic pools.

One mechanism that has been exploited by some intracellular pathogens is to shut down host cell metabolism early in the course of infection and establish conditions where parasite nucleic acids and proteins are produced in an essentially noncompetitive environment. This provides an efficient mechanism for the rapid production of new progeny but it also ensures rapid environmental (host cell) collapse and thus limits the length of time any infected cell supports the growth of the parasite.[1] In contrast to host-cell shut down during the course of chlamydial replication, host cell metabolism proceeds nearly unimpeded for a considerable length of time, and chlamydiae acquire required metabolites at different efficiencies from the host, depending on the particular metabolite. For example, chlamydiae utilize exogenous thymidine pools very poorly,[26] and thymidine must be produced by *Chlamydia* via a salvage pathway from uridine pools. In contrast, *Chlamydia* have apparently equal access with the infected host cell to other deoxynucleoside triphosphate pools,[27] and thus compete with the host cell very effectively for these nucleic acid precursors. In fact, chlamydiae are absolutely dependent upon the host cell for chemical energy stored as ATP and have evolved transport mechanisms that pump ATP in and ADP out,[28] just the reverse process of what is accomplished by the mitochondria.

Competition for essential amino acids between chlamydiae and their host cells appears to be of crucial importance in the interaction between chlamydiae and the host. Hatch[29] showed that isoleucine was essential for the intracellular replication of *C. psittaci*, but isoleucine pools became available for *C. psittaci* protein synthesis only after host cell requirements were met. Thus, under limiting isoleucine concentrations, host metabolism continued, and chlamydial replication was curtailed. The amino acid requirements for various chlamydial strains have been defined in cell culture systems. Karayiannes and Hobson[30] found that an E serovar of *C. trachomatis* required seven amino acids for growth, obtained only poor growth when five other amino acids were omitted, and grew normally in the absence of four amino acids (Table 1). Allan and Pearce[31,32] extensively characterized amino acid requirement for all *C. trachomatis* serovars and representative *C. psittaci* strains and found that requirements for amino acids appeared to be disease-syndrome related. Whether or not this observation relates to availability of amino acids at specific epithelial tissue is at present not known.

What is known is that competition between chlamydiae and their host may be a key factor in defining chlamydial persistence. The work of Hatch[29] stressed this point, and more recent findings concerning persistence in cell culture that was mediated by the addition of exogenous gamma interferon[33] also pertains to this point, at least for some *C. psittaci* strains. The phenomenon of persistence will be discussed in more detail in a subsequent section of this chapter.

D. Exiting the Host Cell: The Cycle is Completed

To successfully complete a cycle of development and be transmitted, chlamydiae must complete four processes. First, the noninfectious RB must redifferentiate to infectious EB. Then the chlamydiae must escape from both the membrane bound inclusion and the host cell, and finally, liberated EB must encounter additional susceptible host cells to initiate a new round of replication.

Table 1
AMINO ACID REQUIREMENTS FOR REPRESENTATIVE *CHLAMYDIA*

Chlamydial Strain	Amino acids[a]																	Ref.
	Arg	Asp	Cys	Glu	Gl	Gly	His	Ileu	Leu	Lys	Meth	Phe	Pro	Thre	Tryp	Ty	Val	
C. *trachomatis* (E)	±	−	+	−	−	−	+	+	+	±	±	+	−	±	±	+	+	30
C. *psittaci* (GPIC)	−	ND	−	ND	−	ND	−	−	+	−	−	+	ND	−	−	−	+	31
C. *psittaci* (GBC)	ND	ND	ND	ND	ND	ND	ND	+	ND	ND	ND	ND	ND	ND	+	ND	ND	29
C. *psittaci* (EAE)	−	ND	−	ND	+	ND	+	±	+	−	−	+	ND	−	+	−	+	32
C. *psittaci* (Cal 10)	−	ND	−	ND	±	ND	+	±	±	−	−	+	ND	−	±	−	+	32
C. *psittaci* (FKC)	−	ND	−	ND	±	ND	+	±	±	−	−	+	ND	−	+	−	+	32
C. *trachomatis* (A)	−	ND	−	ND	±	ND	+	±	±	−	−	±	ND	−	−	±	+	32
C. *trachomatis* (B)	−	ND	−	ND	±	ND	+	±	−	−	−	±	ND	−	−	−	+	32
C. *trachomatis* (C)	−	ND	−	ND	±	ND	+	−	−	−	−	±	ND	−	−	−	+	32
C. *trachomatis* (D)	−	ND	−	ND	±	ND	±	−	±	−	−	±	ND	−	−	−	+	32
C. *trachomatis* (E)	−	ND	−	ND	±	ND	±	−	±	−	−	±	ND	−	−	−	+	32
C. *trachomatis* (F)	−	ND	−	ND	±	ND	+	±	+	−	−	+	ND	−	−	−	+	32
C. *trachomatis* (G)	−	ND	−	ND	±	ND	+	−	+	−	+	+	ND	−	−	−	+	32
C. *trachomatis* (H)	−	ND	−	ND	±	ND	−	−	+	−	±	+	ND	−	−	+	+	32
C. *trachomatis* (I)	−	ND	−	ND	±	ND	−	−	+	−	−	−	ND	−	−	−	+	32
C. *trachomatis* (L₂)	−	ND	−	ND	±	ND	−	−	+	−	−		ND	−	−	−	+	32

[a] (+) Essential; (±) better growth in its presence; (−) not required.

The events that trigger RB to EB differentiation are obscure. Statements concerning depletion of essential nutrients or accumulation of toxic metabolites are often used to explain this morphogensis, but experimental data to support these claims have not been published, although they remain reasonable hypotheses to account for redifferentiation. It is no doubt more complex than simple deprivation of, for example, a single essential amino acid, because under those conditions chlamydiae remain intracellular as nonreplicating RB.[29,33] Since disulfide bonds are essential in maintaining EB rigidity, changes in redox potential together with other chemical signals might be important.

Ultrastructural studies have provided evidence that inclusions, or at least chlamydiae within cytoplasmic vesicles, can be extruded from infected host cells, leaving behind a relatively intact host cell cytoplasm.[34,35] The release of inclusions without disrption of the inclusion membrane or the host cell plasma membrane also has been described,[36] but never confirmed. In cell culture systems observed under inverted phase contrast microscopy, the final stages of *C. psittaci* development are associated with an increase in Brownian motion followed by lysis of the inclusion membrane. Free EB then can be seen in a rapid Brownian frenzy within the cytoplasmic remains. The nucleus is often obscured at this stage. Soon the cytoplasmic membrane lyses and suddenly there is no host cell at all, just a pycnotic residue, and the EB, RB, transition forms between EB and RB, and everything else contained within the bounds of the plasma membrane is released and diluted into the external milieu.[61] Stokes[37] reported that a chlamydiae-specific protease is associated with the terminal stages of infection and this activity may help promote host cell membrane lysis and liberation of EB.

V. CHLAMYDIAL PERSISTENCE AT THE LEVEL OF THE SINGLE CELL

A. Defining Conditions for Persistence

The chlamydial developmental cycle, from entry of EB to the release of progeny EB with the attendant lysis of the host cell has been well studied in cell culture systems. The supposition has always been that similar events occur during natural chlamydial infections. However, if normal cycles of infection, with the obligatory lysis of infected host cells, release of hundreds to several thousand EB, each capable of establishing a new infection, occurred regularly during natural chlamydial infections, it is likely that diseases caused by chlamydiae would be overwhelming, acute, and rapidly progressive. This, in fact, is not the case. A typical chlamydial infection, be it ocular, genital, or pulmonary exhibits a more prolonged course. Even during acute stages, when inflammation is prominent and chlamydiae are readily isolated, great numbers of chlamydiae are never recovered. The two major characteristics that are used to describe essentially all chlamydial infections are the presence of an inflammatory response, at least during acute stages of disease, and the predilection for infections to run a chronic course, i.e., persistence. The former condition may be related to the activation of the alternate complement system by EB,[38] with release of well described chemotactic complement-derived peptides, although some initial chlamydiae-mediated host cell damage no doubt also contributes to elaboration of an inflammatory exudate.

The phenomenon of chlamydial persistence has been a research area that has been investigated sporadically over the past 30 years, but it is only during the past few years that any more than phenomenological observations have been made.

B. Persistence in Cell Culture Systems

Manire and Galasso[39] established persistently infected Hela cell cultures in the late 1950s, thus establishing parameters which demonstrated that chlamydial persistence could be studied in cell culture. Bader and Morgan[40-42] suggested that chlamydial persistence might be related to nutritional deprivation, and Hatch[29] confirmed this hypothesis by demonstrating the deprivation of the amino acid isoleucine was sufficient to retard *C. psittaci* development in L cells. Byrne et al.[33,43-45] provided evidence that the immune response generated against chlamydiae, specifically the production of gamma interferon, served to activate host cells to restrict chlamydial replication without eradicating the organisms and thus established cell culture systems to study immune mediated persistence.

Recently, light has been shed on the mechanism involved in immune-mediated persistence. The activation of the tryptophan catabolic enzyme indoleamine-2,3-dioxygenase (IDO) by gamma-interferon resulted in depletion of intracellular tryptophan pools such that some chlamydial strains could not replicate within gamma-interferon treated uroepithelial cells.[33] De la Maza et al.[46] reported that this mechanism of host cell activation was not involved in the inhibition of an LGV biovar of *C. trachomatis* in gamma-interferon-treated McCoy cells. Immune mediated persistence may therefore be mediated by more than one gamma-interferon-mediated change in the host cells. Those chlamydial strains that require tryptophan for growth (Table 1) would be expected to be more susceptible to changes induced by gamma-interferon-mediated activation of IDO than those strains that do not require tryptophan. The latter strains must therefore be inhibited by an as yet undefined gamma-interferon-mediated mechanism.

Persistence in cell culture has also been studied by Moulder and his colleagues. Initially this group described L cells persistently infected with *C. psittaci*.[47] They found that host cells which survived an initial productive infection were resistant to superinfection[48] and carried a cryptic form of chlamydiae that periodically initiated new productive infectious cycles.[47] Host cells that were resistant to superinfection exhibited altered surface protein composition. A new surface protein with relative molecular mass of 35 kdaltons appeared on resistant cells, and two proteins of 60 and 100 kdaltons that were constituents of normal cells were present at reduced levels within the membrane of resistant cells. When L cells were cured of their persistent infection, the patterns of host cell surface labelling also returned to normal.[49] The maintenance of persistence in the L cell-*C. psittaci* model was also found to be at least partially dependent on the composition of the growth medium. Persistently infected L cells growing in Medium 199® always went through overt cycles of chlamydial replication when redispersed. However, cells plated in minimal essential medium remained persistently infected.[50] The differences between the two media were not characterized with respect to factors required to overcome persistence, but it may be of note that Medium 199® contains twice as much tryptophan (20 μg/mℓ) as minimal essential medium 10 μg/mℓ); and since this amino acid appears to be critical for immune-mediated persistence,[33] it may very well be important in the persistent infections described by Moulder.

Persistently infected L cells were studied by Perez-Martinez and Storz using an ovine abortion strain and an arthropathogenic strain of *C. psittaci*.[51] These authors established persistence with the ovine abortion strain, but were unable to initiate a persistently infected population of L cells with an arthropathogenic strain. The former strain was of a biotype that caused persistence in vivo and the latter strain was not. The authors correlated persistence with the ovine abortion strain to temporary changes in the susceptibility of L cells to infection and by the selection of a persistence-adapted substrain. A similar result was obtained by Lee and Moulder[52] (see below) using a trachoma biovar of *C. trachomatis* and McCoy cells. Perez-Martinez and Storz found no evidence for a cryptic form of chlamydiae, but rather defined persistence on the basis of the host cells cycling through stages where chlamydiae attach, but do not form productive infection. Detection of the production of fibroblast (α + β) interferon as contributing factors proved unsuccessful.

Lee[53] and Lee and Moulder[52] studied persistent infections of a trachoma biovar of *C. trachomatis* in McCoy cells. In this system, persistence was established after an initial productive infection which resulted in destruction of most of the host cells. A fresh population of host cells grew from the few surviving host cells. A small proportion of these cells initially contained inclusions, but host cell growth proceeded more rapidly than chlamydial growth and the fraction of inclusion-bearing cells fell to less than 1% of the total population. With time, however, the proportion of inclusion-containing cells rose and a phase of overt host cell destruction eventually ensued. At no time in these cycles of parasite growth, followed by host cell growth, were the host cells resistant to superinfection and no evidence was obtained for a cryptic *C. trachomatis* developmental form, although data were reported which suggested that a *C. trachomatis* substrain was selected for during prolonged incubation of persistently infected host cells that more easily established the persistent state.

Study of chlamydial persistence in cell culture systems no doubt has its limitations, but the work done thus far has provided clues as to how persistence may result in vivo. Clearly, selection of substrains that retain serovar specific antigenicity may provide important insight concerning parasite-mediated adaptations that lead to persistence. The potential for the presence of a cryptic chlamydial form surviving intracellularly to eventually result in the persistent state remain an intriguing, if unconfirmed, possibility. But in vivo persistence must be more complex than what has been reproduced in cell culture systems. The potential for soluble immune mediators, such as gamma interferon, as contributing factors that alter host cell physiology to promote persistence has been discussed;[33,43-45] but each of these in vitro model systems fall short of the true nature of chlamydial persistence. Until experimental systems are devised that combine selection of chlamydial substrains, host cells in culture that more accurately reflect in vivo host cells, and a more precise idea of immune involvement our understanding of chlamydial persistence will remain incomplete.

C. Parasitizing the Parasite: A Role in Persistence?

One interesting development in the ecology of chlamydial infections that may influence the way in which chlamydiae interact with their host cells has been the discovery of viruses that infect chlamydiae. Chlamydial phages were initially discovered in chlamydiae isolated from clams and oysters from the Chesapeake Bay.[54] Richmond et al.[55] found an avian *C. psittaci* RB infected with phages. The viruses grew in crystalized arrays and eventually disrupted RB, causing release of the viruses which attached to other RB within the same inclusion (Figure 4). Viruses were isolated in cesium chloride gradients from infected monolayers and were found to be polyhedral with a diameter of 22nm. Bevan and Labrain[56] have successfully transferred the virus from one chlamydial strain to another. This work is significant in that it may provide a method that will enable researchers to transfer nucleic acid between chlamydial strains or from other organisms to chlamydiae, a process that heretofore has been lacking and has greatly hampered the study of chlamydial genetics. Another possibility that the discovery of chlamydial phages offers may relate to chlamydial persistence. A role for infection of chlamydiae by phages leading to the development of the persistent state has not been examined, but may help explain the nature of cryptic chlamydiae and the apparent resistance of cryptically infected cells to superinfection.

VI. CHLAMYDIAL VIRULENCE AT THE LEVEL OF THE SINGLE CELL

Determination of virulence factors for any pathogen is a complicated matter, and chlamydiae have proved to be no exception to this axiom. Study of virulence at the cellular level for chlamydiae is very limiting since the pathogenesis of chlamydial diseases is as much related to host responses as it is to any direct effects that may be described for the chlamydiae per se. Clearly, attachment and ingestion are related to virulence, as is inhibition of lysosome fusion. Each of these parameters has been effectively studied at the level of the single cell and essential aspects of these processes have been described above and in previous chapters. The role of immunity in the pathogenesis of chlamydial disease will be considered in other chapters of this volume. There is one additional phenomenon that has been studied both in vivo and in cell culture; that being the induction of an immediate toxic

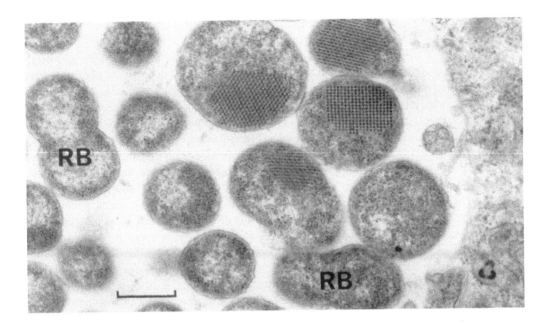

FIGURE 4.   Electron micrograph showing chlamydial reticulate bodes (RB) infected with crystaline arrays of phages. (From Richmond, R. J., Stirling, P., and Ashley, C. R., *FEMS Microbiol. Lett.* 14, 33, 1982. With permission.)

response associated with either injection of large numbers of chlamydiae into susceptible experimental animals or the presence of greater than 100 ID_{50} for host cells in culture systems. No firm correlations exist to conclude that in vivo and in vitro immediate toxicity are the result of similar effects, but both will be described below.

Injection of large numbers of chlamydiae is toxic for mice. Death occurs within 2 to 8 hr after i.v. injection and this toxicity can be prevented by immunization of animals with formalin or phenol killed organisms.[57] Manifestation of chlamydial toxicity are similar to endotoxic shock, and since chlamydiae possess lipopolysaccharide with chemical properties similar to that found in enteric bacteria,[58] it is tempting to conclude that at least in vivo, immediate toxicity is related to chlamydial endotoxic shock. There are, however, two important pieces of information that would argue against this. In the first place, immunization against immediate toxicity is serovar specific, and chlamydial endotoxin is a genus-specific antigen.[58] Secondly, injection of large numbers of *Chlamydia* i.v. into endotoxin responsive (C3H/HeN) and endotoxin insensitive (C3H/HeJ) mice resulted in identical manifestation of immediate toxicity.[59]

Immediate toxicity has also been described in cell culture,[60] and results from these experiments may have relevance concerning the immediate toxic effects observed in vivo. In L cells, very large multiplicities of chlamydiae (500 to 1000 per host cell) resulted in immediate toxicity. The toxic response was related to the ingestion event per se and was independent chlamydial differentiation or replication. In vivo correlates of the events described at the single cell level might include ingestion of chlamydiae and lysis of substantial numbers polymorphonuclear leukocytes, monocytes, or other cell types that normally release factors that act to mediate toxic manifestations similar to those described in response to endotoxin. If in vivo toxicity caused by chlamydiae is related to the ingestion event, the manifestations of chlamydial toxicity would mimic endotoxin shock but would be mediated by an entirely different mechanism.

VII. SUMMARY AND PROSPECTUS

The study of the direct interactions between chlamydiae and their host cells has been instrumental in defining the chlamydial developmental cycle, the dependence of chlamydiae on host cells for ATP, the phenomenon of chlamydial persistence, and understanding parameters related to chlamydial virulence such as attachment and ingestion, inhibition of lysosome fusion, and immediate toxicity. The relationship that chlamydiae have established with their host cells appears to be one whereby chlamydiae subvert normal host cell functions such as endocytosis, production of high energy nucleoside triphosphate, and soluble metabolite pool storage to promote their own survival and at the same time limit host cell damage for as long as possible. Chlamydiae also modify the host cell to better meet their needs. The structure of the chlamydial inclusion itself may be the best example of this, but alteration of host cell membrane fluditiy by insertion of LPS-containing blebs may prove to be an important host-cell modification as well. Even when the host cell responds to limit chlamydial growth, either via immune mediators or by restricting intracellular metabolite pools, *Chlamydia* have adapted such that they remain in the persistent state and then survive intracellularly without replication until conditions more favorable to their survival occur. The subtle checks and balances that exist between chlamydiae and their hosts suggest a long standing evolutionary relationship between these pathogens and the eucaryotic host cell. The ubiquity of chlamydial strains throughout the biosphere also would support this contention.

Study of the interactions between chlamydiae and the host cell can be thought of as an exercise in the study of the life process itself. This little corner of the biosphere can provide clues to a more complete understanding of biology *in toto*. Life, in large part, represents evolutionary steps taken in response to other life forms. The intimate association between *Chlamydia* and their host cells is a fascinating example of the co-evolution of two different life forms and one that can easily be studied within the comfort of our laboratories.

Further increases in our understanding of the interactions between chlamydiae and their host cells will develop from an increased understanding of chlamydial genetics. The means are available and work is underway in several laboratories, but the telling of that story must await further developments.

ACKNOWLEDGMENTS

I thank Caroline Fritsch for typing the manuscript, Dr. J. Carlin, Robin E. Huebner, and Liesa K. Lehmann for helpful discussions, and Dr. Almen Barron for his patience and understanding. Research in my laboratory is supported by Public Health Services grant AI19782 from the National Institutes of Allergy and Infectious Diseases.

REFERENCES

1. **Moulder, J. W.**, Comparative biology of intracellular parasitism, *Microbiol. Rev.*, 49, 298, 1985.
2. **Moulder, J. W.**, Intracellular parasitism: life in an extreme environment, *J. Infect. Dis.*, 130, 300, 1974.
3. **Byrne, G. I.**, Requirements for ingestion of *Chlamydia psittaci* by mouse fibroblasts (L cells), *Infect. Immun.*, 14, 645, 1976.
4. **Byrne, G. I., and Moulder, J. W.**, Parasite specified phagocytosis of *Chlamydia psittaci* and *Chlamydia trachomatis* by L and Hela cells, *Infect. Immun.*, 19, 598, 1978.
5. **Hackstadt, T.**, Identification and properties of chlamydial polypeptides that bind eucaryotic cell surface components, *J. Bacteriol.*, 165, 13, 1986.
6. **Wenman, W. M. and Meuser, R. V.**, *Chlamydia trachomatis* elementary bodies possess proteins which bind to eucaryotic cell membranes, *J. Bacteriol.*, 165, 602, 1986.

7. **Gregory, W. W., Byrne, G. I., Gardner, M., and Moulder, J. W.,** Cytochalasin B does not inhibit ingestion of *Chlamydia psittaci* by mouse fibroblasts (L cells) and mouse peritoneal macrophages, *Infect. Immun.,* 25, 463, 1979.
8. **Hodinka, R. L. and Wyrick, P. B.,** An ultrastructural study on the mode of entry of *Chlamydia psittaci* into L929 cells, *Infect. Immun.,* 54, 855, 1986.
9. **Ward, M. E. and Murray, A.,** Control mechanisms governing the infectivity of *Chlamydia trachomatis* for Hela cells: mechanisms of endocytosis, *J. Gen. Microbiol.,* 130, 1765, 1984.
10. **Winkler, H. H. and Miller, E. T.,** Phospholipase A and the interaction of *Rickettsia prowazekii* and mouse fibroblasts (L929 cells), *Infect. Immun.,* 38, 109, 1982.
11. **Hackstadt, T. and Williams, J. C.,** Biochemical strategum for obligate parasitism of eucaryotic cells by *Coxiella burnetii, Proc. Natl. Acad. Sci. USA,* 78, 3240, 1981.
12. **Chang, K. P. and Dwyer, D. M.,** *Leishmania donovani.* Hamster macrophage interaction *in vitro:* cell entry, intracellular survival and multiplication of amastigotes, *J. Exp. Med.,* 147, 515, 1978.
13. **Hart, P. D., Armstrong, J. A., Brown, C. A., and Draper. P.,** Ultrastructural study of the behavior of macrophage toward parasitic mycobacteria, *Infect. Immun.,* 5, 803, 1972.
14. **Armstrong, J. A. and Hart, P. D.,** Response of cultured macrophages to *Mycobacterium tuberculosis* with observations on fusion of lysosomes with phagosomes, *J. Exp. Med.,* 134, 713, 1971.
15. **Jones, T. C. and Hirsch, J. G.,** The interaction between *Toxoplasma gondii* and mammalian cells. II. The absence of lysosome fusion with phagocytic vacuoles containing living parasite, *J. Exp. Med.,* 136, 1173, 1972.
16. **Friis, R. R.,** Interaction of L cells and *Chlamydia psittaci:* entry of the parasite and host responses to its development, *J. Bacteriol.,* 110, 706, 1972.
17. **Newhall, W. J. V. and Jones, R. B.,** Disulfide-linked oligomers of the major outer membrane protein of chlamydiae, *J. Bacteriol.,* 154, 998, 1983.
18. **Hackstadt, T., Todd, W. J., and Caldwell, H. D.,** Disulfide-mediated interactions of the chlamydial major outer membrane protein: role in the differentiation of chlamydiae?, *J. Bacteriol.,* 161, 25, 1985.
19. **Bavoil, P., Ohlin, A., and Schachter, J.,** Role of disulfide bonding in outer membrane structure and permeability in *Chlamydia trachomatis, Infect. Immun.,* 44, 479, 1984.
20. **Haak, R. A., Collet, B. A., and Newhall, W. J. V.,** Membrane fluidity differences between reticulate bodies and elementary bodies of *Chlamydia trachomatis, J. Bacteriol.,* in press.
21. **Stokes, G. V.,** Cycloheximide-resistant glycosylation in L cells infected with *Chlamydia psittaci, Infect. Immun.,* 9, 497, 1974.
22. **Matsumoto, A.,** Isolation and electron microscope observations of intracytoplasmic inclusions containing *Chlamydia psittaci, J. Bacteriol.,* 145, 605, 1981.
23. **Richmond, S. J. and Sterling, P.,** Localization of chlamydial group antigen in McCoy cell monolayers infected with *Chlamydia trachomatis* or *Chlamydia psittaci, Infect. Immun.,* 34, 561, 1981.
24. **Wilde, C. E., III, Karimi, S. T., and Haak, R. A.,** Cell surface alterations during chlamydial infections, *Microbiology,* in Press.
25. **Weinberg, E. D.,** Iron and infection, *Microbiol. Rev.,* 42, 45, 1978.
26. **Hatch, T. P.,** Utilization of exogenous thymidine by *Chlamydia psittaci* growing in thymidine kinase-containing and thymidine kinase-deficient L cells, *J. Bacteriol.,* 125, 706, 1976.
27. **Hatch, T. P.,** Utilization of L-cell nucleoside triphosphates by *Chlamydia psittaci* for ribonucleic acid synthesis, *J. Bacteriol.,* 122, 393, 1975.
28. **Hatch, T. P., Al-Hassainy, E., and Silverman, J. A.,** Adenine nucleotide and lysine transport in *Chlamydia psittaci, J. Bacteriol.,* 150, 662, 1982.
29. **Hatch, T. P.,** Competition between *Chlamydia psittaci* and L cells for host isoleucine pools: a limiting factor in chlamydial multiplication, *Infect. Immun.,* 12, 211, 1975.
30. **Karayiannis, P. and Hobson, D.,** Amino acid requirements of a *Chlamydia trachomatis* genital strain in McCoy cell cultures, *J. Clin. Microbiol.,* 13, 427, 1981.
31. **Allan, I. and Pearce, J. H.,** Differential amino acid utilization of *Chlamydia psittaci* (strain guinea pig inclusion conjuctivitis) and its regulatory effect on chlamydial growth, *J. Gen. Microbiol.,* 129, 1991, 1983.
32. **Allan, I. and Pearce, J. H.,** Amino acid requriements of strains of *Chlamydia trachomatis* and *C. psittaci* growing in McCoy cells: relationship with clinical syndrome and host origin, *J. Gen. Microbiol.,* 129, 2001, 1983.
33. **Byrne, G. I., Lehmann, L. K., and Landry, G. J.,** Induction of tryptophan catabolism is the mechanism for gamma interferon-mediated inhibition of intracellular *Chlamydia psittaci* replication in T24 cells, *Infect. Immun.,* 53, 347, 1986.
34. **Doughri, A. M., Storz, J., and Altera, K. P.,** Mode of entry and release of chlamydiae in infections of intestinal epithelial cells, *J. Infect. Dis.,* 126, 652, 1972.
35. **de la Maza, L. M. and Peterson, E. M.,** Scanning electron microscopy of McCoy cells infected with *Chlamydia trachomatis, Exp. Molec. Pathol.,* 36, 217, 1982.

36. **Todd, W. J. and Caldwell, H. D.,** The interaction of *Chlamydia trachomatis* with host cells: ultrastructural studies of the mechanism of release of a biovar II strain from Hela 229 cells, *J. Infect. Dis.,* 151, 1037, 1985.

37. **Stokes, G. V.,** Proteinase produced by *Chlamydia psittaci* in L cells, *J. Bacteriol.,* 118, 616, 1974.

38. **Megran, D. W., Stiver, H. G., and Bowie, W. R.,** Complement activation and stimulation of chemotaxis by *Chlamydia trachomatis, Infect. Immun.,* 49, 670, 1985.

39. **Manire, G. P. and Galasso, J.,** Persistent infection of Hela Cells with meningopneumonitis virus, *J. Immunol.,* 83, 529, 1959.

40. **Morgan, H. R.,** Latent viral infection of cells in tissue culture. I. Studies on latent infection of chick embryo tissues with psittacosis virus, *J. Exp. Med.,* 103, 37, 1956.

41. **Bader, J. P. and Morgan, H. R.,** Latent viral infection of cells in tissue culture. VI. Role of amino acids, glutamine, and glucose in psittacosis virus propagation in L cells, *J. Exp. Med.,* 106, 617, 1958.

42. **Bader, J. P. and Morgan, H. R.,** Latent viral infection of cells in tissue culture. VII. Role of water soluble vitamins in psittacosis virus propagation in L cells, *J. Exp. Med.,* 113, 271, 1961.

43. **Byrne, G. I. and Faubion, C. L.,** Lymphokine-mediated microbistatic mechanisms restrict *Chlamydia psittaci* growth in macrophages, *J. Immunol.,* 128, 469, 1982.

44. **Byrne, G. I. and Faubion, C. L.,** Inhibition of *Chlamydia psittaci* in oxidatively active thioglycollate-elicited macrophages: distinction between lymphokine-mediated oxygen-dependent and oxygen-independent macrophage activation, *Infect. Immun.,* 40, 464, 1983.

45. **Byrne, G. I. and Krueger, D. A.,** Lymphokine-mediated inhibition of *Chlamydia* replication in mouse fibroblasts is neutralized by anti-gamma interferon immunoglobulin, *Infect. Immun.,* 42, 1152, 1983.

46. **de la Maza, L. M., Peterson, E. M., Fennie, C. W., and Czarniecki, C. W.,** The anti-chlamydial and anti-proliferative activities of recombinant murine interferon-α are not dependent on tryptophan concentrations, *J. Immunol.,* 135, 4198, 1985.

47. **Moulder, J. W., Levy, N. J., and Schulman, L. P.,** Persistent infection of mouse fibroblasts (L cells) with *Chlamydia psittaci:* evidence for a cryptic chlamydial form, *Infect. Immun.,* 30, 874, 1980.

48. **Moulder, J. W., Levy, N. J., Zeichner, S. L., and Lee, C. K.,** Attachment defect in mouse fibroblasts (L cells) persistently infected with *Chlamydia psittaci, Infect. Immun.,* 34, 285, 1981.

49. **Moulder, J. W., Zeichner, S. L., and Levy, N. J.,** Association between resistance to superinfection and patterns of surface protein labeling in mouse fibroblasts (L cells) persistently infected with *Chlamydia psittaci, Infect. Immun.,* 35, 834, 1982.

50. **Moulder, J. W.,** Inhibition of onset of overt multiplication of *Chlamydia psittaci* in persistently infected mouse fibroblasts (L cells), *Infect. Immun.,* 39, 898, 1983.

51. **Perez-Martinez, J. A. and Storz, J.,** Persistent infection of L cells with an ovine abortion strain of *Chlamydia psittaci, Infect. Immun.,* 50, 453, 1985.

52. **Lee, C. K. and Moulder, J. W.,** Persistent infection of mouse fibroblasts (McCoy cells) with a trachoma strain of *Chlamydia trachomatis, Infect. Immun.,* 32, 822, 1981.

53. **Lee, C. K.,** Factors affecting the rate at which a trachoma strain of *Chlamydia trachomatis* establishes persistent infections in mouse fibroblasts (McCoy cells), *Infect. Immun.,* 33, 954, 1981.

54. **Harshbarger, J. C., Chang, S. C., and Otto, S. V.,** Chlamydiae (with phages), mycoplasmas, and rickettsia in Chesapeake Bay bivalves, *Science,* 196, 666, 1977.

55. **Richmond, S. J., Stirling, P., and Ashley, C. R.,** Virus infecting the reticulate bodies of an avain strain of *Chlamydia psittaci, FEMS Microbiol. Lett,* 14, 31, 1982.

56. **Bevan, B. J. and Labrain, J.,** Laboratory transfer of a virus between isolates of *Chlamydia psittaci, Vet. Rec.,* 112, 280, 1983.

57. **Bell, S. D., Jr., Snyder, J. C., and Murray, E. S.,** Immunization of mice against toxic doses of homologous elementary bodies of trachoma, *Science,* 130, 626, 1959.

58. **Nurminen, M., Leinonen, M., Saikku, P., and Mäkelä, P. H.,** The genus specific antigen of *Chlamydia:* resemblance to the lipopolysaccharide of enteric bacteria, *Science,* 220, 1279, 1983.

59. **Ivins, B. E. and Wyrick, P. B.,** Response of C3H/HeJ and C3H/HeN mice and their peritoneal macrophages to the toxicity of *Chlamydia psittaci* elementary bodies, *Infect. Immun.,* 22, 620, 1978.

60. **Moulder, J. W., Hatch, T. P., Byrne, G. I., and Kellogg, K. R.,** Immediate toxicity of high multiplicities of *Chlamydia psittaci* for mouse fibroblasts (L cells), *Infect. Immun.,* 14, 277, 1976.

61. **Byrne, G. I.,** unpublished data.

Chlamydia as Pathogens

Chapter 8

OVERVIEW OF HUMAN DISEASES

Julius Schachter

TABLE OF CONTENTS

I. TRACHOMA

A. History and Epidemiology

Trachoma is one of the oldest known human diseases. It was described in ancient Egyptian and Chinese writings. Once common throughout the world, it is now a major problem only in certain developing countries.[1] In the 19th century trachoma was endemic in much of western Europe and the U.S. It disappeared with improved standards of living and hygienic conditions, rather than as a result of antitrachoma programs. Trachoma is a disease of poverty and it is most important in areas where environmental sanitation and personal hygiene are poor. Today it is the world's leading preventable cause of blindness, with over 600 million people affected and millions blinded. Parts of the world where trachoma is a particularly large-scale problem include northern and southern subSaharan Africa and parts of Southeast Asia.

Unfortunately, in the endemic area the economics of the involved nations often preclude massive efforts at improving the standard of living. In many of these countries trachoma remains a relatively low priority problem because there are other life threatening or debilitating conditions that are more immediate.

In the hyperendemic area, active trachoma is a disease of young children and virtually all will be infected before they reach 2 years of age.[2] The condition often occurs in familial clusters. Even in the poor environment where trachoma flourishes, it is possible to show an association of trachoma severity with gradation of poverty within the village environment.

Blindness generally does not occur until many years after the active disease has waned. Occasionally children are blinded from corneal damage as a result of secondary bacterial infections. More typically, blindness results from mechanical abrasion of the cornea by the eyelashes that in-turn as a result of eyelids deformed by conjunctival scarring. In hyperendemic areas age-specific rates of blindness often are in excess of 20% above age 60.

The etiologic agent of trachoma was first demonstrated in conjunctival cells of baboons that had been inoculated with scrapings from human trachoma cases. Although it was possible to perform human volunteer inoculations and infect the conjunctivae of subhuman primates, the trachoma organism was not isolated until 1956 when Chinese researchers succeeded in growing the organism in the yolk sac of an embryonated hen's egg.[3] In 1970 a useful serotyping method was developed based on an indirect fluorescent antibody test.[4] This test demonstrated the presence of 15 serovars of *Chlamydia trachomatis*. It was confirmed that only three serotypes (as had been identified in previous mouse toxicity prevention tests) were associated with endemic trachoma.

B. Clinical Description

Although onset can be insidious, the disease often begins as a mucopurulent conjunctivitis, developing into a follicular keratoconjunctivitis. Over time some of the follicles necrose, resulting in scarring of the conjunctivae. The cornea may be involved with keratitis and pannus (neovascularization of the corneal limbus) developing. The only pathognomonic signs of the disease are Herbert's peripheral pits, which results from necrosis of limbal follicles. The scars within the conjunctiva may slowly contract, distorting the eyelid, and causing an in-turning of the eyelashes so that they abrade the cornea. This complex represents the blinding lesions of trachoma, called entropion and trichiasis. Severe lid deformity may not develop until 20 or 30 years after active disease has waned (typically between 5 and 10 years of age). Trachoma often occurs in areas of the world where seasonal outbreaks of bacterial conjunctivitis contribute to ocular pathology and make the disease worse. It is likely that a single chlamydial infection is inadequate to cause severe ocular damage. However, in a trachoma endemic area, pressure of reinfection is great and subsequent infections probably result in enhanced hypersensitivity, resulting in worse disease.

C. Diagnosis

Clinical diagnosis of trachoma is based mainly on findings within the conjunctiva.[5] It focuses on identification of follicular reaction, papillary hypertrophy, scars within the conjunctiva, and on vascularization (pannus) of the cornea. Laboratory diagnosis is often based on demonstration of typical chlamydial inclusions or elementary bodies in stained conjunctival smears. Fluorescent antibody methods are more sensitive than the classical Giemsa stain for this purpose. *Chlamydia trachomatis* may be readily isolated from active cases by use of cell culture systems. This is the most sensitive laboratory test. It is difficult to isolate chlamydiae in the late chronic stages of the disease.

Serologic tests are not useful for diagnosing individual cases. In hyperendemic areas, the great majority of individuals have serum antibodies to *C. trachomatis*. Antibodies in tears correlate with intensity of disease.[6]

D. Prevention and Treatment

Experimental vaccines have induced a short-lived immune response, but some vaccines developed more severe disease, suggesting hypersensitivity to the organism.[1,7] These early vaccines used relatively crude, purified elementary body suspensions as antigen. Currently, no vaccine trials are in progress and there are no commercially available vaccines. Trachoma control is currently based on mass treatment with topical tetracycline ointment of all affected individuals within a village setting and on surgical intervention to correct lid deformities. These efforts are aimed at preventing blindness not at trachoma eradication. Improved living conditions provide the most effective trachoma control.

II. GENITAL TRACT INFECTIONS

A. History and Epidemiology

Sexually transmitted *C. trachomatis* infections were first recognized shortly after the discovery of the trachoma agent in 1907.[1] Ophthalmologists had recognized a form of nongonococcal ophthalmia neonatorum and found characteristic intracytoplasmic chlamydial inclusions in the conjunctivae of infants with that condition. The disease was called inclusion blennorrhea, or inclusion conjunctivitis of the newborn. Similar cytologic studies were performed on genital tract specimens from the parents of affected infants and then on men with nongonococcal urethritis and the cervices of their contacts. The same type of intracellular inclusions were found in urethral and cervical epithelium. Thus, much of the epidemiology and clinical manifestations of sexually transmitted chlamydiae were elucidated by ophthalmologists.[8]

Unfortunately, these cytologic tests for diagnosis were relatively insensitive and quite time consuming, precluding the testing of large numbers of patients. Thus, it was not until the trachoma agent was isolated in 1957 in the yolk sac of embryonated hen's eggs, and tissue culture isolation methods were developed in 1965, that it became possible to test large numbers of patients.[3,9] With the introduction of a serologic test that proved to be a useful epidemiologic tool to identify populations deserving further study, the clinical spectrum of sexually transmitted *C. trachomatis* rapidly expanded.[4,6]

In industrialized societies, *C. trachomatis* is now considered to be the most common sexually transmitted pathogen.[11] In the U.S. it is estimated that more than 3 million new infections occur each year (this may be compared with 1.8 million gonococcal infections and approximately 500,000 genital infections with *Herpesvirus hominis*)[12] Most prevalence and clinical studies have been done in the U.S. and in western Europe. More information is becoming available on the importance of sexually transmitted chlamydiae in the developing world and it appears that these infections are just as important in many developing countries.

In industrialized societies sexually transmitted chlamydial infections are widely distributed among the population. These infections are found at relatively high rates in populations that are at low risk for gonorrhea. For example, in higher socioeconomic class populations gonorrhea is seldom encountered, but chlamydial infections are common. In these settings, as exemplified by students being screened at private colleges, chlamydial prevalence often exceeds the gonococcal infection rate by a factor of 5- to 10-fold in symptomatic men or in asymptomatic women having routine pelvic examinations.[13] In general, it is only in the urban center venereal disease clinics that chlamydial and gonococcal infections occur at the same rate or where, perhaps, gonococcal infections may be more common. In screening studies of women attending family planning clinics for birth control advice, chlamydial infection rates often exceed gonococcal recovery rates by a factor of ten.[14] Lower socio-economic classes have higher infection rates, although the ratio of chlamydial to gonococcal infection rates in these populations tends to be lower than in the more affluent.

Age is the most important risk factor for *C. trachomatis* infection. Younger women have the highest infection rates. A number of studies have shown that approximately 1 in 6 sexually active teenagers attending university adolescent clinics yielded *C. trachomatis* from their cervices during routine pelvic examinations.[15] Other risk factors have been identified. Use of oral contraceptives is associated with higher chlamydial infection rates. Race (which may be a proxy for socioeconomic status) is also a risk factor and black women have been found to have higher infection rates. Sexual preference is another risk factor: *C. trachomatis* is a more common cause of urethritis in heterosexual males than in homosexual males.

B. Clinical Description

In the male, *C. trachomatis* is recognized as the most common cause of nongonococcal urethritis (NGU), being responsible for 35 to 50% of cases.[1,16,17] The condition cannot be diagnosed (i.e., differentiated from gonococcal urethritis) on clinical grounds. The spectrum of chlamydial infection in the male urethra is broad, ranging from producing a frankly purulent discharge to being inapparent. Chlamydial urethritis has an incubation period of 1 to 3 weeks and is usually seen as a mucopurulent discharge in a man suffering from dysuria. Management of this condition does not require an etiologic diagnosis. It can be based on use of Gram stain (exclusion of gonorrhea by failure to demonstrate intracellular diplococci in polymorphonuclear leukocytes) of a urethral smear.

Postgonococcal urethritis is a special subset of NGU, which is found in men who have been successfully treated for gonococcal infection and either develop symptoms shortly after therapy or remain symptomatic, although their gonococcal infections are cured. *C. trachomatis* is the leading cause of this condition, being responsible for 70 to 90% of the cases.[18] Approximately 20% of men with gonorrhea have concomitant chlamydial infections. Betalactam drugs in dosages used to treat gonorrhea are largely ineffective in treating chlamydial infections. Thus, these men have had adequate therapy for their gonococcal infection but will develop a symptomatic chlamydial urethritis. As a result of these findings the Centers for Disease Control in the U.S. has recommended that all cases of gonorrhea in heterosexual males should be treated presumptively for chlamydial infection.[19] Thus, today the current treatment of choice for gonorrhea is a beta-lactam (ampicillin or amoxycillin taken orally with benemid) followed by a week of tetracycline.

Although NGU was long considered to be a trivial condition, it is now recognized that serious complications can result. Ascending infections can occur, resulting in epididymitis.[20] *C. trachomatis* is the leading cause of epididymitis in sexually active young men. Although serologic studies indicate that homosexual males have a high rate of exposure to *C. trachomatis*, they are less likely to develop chlamydial urethritis as diagnosed by culture. Rectal infections are not uncommon. Screening studies in gay men attending venereal disease clinics found approximately 6% of asymptomatic men yielded *Chlamydia* from rectal swabs, as compared to a 12% recovery rate from men with proctitis.[21] *C. trachomatis* can be recovered from the pharynx of sexually active males or females at risk for genital tract infection. The role of this agent in causing pharyngitis is uncertain.

In the female the most commonly affected site is the cervix, where the organism can cause a mucopurulent endocervicitis.[22] This condition is characterized by a mucopurulent endocervical discharge, often together with friability and edema within a zone of ectopy.[23,24] Unfortunately, the clinical spectrum of chlamydial infection in the cervix is quite wide. Asymptomatic and inapparent infections are common. In screening studies of women having routine pelvic examinations for cervical cytology or birth control advice, approximately 70% of the infections detected by isolation of the agent were not associated with any abnormal clinical findings.

Because women with gonorrhea who have a mucopurulent cervical discharge tend to maintain that discharge if they have a chlamydial infection and are treated with beta-lactam drugs, the same rationale for treatment of women with gonorrhea has been applied, as was discussed above for men with gonococcal infection. The same regimens are used. It is even more important to treat women with gonorrhea for chlamydial infection on a presumptive basis. Women appear to have double infection rates approximately twice as high as are observed with men (35 to 45% of women with gonorrhea have a concomitant chlamydial infection) and eradication of the chlamydiae will reduce the subsequent development of salpingitis.[17,25]

Efforts have been made to establish clinical criteria for diagnosing chlamydial cervicitis. Swab tests can be used to demonstrate a purulent discharge (a white swab inserted into the endocervical canal will be colored yellow in the presence of a purulent discharge) or friability (a swab rubbed against the endocervical wall will be reddened by bleeding caused by pressure if the cervix is edematous and friable).[24] These tests have a predictive value of 30 to 50% for *C. trachomatis* infection in different populations and have been used as a guideline for presumptive therapy.

Urethral infections also occur in women and chlamydial infection has been associated with sterile pyuria in some female populations.[26] This association of chlamydial infection with the urethral syndrome was shown in college-age women. It is probably unwise to generalize those findings to other populations of women.

Ascending genital infection is common. *C. trachomatis* is found in the endometrium or fallopian tubes of approximately 25% of women with acute salpingitis in the U.S. and at higher rates in western Europe.[27-29] As expected, younger women are at higher risk. Chlamydial salpingitis tends to be clinically milder than salpingitis associated with gonococcal or mixed anaerobic infections. Chlamydia-infected women usually have a longer prodrome before being admitted to the hospital.[28] *C. trachomatis* is also associated with complications of salpingitis, such as the Fitz-Hugh-Curtis syndrome (perihepatitis). Chlamydiae are important causes of tubal factor infertility and ectopic pregnancy as a result of tubal damage following salpingitis.[30] Unfortunately, because chlamydial salpingitis can be clinically mild, or even inapparent, evidence of chlamydial infection is often first obtained retrospectively by serologic tests performed during evaluation of infertility.

C. Diagnosis

Definitive diagnosis is made by isolation of the agent in cell culture systems.[31] Serology does not play a role in diagnosing uncomplicated genital tract infections, although the higher antibody titers seen by the micro-immunofluorescence (micro-IF) test in complications (epididymitis, salpingitis, Fitz-Hugh-Curtis syndrome, etc.) may provide some support for an etiologic diagnosis. Nonculture methods for chlamydial diagnosis are available. Antigen detection based on use of fluorescein-conjugated monoclonal antibodies or enzyme immunoassay tests may provide greater access to chlamydial diagnosis.[32,33] However, these tests, while being simpler and less expensive than cell culture systems, appear to be less sensitive than the tissue culture methods. It should be stressed that some of chlamydial diseases can be managed without specific diagnosis.[19] For example, urethritis in men can be managed easily based on Gram stain. Women with mucopurulent endocervicitis or salpingitis should be automatically treated for chlamydial infection. Treatment of sex partners is indicated in all of these conditions.

D. Prevention and Treatment

Use of barrier contraceptives will reduce transmission. Treatment of uncomplicated genital tract infection is relatively easy with 2 g tetracycline per day for 7 days, resulting in cure rates in excess of 95%. Upper genital tract infections call for longer courses of therapy. Short-term therapy plays no role in management of chlamydial infections.

Chlamydia control programs are in their genesis. In the past few years the Centers for Disease Control has recommended treatment guidelines for the management of chlamydial infections. With the broader availability of diagnostic tests, the recommendation has been made that chlamydial infections be made reportable. Reporting of nongonococcal urethritis and pelvic inflammatory disease (the former being the best proxy for chlamydial infections and the latter being the most important complication) should provide a firmer data base and the prevalence and incidence of these conditions can be established. When high risk populations are better defined, control efforts can be implemented.

III. NEONATAL INFECTIONS

A. History and Epidemiology

The history of chlamydial infections in infants has established a pattern which has been repeated with many of the chlamydial diseases that have been recognized in the past century. Shortly after the identification of the gonococcus and the introduction of Credé prophylaxis to prevent gonococcal ophthalmia neonatorum, it was recognized that a nongonococcal (or "abacterial") form of ophthalmia neonatorum existed.[1] Epidemiology of this condition was clarified in the early part of the 20th century by ophthalmologists using the cytologic techniques for diagnosis of trachoma. In the 1950s inclusion conjunctivitis of the newborn (ICN) was considered to be a relatively uncommon condition occurring at a rate of one to four cases per five thousand live births. This rate was based on chart reviews identifying severe cases of conjunctivitis who were referred to ophthalmologists for diagnosis. With the advent of tissue culture techniques it became possible to screen the maternal cervix for chlamydial infection and determine true incidence and attack rates for this condition. In two prospective studies, 2 and 6% of all newborns in the study population were found to develop ICN.[10,34] During one of these prospective studies, the potential for *C. trachomatis* to cause pneumonia in infants was recognized.[35] Beem and Saxon then described a characteristic pneumonia syndrome attributable to chlamydial infection.[36] This entity had probably been described in the earlier literature as eosinophilic pertussoid pneumonia.

Approximately 60 to 70% of infants born through a Chlamydia infected birth canal will acquire the infection.[37,38] About one in three of the exposed infants develop ICN, while approximately one in six develops pneumonia. Vaginal and gastrointestinal tract infections also occur, but have no known clinical consequences.

B. Clinical Description

Approximately 5 to 21 days after birth, the infant develops a mucopurulent conjunctivitis. Hyperemia and discharge are the most prominent findings. Follicles are not seen unless the condition persists for longer than a month. ICN is usually self-limiting and will resolve in a few months without treatment. ICN is not considered to be a sight-threatening condition. Corneal damage is minimal, although some keratitis and micropannus can develop. Conjunctival scarring is relatively uncommon, although sheet scarring may follow the disease in infants who develop pseudomembranes. These scars do not result in lid deformity. Occasional cases persist and severe disease can rarely develop, which may threaten vision.

The incubation period for chlamydial pneumonia of infants is usually between 2 and 12 weeks.[38-40] The infants will often have a prodrome of rhinitis and many will have had conjunctivitis. Affected infants are usually afebrile, are markedly tachypneic and occasionally apneic and have a staccato cough. They are hypergammaglobulinemic, particularly in the IgM class. A relative eosinophilia occurs in approximately one-third of affected infants. Radiographs usually show hyperinflation.

The spectrum of chlamydial respiratory involvement in infants is quite broad. Nasopharyngeal infections are quite common.[40] Some infants develop a severe rhinitis, without lower respiratory involvement, that can occasionally interfere with respiration.[41] The pathogenesis of Chlamydia pneumonia probably reflects descending infection and bronchiolitis can occur.[42] Infants with Chlamydia pneumonia typically fall into the category of infants with failure to thrive and often have only mild respiratory distress, with tachypnea being the prominent finding.[36,40] Occasionally these infants have severe respiratory problems and may become apneic and require respiratory assistance. In our experience, approximately three-quarters of the infants can be managed on an outpatient basis. The other infants are severely ill and must be treated in an intensive care nursery.

C. Diagnosis

Conjunctivitis may be diagnosed readily by any of the cytologic tests.[31] Giemsa stain is adequate in diagnosing severe cases of conjunctivitis, while the fluorescent-antibody techniques are quite sensitive. The agent may be easily isolated.

A specific diagnosis for pneumonia may be more difficult because of sampling problems, but the organism can often be isolated from the nasopharynx or tracheobronchial aspirates.[36] Serology may be the test of choice in diagnosing chlamydial pneumonia because of the sampling problems. Infants with chlamydial pneumonia almost always develop high IgM antibody levels and because of their defined exposure (at birth) the diagnosis may be readily established on the basis of a single point titer of specific antichlamydial IgM antibodies > 1:32 in the micro-IF test.[43]

D. Prevention and Treatment

Chlamydial infection in the infant calls for systemic therapy with erythromycin (50 mg/ kg in divided doses each day). Conjunctivitis will respond to 7 to 10 days of therapy, while pneumonia should be treated for 14 to 21 days.[19] Topical therapy is not recommended for ICN because of relatively high failure rates and the fact that systemic therapy will prevent subsequent development of pneumonia.

Ocular prophylaxis with erythromycin given soon after birth appears to prevent the development of ICN, but will not prevent the development of pneumonia.[44] C. trachomatis is a far more common pathogen in the U.S. than is N. gonorrhoeae. Thus it seems reasonable to recommend routine ocular prophylaxis be based on the use of erythromycin ointment, which appears to be effective against N. gonorrhoeae and C. trachomatis, instead of silver nitrate (standard Credè prophylaxis), which is active against the gonococcus but not Chlamydia. This recommendation, however, should not carry over into areas where antibiotic resistant Neisseria are a problem, until further studies determine efficacy.

Pregnant women can be screened for chlamydial infection and those found to be infected, treated with erythromycin. This will prevent perinatal transmission.[45] Because attack rates have been fairly consistent in most studies, the prevalence of chlamydial infection in pregnant women will determine the cost-benefit relationship of this stratagem.[38] A number of studies have found that more than 20% of infected women in specific settings have chlamydial infection and such prenatal clinics would clearly be appropriate sites for screening and treatment.

IV. INCLUSION CONJUNCTIVITIS IN ADULTS

A. History and Epidemiology

Adult inclusion conjunctivitis was recognized in the decades following the demonstration of inclusion conjunctivitis in infants.[1] The condition was seen in young adults in western Europe and many of the early cases were considered to be a form of "swimming pool conjunctivitis". It is now recognized that these cases result from exposure to infected genital tract discharges.

The condition has an incubation period of approximately 1 to 3 weeks. It is typically seen in sexually active young adults.

B. Clinical Description

Inclusion conjunctivitis in adults is an acute follicular conjunctivitis which must be differentiated from adenovirus keratoconjunctivitis by microbiologic tests. It is quite similar in presentation to early trachoma, although it is not usually associated with chronic disease and is not considered to be a sight-threatening condition. It can become chronic, although most cases will clear spontaneously after several months, if not treated.

C. Diagnosis

In the early stages of the disease almost any of the cytologic methods are highly efficient diagnostic tools. As the condition progresses into the more chronic stages prior to resolution it is more difficult to demonstrate the organism. Isolation is the diagnostic test of choice.

D. Prevention and Treatment

This is a sexually transmitted disease and prevention will be based on reducing the genital tract reservoir for the agent. Treatment is with systemic tetracycline or sulfonamides. The ocular infection may be more difficult to treat than uncomplicated genital tract infections and 3 week courses of therapy are recommended.

V. LYMPHOGRANULOMA VENEREUM

A. History and Epidemiology

Although this condition has been known for 200 years and is worldwide in distribution, it is not a well-studied entity. It is a more common disease in some parts of the world (Southeast Asia, India, Africa) than in others. In certain countries lymphogranuloma venereum (LGV) can be responsible for a fairly large proportion of clinic visits. Some censuses show 2 to 6% of all patients attending venereal disease clinics may have a diagnosis of LGV.[46]

B. Clinical Description

The incubation period is variable (usually 1 to 3 weeks, but may be much longer). The first manifestation of the disease appears to be a primary lesion — a painless superficial ulcer or vesicle on the genitals. In temperate climes this lesion is often not apparent.[1] In tropical countries, the ulcerative form of the disease appears to be quite important.[47] Within 1 to 3 weeks after the primary lesion appears, regional lymphadenopathy develops. Bubo development is known as the secondary stage of the disease and is typically seen in young men. Because women probably have primary implantation of the organism within the vagina (where the draining lymph nodes will be retroperitoneal rather than inguinal), they usually do not develop inguinal lymphadenopathy. The lymph nodes will ultimately heal, often with some scarring, but the infection can persist and cause late destructive lesions, involving the gastrointestintal tract and genitalia. Scarring can result in obstruction, and fistulae are common. Primary implantation within the rectum can result in a severe proctocolitis. This is not an uncommon condition among homosexual men in some parts of the U.S.[21,46]

C. Diagnosis
LGV results in relatively high complement fixing (CF) antibody levels and the CF test may be used to support a diagnosis.[31] The micro-IF test can be used the same way, and very high titers of broadly reactive antibody are usually found. Chlamydiae can be isolated from ulcers, lymph node aspirates, rectal swabs, or biopsies. In the past, diagnosis was based on the use of a delayed hypersensitivity skin test (Frei test). This test is not recommended because of problems with sensitivity and specificity.[48] It is not available in the U.S. but is still available in some other countries.

D. Prevention and Treatment
There are no data available from controlled treatment trials. Tetracyclines and sulfonamides have been used and the treatment of choice appears to be at least 2 weeks of tetracycline at 2 g/day. In some instances, several courses of therapy appear to be necessary.

VI. *CHLAMYDIA PSITTACI*

A. History and Epidemiology
Psittacosis was first described in Switzerland in the 1870s.[1] In a few decades the disease had been reported from a number of countries in Europe and the association with exotic or psittacine birds was well described. A world-wide outbreak occurred in 1929 to 1930 and focused much attention on the condition because of the approximate 20% fatality rates seen in this preantibiotic pandemic. Although *C. psittaci* infections in nonpsittacine birds were known, it was not until the 1950s that the importance of ornithosis in poultry was recognized, and human psittacosis was described as an important occupational disease in workers in poultry processing plants.[49]

Human psittacosis is a disease contracted from exposure to any infected avian species. *C. psittaci* is ubiquitous among avian species and infection in the birds is usually of the intestinal tract. The organism is shed in the feces, contaminates the environment, and is spread by aerosol.

B. Clinical Description
Human psittacosis usually occurs in one of two forms. Respiratory disease can be a mild influenzal disease or a severe and fatal (if untreated) pneumonia can develop. The incubation period is typically 1 to 3 weeks. Fever, chills, and severe headache usually occur. Radiographs may show more extensive lung involvement than is expected on the basis of respiratory difficulty. Pulse rate may be lower than expected.

The other form of the disease is more typhoidal and involves a general toxic febrile state without respiratory findings. Person to person transmission is uncommon, but has occurred.

C. Diagnosis
Clinical signs are not pathognomonic, although a relatively low pulse associated with a high fever and severe headache can be suggestive. A clinician's index of suspicion (asking questions about potential exposure to birds) is usually crucial to arriving at a diagnosis. Serodiagnosis is generally considered to be the method of choice because isolation of the agent is seldom achieved. Rising antibody levels can be demonstrated by CF or micro-IF tests.

D. Prevention and Treatment

Tetracycline at 2 g/day for at least 2 weeks is considered the treatment of choice, with erythromycin being the alternative drug. Human psittacosis is considered an occupational hazard for those in the poultry or pet bird industries.

Traditionally, the administrative method of controlling psittacosis derived from exotic birds has been embargo. Importation of psittacine birds has been prohibited in many countries. Screening to select psittacosis-free birds for breeding has been used to establish uninfected flocks of those small birds that can be bred in captivity. Chemoprophylaxis for exotic birds has been developed.[50] If birds are held in quarantine and appropriately treated, that can be cleared of *C. psittaci* infection before they are introduced into normal distribution channels. When treated birds are introduced into commerce, clean premises can be maintained by keeping closed premises and avoiding introduction of any untreated birds. Unfortunately, this approach will not work in the poultry industry because of potential contamination of premises by feral birds.[51]

VII. OTHER HUMAN *C. PSITTACI* INFECTIONS

A. History and Epidemiology

C. psittaci is common in domestic mammals. In some parts of the world these infections have important economic consequences, as *C. psittaci* is a cause of a number of systemic and debilitating diseases in domestic mammals and most importantly, can cause abortions (see Chapter 9).[52] Human chlamydial infections resulting from exposure to infected domestic mammals are known, but are relatively uncommon.

During trachoma studies performed in Taiwan and Iran, some *C. psittaci* strains were recovered from conjunctival swabs. Seroepidemiologic studies have suggested that infections with these strains (currently designated as TWAR) are common in many parts of the world.[53,54] Age-specific prevalence rates suggest that transmission occurs in childhood and peaks early in adult life. It has been suggested that these strains are representative of *C. psittaci* circulating among humans without an avian reservoir.

B. Clinical Description

C. psittaci infection contracted from mammalian isolates appeared to be relatively inocuous, as seroconversion in the absence of disease has been demonstrated in veterinarians, ranchers, and workers in slaughterhouses. Occasionally, however, severe disease may develop. There have been a number of instances when *C. psittaci* (apparently derived from aborting ewes) infected pregnant women and resulted in abortion and even life-threatening disease.[55]

The TWAR strains' seroepidemiologic patterns suggest that inapparent infections are quite common and it is likely that mild respiratory disease can result from these infections. These organisms appear to be capable of causing conjunctivitis in humans and have been associated with a broad spectrum of respiratory disease in the few studies that have been reported. There are serologic data suggesting that TWAR strains caused an outbreak of mild pneumonia in Finland, and in Canada they have been associated with a wide spectrum of disease, including fatal pneumonia cases.[56,57] There are relatively few isolates of these strains available, but recent studies in Seattle have reported isolation of this organism from throat swabs collected from college students with pneumonia.[53]

C. Diagnosis

This field is still in its infancy and the diagnostic methods of choice have yet to be determined. What has been learned to date has been based on specific antibody detection by the micro-immunofluorescence test using the TWAR antigen. Specific complement fixing antibodies appear to accompany many of the TWAR infections. The agent can be isolated, with considerable difficulty, in tissue culture systems (HeLa or cycloheximide-treated McCoy cells, as used in *C. trachomatis* isolation attempts), but the inclusions will not stain by the iodine stain and the Giemsa stain is a difficult one to read. Inclusions would be best demonstrated by use of fluorescein-conjugated homologous monoclonal antibodies or perhaps by genus-specific monoclonals. Antibodies could be used to demonstrate elementary bodies in a direct specimen test using smears prepared from the involved sites.

D. Prevention and Treatment

There currently are no modalities available for prevention of these infections. The *C. psittaci* coming from lower mammals are shed in extraordinary quantities and contaminate the environment, and those with environmental exposure could be infected. With the TWAR strains the epidemiology of these strains must be elucidated before any strategies for prevention can be proposed.

There are relatively few case reports so that one can only speculate on treatment of these infections. On the basis of what is known about other chlamydial strains, it is likely that high dose tetracycline or erythromycin regimens will be effective.

REFERENCES

1. **Schachter, J. and Dawson, C. R.,** *Human Chlamydial Infections,* John Wright-PSG, Littleton, Mass., 1978, 273.
2. **Dawson, C. R., Daghfous, M., Messadi, M., Hoshiwara, I., and Schachter, J.,** Severe endemic trachoma in Tunisia, *Br. J. Ophthalmol.,* 60, 245, 1976.
3. **T'ang, F-F., Chang, H.-L., Huang, Y-T., and Wang, K-C.,** Trachoma virus in chick embryo, *Natl. Med. J. China,* 43, 81, 1957.
4. **Wang, S-P. and Grayston, J. T.,** Immunologic relationship between genital TRIC, lymphogranuloma venereum, and related organisms in a new microtiter indirect immunofluorescence test, *Am. J. Ophthalmol.,* 70, 367, 1970.
5. **Dawson, C. R., Jones, B. R., and Tarizzo, M.,** *Guide to Trachoma Control in Programmes for the Prevention of Blindness,* World Health Organization, Geneva, 1982, 62.
6. **Treharne, J. D., Dwyer, R. St. C., Darougar, S., Jones, B. R., and Daghfous, T.,** Antichlamydial antibody in tears and sera, and serotypes of *Chlamydia trachomatis* isolated from school children in southern Tunisia, *Br. J. Ophthalmol.,* 62, 509, 1978.
7. **Grayston, J. T. and Wang, S-P.,** New knowledge of chlamydiae and the diseases they cause, *J. Infect. Dis.,* 132, 87, 1975.
8. **Lindner, K.,** Gonoblennorrhoe, einschlussblennorrhoe, und trachoma, *Graefe's Arch. Ophthalmol.,* 78, 380, 1911.
9. **Gordon, F. B. and Quan, A. L.,** Isolation of the trachoma agent in cell culture, *Proc. Soc. Exp. Biol. Med.,* 118, 354, 1965.
10. **Schachter, J.,** Chlamydial infections, *N. Engl. J. Med.,* 298, pp. 428, 490, 540, 1978.
11. **Schachter, J., Hanna, L., Hill, E. C., Massad, S., Sheppard, C. W., Conte, J. E., Jr., Cohen, S. N., and Meyer, K. F.,** Are chlamydial infections the most prevalent venereal disease?, *J. Am. Med. Assoc.,* 231, 1252, 1975.
12. Prepared by the participants in a NIAID symposium, *Sexually Transmitted Disease: 1980 Status Report,* Department of Health and Human Services Publ. No. 81-2213, U.S. Government Patent Office, Washington, D.C., 1981.
13. **McCormack, W. M., Evard, J. R., Laughlin, C. F., Rosner, B., Alpert, S., Crockett, V. A., McComb, D., and Zinner, S. H.,** Sexually transmitted conditions among women in college, *Obstet. Gynecol.,* 139, 130, 1981.
14. **Schachter, J., Stoner, E., and Moncada, J.,** Screening for chlamydial infections in women attending family planning clinics: evaluations of presumptive indicators for therapy, *West. J. Med.,* 138, 375, 1983.
15. **Shafer, M. A., Blain, B., Beck, A., Dole, P., Irwin, C. E., Sweet, R., and Schachter, J.,** *Chlamydia trachomatis:* important relationships to race, contraception, lower genital tract infection, and papanicolaou smears, *J. Pediatr.,* 104, 141, 1984.

16. **Holmes, K. K., Handsfield, H. H., Wang, S-P., Wentworth, B. B., Turck, M., Anderson, J. B., and Alexander, E. R.,** Etiology of nongonococcal urethritis, *N. Engl. J. Med.,* 292, 1199, 1975.

17. **Oriel, J. D. and Ridgway, G. L.,** *Genital Infection by Chlamydia trachomatis,* Edward Arnold, London, 1982, 144.

18. **Oriel, J. D., Ridgway, G. L., Reeve, P., Beckingham, D. C., and Owen, J.,** The lack of effect of ampicillin plus probenecid given for genital infections with *Neisseria gonorrhoeae* on associated infections with *Chlamydia trachomatis, J. Infect. Dis.,* 133, 568, 1976.

19. **Centers for Disease Control,** Division of Sexually Transmitted Diseases, 1985 STD Treatment Guidelines, *Morbidity and Mortality Weekly Report,* 34(4S), 1985.

20. **Berger, R. E., Alexander, E. R., Monda, G. D., Ansell, J., McCormick, G., and Holmes, K. K.,** *Chlamydia trachomatis* as a cause of acute "idiopathic" epididymitis, *N. Engl. J. Med.,* 298, 301, 1978.

21. **Quinn, T. C., Goodell, S. E., Mkrtichian, E., Schuffler, M. D., Wang, S-P., Stamm, W. E., and Holmes, K. K.,** *Chlamydia trachomatis* proctitis, *N. Engl. J. Med.,* 305, 195, 1981.

22. **Rees, E., Tait, I. A., Hobson, D., and Johnson, F. W. A.,** *Chlamydia* in relation to cervical infection and pelvic inflammatory disease, in *Nongonococcal Urethritis and Related Infections,* Holmes, K. K. and Hobson, D., Eds., American Society for Microbiology, Washington, D.C., 1977, 67.

23. **Paavonen, J., Brunham, R., Kiviat, N., Stevens, C., Kuo, C-C., Stamm, W. E., and Holmes, K. K.,** Cervicitis — etiologic, clinical, and histopathologic findings, in *Chlamydial Infections,* Mardh, P-A., Holmes, K. K., Oriel, J. D., Piot, P., and Schachter, J., Eds., Elsevier, Amsterdam, 1982, 141.

24. **Brunham, R. C., Paavonen, J., Stevens, C. E., Kiviat, N., Kuo, C-C., and Holmes, K. K.,** Mucopurulent cervicitis — the ignored counterpart in women of urethritis in men, *N. Engl. J. Med.,* 311, 1, 1984.

25. **Stamm, W. E., Guinan, M. E., Johnson, C., Starcher, T., Holmes, K. K., and McCormack, W. M.,** Effect of treatment for *Neisseria gonorrhoeae* on simultaneous infection with *Chlamydia trachomatis, N. Engl. J. Med.,* 310, 545, 1984.

26. **Stamm, W. E., Wagner, K. F., Amsel, R., Alexander, E. R., Turck, M., Counts, G. W., and Holmes, K. K.,** Causes of the acute urethral syndrome in women, *N. Engl. J. Med.,* 303, 409, 1980.

27. **Mardh, P-A., Ripa, T., Svensson, L., and Westrom, L.,** *Chlamydia trachomatis* infection in patients with acute salpingitis, *N. Engl. J. Med.,* 296, 1377, 1977.

28. **Westrom, L.,** Incidence, prevalence and trends of acute pelvic inflammatory disease and its consequences in industrialized countries, *Am. J. Obstet. Gynecol.,* 138, 880, 1980.

29. **Sweet, R. L., Schachter, J., and Robbie, M.,** Failure of beta lactam antibiotics to eradicate *Chlamydia trachomatis* in the endometrium despite apparent clinical cure of acute salpingitis. *J. Am. Med. Assoc.,* 250, 2641, 1983.

30. **Cates, W., Jr.,** Sexually transmitted organisms and infertility: the proof of the pudding, *Sex. Transm. Dis.,* 11, 113, 1984.

31. **Schachter, J.,** Chlamydiae (Psittacosis-Lymphogranuloma Venereum-Trachoma Group), in *Manual of Clinical Microbiology,* 4th ed., Lennette, E. H., Ed., American Society for Microbiology, Washington, D.C., 1985, 856.

32. **Tam, M. R., Stamm, W. E., Handsfield, H. H., Stephens, R., Kuo, C-C., Holmes, K. K., Ditzenberger, K., Crieger, M., and Nowinski, R. C.,** Culture-independent diagnosis of *Chlamydia trachomatis* using monoclonal antibodies, *N. Engl. J. Med.,* 310, 1146, 1984.

33. **Herrmann, J. E., Howard, L. V., Armstrong, A. S., and Craine, M. C.,** Immunoassay for detection of *Neisseria gonorrhoeae* and *Chlamydia trachomatis* in samples from a single specimen, in Programs and Abstracts of the International Society for STD Research 5th International Meeting, August 1-3, 1983, Seattle, Wash., (abstr. 44) 1983, 76.

34. **Chandler, J. W., Alexander, E. R., Pheiffer, T. A., Wang, S-P., Holmes, K. K., and English, M.,** Ophthalmia neonatorum associated with maternal chlamydial infections, *Trans. Am. Acad. Ophthalmol. Otolaryngol.,* 83, 302, 1977.

35. **Schachter, J., Lum, L., Gooding, C. A., and Ostler, B.,** Pneumonitis following inclusion blennorrhea, *J. Pediat.,* 87, 779, 1975.

36. **Beem, M. O. and Saxon, E. M.,** Respiratory tract colonization and a distinctive pneumonia syndrome in infants infected with *Chlamydia trachomatis, N. Engl. J. Med.,* 293, 306, 1977.

37. **Schachter, J., Holt, J., Goodner, E., Grossman, M., Sweet, R., and Mills, J.,** Prospective study of chlamydial infection in neonates, *Lancet,* 2, 377, 1979.

38. **Schachter, J. and Grossman, M.,** Chlamydial infections, *Annu. Rev. Med.,* 32, 45, 1981.

39. **Harrison, H. R., English, M. G., Lee, C. K., and Alexander, E. R.,** *Chlamydia trachomatis* infant pneumonitis (comparison with matched controls and other infant pneumonitis), *N. Engl. J. Med.,* 288, 702, 1978.

40. **Beem, M. O. and Saxon, E. M.,** *Chlamydia trachomatis* infections in infants, in *Chlamydial Infections,* Mardh, P-A., Holmes, K. K., Oriel, J. D., Piot, P., and Schachter, J., Eds., Elsevier, Amsterdam, 1982, 199.

41. **Cohen, S. D., Azimi, P. H., and Schachter, J.,** *Chlamydia trachomatis* associated with severe rhinitis and apneic episodes in a one-month-old infant, *Clin. Pediat.*, 21, 498, 1982.

42. **Arth, C., Von Schmidt, B., Grossman, M., and Schachter, J.,** Chlamydial pneumonitis, *J. Pediat.*, 93, 447, 1978.

43. **Schachter, J., Grossman, M., and Azimi, P. H.,** Serology of *Chlamydia trachomatis* in infants, *J. Infect. Dis.*, 146, 530, 1982.

44. **Hammerschlag, M. R., Chandler, J. W., Alexander, E. R., English, M., Chiang, W-T., Koutsky, L., Eschenbach, D. A., and Smith, J. R.,** Erythromycin ointment for ocular prophylaxis of neonatal chlamydial infection, *J. Am. Med. Assoc.*, 244, 2291, 1980.

45. **Schachter, J., Sweet, R. L., Grossman, M., Landers, D., Robbie, M., and Bishop, E.,** Experience with the routine use of erythromycin for chlamydial infections in pregnancy, *N. Engl. J. Med.*, 314, 276, 1986.

46. **Schachter, J. and Osoba, A. O.,** Lymphogranuloma venereum, *Br. Med. Bull.*, 39, 151, 1983.

47. **Piot, P., Ballard, R. C., Fehler, H. G., Van Dyck, E., Ursi, J. P., and Meheus, A. Z.,** Isolation of *Chlamydia trachomatis* from genital ulcerations in southern Africa, in *Chlamydial Infections*, Mardh, P-A., Holmes, K. K., Oriel, J. D., Piot, P., and Schachter, J., Eds., Elsevier, Amsterdam, 1982, 115.

48. **Schachter, J., Smith, D. E., Dawson, C. R., Anderson, W. R., Deller, J. J., Jr., Hoke, A. W., Smartt, W. H., and Meyer, K. F.,** Lymphogranuloma venereum. I. Comparison of Frei test, complement fixation test and agent isolation, *J. Infect. Dis.*, 120, 372, 1969.

49. **Meyer, K. F.,** Ornithosis, in *Diseases of Poultry*, 5th ed., Biester, H. E. and Schwarte, L. H., Eds., Iowa State University Press, Ames, Iowa, 1965, 675.

50. **Arnstein, P., Eddie, B., and Meyer, K. F.,** Control of psittacosis by group chemotherapy of infected parrots, *Am. J. Vet. Res.*, 29, 2213, 1968.

51. **Grimes, J. E., Owens, K. J., and Singer, J. R.,** Experimental transmission of *Chlamydia psittaci* to turkeys from wild birds, *Avian Dis.*, 23, 915, 1979.

52. **Storz, J.,** *Chlamydia and Chlamydia-induced Diseases*, Charles C. Thomas, Springfield, Ill. 1971.

53. **Grayston, J. T., Kuo, C-C., Wang, S-P., and Altman, J.,** A new *Chlamydia psittaci* strain, TWAR, isolated in acute respiratory tract infections, *N. Engl. J. Med.*, 315, 161, 1986.

54. **Forsey, T., Darougar, S., Treharne, J. D.,** Prevalence in human being of antibodies to Chlamydia IOL-207, an atypical strain of Chlamydia, *J. Infect.*, 12, 145, 1986.

55. **Schachter, J.,** Human *Chlamydia psittaci* infection, in *Chlamydial Infections*, Oriel, D., Ridgway, G., Schachter, J., Taylor-Robinson, D., Ward, M., Eds., Cambridge, University Press, Cambridge, 1986, 311.

56. **Saikku, P., Wang, S-P., Kleemola, M., Brander, E., Rusanen, E., Grayston, J. T.,** An epidemic of mild pneumonia due to an unusual strain of *Chlamydia psittaci*, *J. Infect. Dis.*, 151, 832, 1985.

57. **Marrie, T. J., Wang, S-P., Kuo, C-C., and Grayston, J. T.,** Chlamydia pneumonia, in Program and Abstracts of the 25th Interscience Congress of Antimicrobial Agents and Chemotherapy, American Society for Microbiology, Washington, D.C., (abstr. 864), 1985, 252.

Chapter 9

OVERVIEW OF ANIMAL DISEASES INDUCED BY CHLAMYDIAL INFECTIONS

Johannes Storz

TABLE OF CONTENTS

I. INTRODUCTION

Chlamydiae are of medical interest because of their characteristic developmental cycle with the associated features of intracellular parasitism, and the diverse diseases they cause in man and animals. Currently, two recognized species compose the genus *Chlamydia*.[1] Man is the primary host of *Chlamydia trachomatis*, with the exception of the mouse biotype. Animals susceptible to *C. psittaci* infections are widely distributed in the animal kingdom, ranging from ectothermic vertebrates, wild and domesticated birds, and mammals to man.[2,3] The goals of this discourse are to describe the diverse diseases caused by chlamydiae in different animal species and to illustrate pathogenetic events and lesion development following chlamydial infections.

II. HOST RANGE AND DISEASE DIVERSITY

The pathogenic role of chlamydiae in several diseases is well established. Depending on such factors as virulence of the agent, host species, age and sex of the animal, environment, management practices, as well as ecological and physiological conditions, the following syndromes may be elicited: intestinal infection and diarrhea, pneumonia, abortion, urogenital infections, mastitis, polyarthritis-polyserositis, encephalomyelitis, hepatitis, and conjunctivitis.[2,3] Avian chlamydial infections may lead to pneumonia and airsacculitis, pericarditis, conjunctivitis, encephalitis, as well as intestinal infections and diarrhea. Chlamydial infections often establish a balanced host-parasite relationship in most animals known as natural hosts.[3] Clinically inapparent intestinal infections of animals lead to prolonged intestinal shedding of chlamydiae, which is one important factor for successfully perpetuating and maintaining this infection in animal populations.

Chlamydiae were isolated from different organ sites and disease conditions of ruminants, horses, swine, cats, dogs, rodents, rabbits, ferrets, opossums, koala bears, many avian species, and frogs. Additionally, serologic findings indicate that monkey, Fallow, roe, red- and white-tailed deer, reindeer, pronghorn antelope, wild boar, and hedgehogs have antibody levels reflecting chlamydial infections.[3] Common reservoirs of chlamydiae include seagulls, ducks, herons, egrets, pigeons, blackbirds, grackles, sparrows, and killdeer. The bird species of the economically important poultry industries are also natural and highly susceptible hosts for chlamydial infections.[4]

In contrast to the host and tissue specificity of *C. trachomatis*, members of the species *C. psittaci* are less host and tissue specific. Continuing interspecies transfer of chlamydiae among wild birds and poultry is considered an explanation for epornitics in domestic birds.[5] Strains isolated from cases of chlamydial abortions of sheep produced mastitis and abortion in cattle, pneumonia in calves and piglets, and airsacculitis in pigeons, turkeys, and sparrows. Isolates from sheep with polyarthritis caused signs of chlamydiosis in budgerigars, turkeys, and sparrows, and arthritis in turkeys. Strains from pigeons produced chlamydiosis in sparrows, turkeys, and budgerigars, and abortion in sheep. Turkey chlamydial isolates induced penumonia in calves and abortion in sheep. Chlamydiae isolated from feces of sheep or cattle caused abortion in these species. Unequivocal differentiation of properties of *C. psittaci* strains defining host and disease specificity remains a challenge.

III. ANTIGENIC AND PATHOGENIC DIVERSITY OF *C. PSITTACI*

The current classification scheme and chlamydial species designations are generally accepted by microbiologists, but significant biological differences between strains within each species have been observed.[1] The development of the indirect micro-immunofluorescence (IMIF) technique by Wang[6] provides a powerful tool for antigenic analysis of chlamydial strains. This method of chlamydial antigenic differentiation was applied to strains of *C. psittaci* of mammalian origin. The IMIF test differentiated 25 strains of *C. psittaci* of mammalian origin into 9 immunotypes.[7] A good correlation was observed between immunotypes and biotypes, indicating that strains with unique pathogenic properties also have unique antigenic compositions. This proposed immunotyping scheme should be expanded to include other chlamydial strains, especially those of avian origin. Although none of the *C. psittaci* immunotypes identified in this study cross reacted significantly with the mouse pneumonitis strain, antigenic relationships with other strains of *C. trachomatis* should also be explored.[7] Good agreement existed with previous reports on the antigenic and genomic relationships of limited numbers of ovine strains of *C. psittaci*. Six of these strains were differentiated with the IMIF test from the ovine arthropathogenic strain LW-679.[8] DNA restriction endonuclease analysis identified unique DNA fragments common to eight ovine chlamydial abortion strains which differed to some extent from those of a single isolate from polyarthritic lambs in Scotland.[9]

The different immunotypes identified in our investigation apparently have some degree of disease and host specificity. The intestinal mucosa emerges as a common site of infection. Immunotype 1 includes strains isolated from ruminants affected with abortions, seminal vesiculitis, pneumonia, and clinically inapparent intestinal infections. The pathogenic potential of most of these isolates had been confirmed experimentally. Immunotype 2 also includes strains isolated from ruminants, but the disease association of this group of strains is conjunctivitis, polyarthritis, encephalitis, and enteritis, under natural and experimental conditions. Immunotypes 3 and 9 are represented by chlamydial isolates which appear to be part of the intestinal flora of cattle and sheep.[7]

Immunotypes 4, 5, and 6 represent porcine strains associated with polyarthritis or generalized infections, clinically inapparent intestinal infections, and abortion or pneumonia, respectively. A single isolate from bovine pneumonia is also identified as immunotype 6. The pathogenicity of these strains has not been studied experimentally.

Immunotype 7 includes the chlamydial strain used in the live feline pneumonitis vaccine, and an isolate presumably cultured from a calf with pneumonia. The isolates from guinea pig inclusion conjunctivitis comprised immunotype 8.

While some of the immunotypes identified induced antibodies which reacted only with the homologous strains, other immunotypes induced cross-reacting antibodies. Notably, immunotype 2 strains induced antibodies that reacted with antigens of other immunotypes.[7]

IV. CHLAMYDIOSIS OF WILD AND DOMESTIC BIRDS

A. Historical and Epidemiological Connotations

Avian chlamydiosis is an infection or disease in domesticated and wild birds and is caused by *C. psittaci*. This infection of birds remains of importance because of the hazards to public health and because of the economic losses that result in the poultry industry. Meyer described naturally occurring chlamydial infections in 130 species of birds belonging to 10 orders.[10] Extensive chlamydial epornitics in turkeys appeared in the U.S. during the 1950s and early 1960s. Simultaneously, severe epornitics in ducks, geese, chickens, and turkeys occurred in Europe, particularly in eastern European countries. Public health interest was high because the disease occurred concurrently in workers involved in handling or processing of infected fowl. Acute awareness of this disease problem, improved methods of diagnosis, effective treatment, and management, as well as quarantine procedures and effective hygienic measures during poultry processing, reduced the losses resulting from this disease in the 1960s. During recent years larger-scale epornitics in turkeys and geese reappeared in the U.S. and Poland.[11,12] Chlamydial infections in pet birds and pigeons remain a hazard to human health.[13] The reasons for the apparent reemergence and increase of avian chlamydial infections and their transmission to man may be found in the relaxation or breakdown of any one of the measures established in the 1960s to control this infectious disease problem.[11]

Page characterized the subtle but extensive interplay between chlamydiae and wild birds and turkey flocks during an epornitic in Texas.[14] Wild birds and feral and domestic mammals in the vicinity of sick and dying turkeys were tested serologically and by chlamydial isolation attempts. Chlamydial antibodies were detected in 65% of the blackbirds, 44% of the killdeer, 27% of the sparrows, 43% of the cattle, and in all goats tested. None of the rats, mice, squirrels, and gophers were positive, but chlamydiae were isolated from an opossum and a house cat that scavenged dead turkeys.

The respiratory tract seems to be the most important route of infection of birds. Transmission among birds occurs predominantly by inhalation of infected dust. Chlamydiae are shed in infected birds in their droppings and nasal secretions. Dried excrement may remain infectious for several months, thus providing a source of infection for other birds.[15]

B. Signs, Morbidity, and Mortality in Avian Chlamydiosis

The acute disease in parrots, parakeets, and other cage birds is characterized by diarrhea, anorexia, and droopiness. A common sign in birds is conjunctivitis or keratoconjunctivitis, which sometimes may be the only symptom. Turkeys develop similar signs. Egg production drops precipitously in affected hens to 10 to 20% or less, depending on the virulence of the infecting chlamydial strain. The birds develop signs of generalized infection, diarrhea with yellow-green gelatinous droppings, and respiratory distress. There may be 50 to 80% morbidity and 10 to 30% mortality in a flock infected with highly virulent strains. Predominantly respiratory signs — such as rhinitis, bronchitis, and air sacculitis — characterize chlamydiosis in racing pigeons. Ducks characteristically have serous or purulent nasal and ocular discharges, whereby the feathers around the nostrils and eyes become encrusted.[10-13]

C. Pathogenesis and Pathological Changes

The events leading to disease were studied in detail during chlamydial infections of orally or aerosol-exposed turkeys.[15] Small numbers of chlamydiae had reached the abdominal air sacs and mesentery within 4 hr after infection, and larger numbers were present in the lungs and thoracic air sacs. The infectivity in lungs and air sacs increased dramatically during the following 24 hr. Chlamydiae had reached the blood, spleen, and kidneys by 48 hr after exposure. They were present in large numbers in the turbinates and colon contents within 72 hr, while the highest infectivity levels were in the pericardium on the 4th day. Chlamydiae were shed abundantly through nasal excretions and in feces or diarrhea fluid. Infection persisted in pericardial membranes, kidneys, or livers for almost 2 months. The incubation period in chlamydial infections ranged from 5 to 10 days, depending on degree of exposure and virulence of the strain used.

Pathological changes of chlamydiosis in poultry and other avian species are basically similar. The lungs are edematous and hyperemic and have thick, cloudy, and edematous air sacs. Fibrinous exudates often cover the serous surfaces of the air sacs, pericardium, liver, and intestine. The livers are usually swollen, hemorrhagic, mottled, and off-color. Greyish, small necrotic foci can be found in livers of affected pigeons. Enlargement of the spleen is a frequently seen lesion in psittacine birds and pigeons. The intestinal tract may be acutely inflamed in cases with diarrhea.

Chlamydiosis of birds is characterized histologically by necrotizing and proliferative changes. The lesions in the respiratory system consist of lung consolidation resulting from focal inflammatory cell infiltration, edema, congestion, and hemorrhage. Histiocytic and lymphatic cells accumulate in the interalveolar septa and in the propria of large bronchioli. The tracheal and other mucous membranes of the respiratory system are infiltrated diffusely or focally with mononuclear cells, lymphocytes, and heterophils. Similar changes occur in the heart, kidneys, spleen, brain, and reproductive organs. The livers of affected birds may have areas of eosinophilic necrosis. Chlamydial infection of birds thus may lead to pneumonia, air-sacculitis, tracheitis, hepatitis, myocarditis, splenitis, nephritis, orchitis, enteritis, and encephalitis.[10,11]

V. INTESTINAL INFECTIONS AND ENTERITIS IN RUMINANTS

A. The Intestinal Habitat

Intestinal chlamydial infections are frequently overlooked, yet they may cause diarrhea in young animals, initiate events in the pathogenesis of other chlamydial disease, and they represent an important epidemiological factor in perpetuating and spreading this infection in animal populations.

Chlamydiae were first isolated from feces and diarrhea fluid of calves, and intestinal chlamydial infections of cattle were found to be worldwide.[16,17] Clinically normal goats and sheep harbored chlamydial agents in the intestinal tract and excreted these organisms in readily detectable amounts. Chlamydiae were also isolated from feces of dogs and from pigs.[3,18] Evidently, the intestinal chlamydial infection of mammals is comparable to the infection of acutely, as well as persistently, infected birds which are known to excrete chlamydiae in the feces or diarrhea fluid.

Colostrum-deprived 24-hr-old calves developed diarrhea a day after oral inoculation with chlamydiae. They also developed fever and leukocytosis. The majority of these calves died within 17 days of exposure. Calves inoculated orally with chlamydiae isolated from polyarthritic joints became stiff and reluctant to move.[19]

B. Cytopathic and Cytocidal Chlamydial Functions and Intestinal Infections

The events of chlamydial infections leading to enteritis and diarrhea were studied in newborn calves. Chlamydiae were isolated from mucosal scrapings of the abomasum, duodenum, jejunum, ileum, caecum, and colon at various times after oral inoculation.[20] Fluorescent antibodies revealed that the epithelial cells on the tips of the villi and in the intervillous zones as well as some cells in the crypts of Lieberkühn and the transition zones were infected.[21] Infection of intestinal cells occurred predominantly through the brush border, a statement derived from ultrastructural analysis. The host cell range included absorptive epithelial cells at the tips of the villi and the intervillous zones, follicle-associated M cells, enterochromaffin and goblet cells of the villi, and undifferentiated cells in the crypts, as

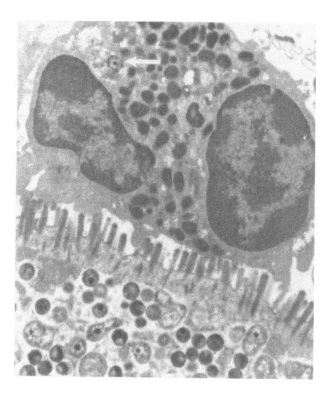

FIGURE 1.   Leukocyte interacting with infected absorptive intestinal epithelial cell with elementary bodies, condensing and reticulate chlamydial forms in supranuclear cytoplasmic chlamydial inclusion. The leukocyte extends pseudopods into the brush border and contains a degenerating chlamydial elementary body (arrow), × 8000. Courtesy of Dr. A. M. Doughri.

well as macrophages and fibroblasts in the lamina propria and endothelial cells in the central lacteals.[22] Neutrophilic leukocytes rarely contained chlamydiae, but leukocytes were pavementing the vicinity of chlamydia-infected epithelial cells (Figure 1).

Morphological features of sequential stages of chlamydial development were correlated with ultrastructural lesions in infected intestinal cells.[22-24] During the early stage of chlamydial infection, the plasmalemma at the site of adsorption developed an electron-dense zone opposite the adsorbed elementary body. An early ultrastructural change was the space-occupying presence of the endosome containing the dispersing chlamydial form within deeper regions of the cytoplasm or in the Golgi region. The rough endoplasmic reticulum remained near the inclusion, but the lysosomes appeared to migrate away and accumulated close to the cytoplasmic periphery. The number and prominence of free ribosomes and polysomes in the cytosol diminished in all types of cells infected. In highly specialized infected cells, such as goblet cells, the mucus droplets became depleted. The reticulate bodies were located at the secreting face of the Golgi stacks and little mucus was formed. The specific granules of infected enterochromaffin cells, which supported chlamydial multiplication, decreased in number and electron density.

The appearance of condensing forms and elementary bodies in infections of all cell types studied coincided with the beginning of severe degenerative changes characterized by cytoplasmic vesiculation, fragmentation of membranes, and lysis of cellular organelles. The network of the smooth and rough endoplasmic reticulum dilated, became ill-defined, and vacuolated. Golgi complexes became dilated and vesiculated. The nuclei were pyknotic or

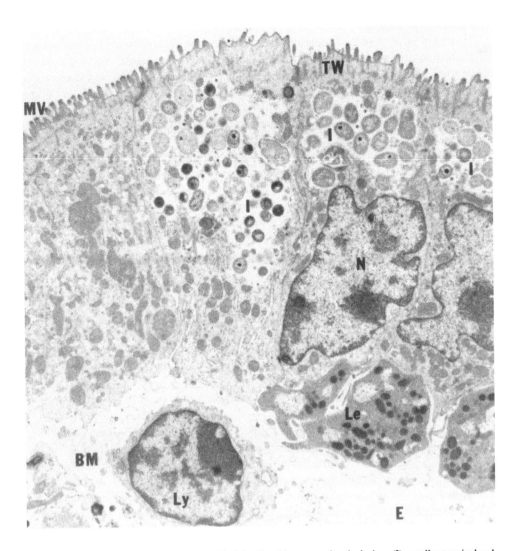

FIGURE 2. Infected absorptive intestinal epithelial cells with supranuclear inclusions (I), swollen terminal web (TW), loss of microvilli, and their rootlets. The cells detached from the festooning basal membrane (BM) in intercellular edema (E) with invading leukocytes (Le) and lymphocytes (Ly), × 7500. Courtesy of Dr. A. M. Doughri.

karyolytic after an initial loss of chromatic pattern and nucleoli, and separation of the membranes of the nuclear envelope. The chlamydial inclusion membranes ultimately fragmented and ruptured, followed by the lysis of the plasmalemma and liberation of chlamydiae.

Specialized organelles of infected cells such as microvilli and terminal webs of enterocytes were also affected (Figure 2). The microvilli lost their uniform, regular spacing and became shorter, irregularly shaped, and vesiculated. Their rootlets were lost and the terminal web became ill-defined. Lateral junctional complexes were displaced and fragmented. The desmosomes persisted, and converging fibrils remained in their vicinity. As the cells degenerated, the number of multivesiculated bodies, myeline figures, fat droplets, membrane-bound, lysosome-like structures, and dense bodies increased. The infection was ultimately cytocidal. Diarrhea most likely resulted from the enterocytes' loss of absorptive and digestive functions, interference with cellular transport and energy generation, accelerated death of infected intestintal epithelial cells, edema, and cellular infiltration of the lamina propria mucosae.

C. Pathological Lesions

The serosal surfaces of the abomasum and small intestine of calves were dull and granular. Paint-brush hemorrhages were present in the duodenal and cecal serosa. The mucosa of the entire small intestine was congested and petechiated. The jejunal wall was thickened by edema. The ileocecal valve was congested and often had numerous petechiae, and the cecal mucosa was hyperemic. Rugae of the colon were congested, and their free edges were slightly eroded. The mesenteric lymph nodes were enlarged and edematous, and the lymphatics of the mesentary were abnormally dilated.[25]

Histologically, the intestinal villi had a bulbous appearance caused by dilation of capillaries and central lacteals, edema, and infiltration of mononuclear cells. The inflammatory process spanned from the mucosal to the serosal surfaces in some instances. The epithelial cells desquamated extensively at many levels of the small intestine. Numerous crypts of Lieberkühn were distended by leukocytes and degenerated epithelial cells. The remaining epithelial cells lining such crypts were cuboidal rather than columnar. The Peyer's patches had lost their follicular architecture and contained necrotic centers, a feature also seen in the tonsils.[25]

VI. RESPIRATORY INFECTIONS AND PNEUMONIA

A. Pneumopathogenic Potential

Chlamydia-induced penumonia of different animal species occurs worldwide.[2,3] An inapparent respiratory chlamydial infection that can be activated to clinical pneumonia was first detected in mice.[26,27] The chlamydial agent involved was subsequently identified as the murine biotype of *C. trachomatis*.[1] The etiologic role of strains of *C. psittaci* as a cause of pneumonia is clearly established in cats, sheep, goats, cattle, swine, horses, and domestic and wild rabbits.[2,3,28-30]

The mouse pneumonitis agent (MoPn) was isolated by intranasal inoculation of lung material from infected, apparently normal mice into susceptible mice through blind passages to activate the infection. Affected mice were reluctant to move, assumed a hunched posture, and had a ruffled fur coat. Breathing became labored and was accompanied by a wet, clicking sound. The ears and tails were cyanotic, and the tails assumed a beaded appearance.[26,27]

Cats affected with chlamydial pneumonia sneeze, cough, have fever, and are depressed and anorectic. Mucopurulent discharge and excessive fluid flow form the eyes and nostrils because of conjunctivitis, rhinitis, and pharyngitis.[28,31] The cats recover in 2 to 4 weeks after infection, but they frequently remain asymptomatic carriers. Under adverse circumstances, periodic relapses can occur affecting cats of all ages. The disease usually is not fatal, but very young or elderly cats may die of severe pneumonia.

Calves with pneumonia and positive chlamydial cultures from respiratory tract samples have serous, mucous, or mucopurulent nasal discharge. They usually are febrile and appear depressed, while a dry cough may be associated with dyspnea. Most cases are of moderate severity, and affected calves recover, but some remain chronically debilitated.[2,3]

B. Pathogenetic Events and Pathological Changes

The sequence of events in the genesis of chlamydial penumonia was studied in mice, in sheep, and in calves. After intranasal inoculation of mice with purified elementary bodies of the murine biotype of *C. trachomatis*, inflammation was not observed in the alveoli during the entire period of the initial developmental cycle. The first observable reaction, characterized by an accumulation of heterophils in the alveoli, was initiated by the release of chlamydial elementary bodies. Mobilization of lung macrophages and an outpouring of phagocytes from the blood vessels into the alveoli preceded intense cellular infiltration in the periphery of bronchioli and the interstitial tissue. Cellular proliferation was seen at the periphery of the inflammatory foci, creating consolidation through local spread of the infection.[32]

Pulmonary infections of lambs with *C. psittaci* of ovine origin most likely are initiated in the bronchioles and spread from there to the lung parenchyma. Lung involvement was most severe on the 5th day after inoculation, when signs of regression became evident. Lungs were virtually normal after 30 days.[32] Consolidation of the anterior lobes and of lung in the hilus region is observed in naturally occurring cases of chlamydia-induced pneumonia of sheep and goats. Mature lesions consisted of irregular but sharply defined areas of consolidation with a dull, grey-pink color. The consolidated areas appeared elevated, felt lumpy, and slightly opaque mucus appeared in opened bronchioli. During the evolving and resolving stages, the lung lesions had streaky, irregular, dark-red bands that consisted of atelectatic regions.

Dungworth and Cordy[33] classified the chlamydia-induced pneumonia of sheep as an interstitial bronchopneumonia. Since they found alveolar epithelialization, the pulmonary lesions were also considered as hyperplastic pneumonia. There was an early acute inflammatory response in terminal bronchioles and adjacent alveoli, followed by proliferation of alveolar cells and accumulation of macrophages within alveoli. Extensive alveolar epithelialization was present at the height of the pulmonary reaction 5 days after experimental infection. Reticulin stains revealed well-preserved alveolar architecture that accounted, at least in part, for the ease with which resolution occurred.

VII. PLACENTAL AND FETAL CHLAMYDIAL INFECTIONS AND ABORTIONS

A. Importance and Signs

In 1950 when Stamp and co-workers[34] identified chlamydiae as the cause of abortions in ewes, they detected another important pathogenic potential of these infections. Placental and fetal infections with chlamydiae with ensuing abortion or the birth of weak offspring are now recognized as a significant cause of reproductive failure in sheep, cattle, goats, and other domestic animals.[2,3]

Chlamydial infection of pregnant ewes result in abortion in the last month of gestation, but abortions may occur as early as the 100th day of gestation. The incidence of abortions may reach 30% of ewes in flocks exposed to this infection for the first time. More commonly, 1 to 5% of the pregnant ewes in enzootically infected herds abort, and losses, occur year after year. Experimentally infected ewes or goats respond with fever 1 to 2 days after parenteral exposure. The fever lasts 3 to 5 days, but normal behavior is not noticeably influenced.[35,36] Following inoculation with chlamydiae isolated from aborted bovine fetuses, pregnant and nonpregnant cows invariably developed fever and a marked leukopenia for the next 3 to 5 days.[37]

Pregnant cows became infected following intravenous, intramuscular, subcutaneous, intracutaneous, and intracysternal inoculation at 3 to 8 months of gestation, and cows of all ages are susceptible. Cows inoculated by the intravenous route aborted within 5 to 36 days, while those inoculated by other routes in the 2nd and 3rd trimesters aborted or had weak calves 33 to 126 days later. Chlamydia-induced abortions were detected under field conditions as early as the 5th month of gestation, but most occurred during the last trimester. Retained placentas were observed in both experimentally induced and naturally occurring abortions. Chlamydial abortions occur sporadically, although in some herds as many as 20% of the pregnant cows abort.[37,38]

B. Pathogenesis of Placental and Fetal Chlamydial Infection

Chlamydiae were isolated from the blood after parenteral inoculation of pregnant ewes and cows. Chlamydiae were cleared rapidly from the blood after intravenous inoculation of ewes. Following initial clearance, chlamydial infectivity could not be recovered from the blood and organs of the ewes for a period lasting at least 18 hr. Chlamydial infectivity reappeared in the somatic organs of the ewes, where it multiplied to establish a secondary chlamydemia during the ensuing 72 hr.[35,39]

The placental junction is breached during the phases of chlamydemia. With localization of chlamydiae in the placenta, the infectious events *in utero* proceed independently of those in the dam.[35] Chlamydial infectivity is virtually eliminated from the somatic organs of the dam while the infection progresses in placenta and fetus. Kwapien and co-workers[40] found the earliest histologic lesions in the interplacentomal areas of experimentally inoculated cows. Lesions were found at the marginal zones neighboring areas of interplacentomal involvement. The chlamydial infection localized first in the endometrial epithelium of intercotyledonary areas where epithelial ulceration and direct contact with the intercotyledonary chorion established placental infection. Chlamydiae had a predilection for the chorionic epithelium in the hilar region of the placentomes of ewes.[41] Chlamydial inclusions were first detected in the chorionic epithelial cells near the hematomas, and their presence stimulated leukocytic reactions in the septal tips.

The earliest infection of organs of fetal lambs was detected 3 to 5 days after inoculation of the ewe.[35] Reed and co-workers[42] found chlamydial infection in liver and spleen of bovine fetuses as early as 6 days after maternal inoculation. Ovine and bovine fetuses from chlamydia-inoculated dams had infection of the liver, spleen, kidney, lung, thymus, brain, regional lymph nodes, and intestine. The progression of the infection *in utero* strongly indicates that the fetus becomes infected hematogenously by chlamydial invasion of the fetal circulation. Oral, conjunctival, or respiratory infections of the fetus are possible but may play a minor role. The uterine mucosa of ewes and cows did not harbor chlamydiae when tested 3 to 6 months after abortion. Infertility problems were observed in dairy cows following chlamydial abortions.

Abortion of a conceptus is a pathological phenomenon of considerable complexity. It is known that placentitis is induced by chlamydial infection, and this injury probably can be the cause of abortion; however, placentitis is often localized. The fetuses become infected, and chlamydial multiplication in fetal organs produces lesions leading to fetal disease and to death.[37,40,42,43]

C. Reaction of Placenta to Infection

The prominent pathologic change in chlamydial abortions in ewes is placentitis.[34,40,41] Various numbers of cotyledons are necrotic, and the periplacentome of affected cotyledons often is thickened and has an opaque, yellow-pink color. The uterine surface of the intercotyledonary chorion has a tough, granular consistency, a turbid, yellow-pink color, and yellowish flaky material on the surface. The margins of the placental lesions consists of zones of hyperemia and hemorrhage (Figure 3).

Bovine placentas were examined in the 2nd and 3rd trimesters of gestation 5 to 49 days after intravenous chlamydial inoculation of pregnant cows. Macroscopic placental lesions were not detected within 5 to 12 days after inoculation.[40] Later lesions consisted of various amounts of yellow to yellow-brown exudate in the interplacentome and periplacentome, and edema of the chorio-allantois. Cows that aborted during the later stages of gestation, 50 to 126 days after inoculation, had placental lesions resembling those described in chlamydial abortions of ewes. Often only parts of these placentas had lesions, while remaining portions appeared normal.[44] The affected parts of the chorion were slightly elevated and could not be scraped off. The margins of the affected cotyledons had small, round focal areas of soft necrosis.[45]

Focal ulceration of the epithelium of the endometrium in intercaruncular areas was observed in cows 5 to 7 days after inoculation.[40] Interplacentomal inflammatory exudate and periglandular infiltration with mononuclear and polymorphonuclear cells in the subepithelial

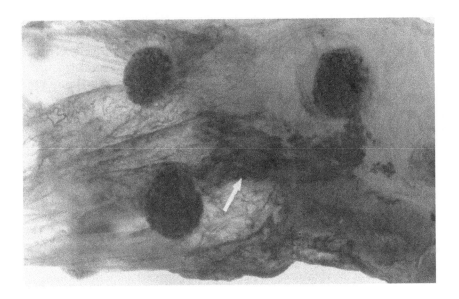

FIGURE 3.   Chlamydia-infected ovine placenta from experimentally induced abortion. Notice zone of hyperemia (arrow) demarkating fibrotically thickened chorion with three cotyledons.

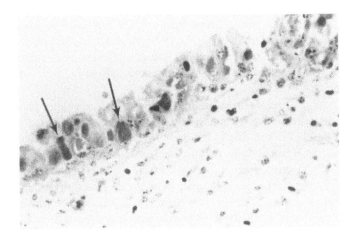

FIGURE 4.   Numerous chlamydial inclusions in epithelial cells of chorion in periplacentomal area of a cow. Gimenez stain, × 400.

stroma of the endometrium were also found. Cows examined 12 to 20 days after inoculation had involvement of the arcade zone of the placentome and acute inflammatory changes in the maternal septal tips of the placentome. Necrosis of the trophoblastic epithelium, infiltration with leukocytes, and angiitis were observed consistently in the intercotyledonary chorion at this stage. Fibrinopurulent exudate was observed between uterine and chorionic surfaces. Ulcerative endometritis and leukocytic infiltration of the subepithelial uterine stroma were present. Cytoplasmic chlamydial inclusions were present in the chorionic cells of the intercotyledonary region and in cells in the endometrium. The microscopic placental lesions were more extensive and severe in the group of cows investigated 27 to 49 days after inoculation. Chlamydial inclusions were found at this stage in trophoblastic epithelial cells of the intercotyledonary chorion (Figure 4).

Microscopic lesions in placentas of ewes examined 20 days after inoculation were found predominantly in the placentome.[41] The chorionic epithelial cells in the hilar region were enlarged and contained chlamydial inclusions. Some arcades were denuded of epithelium which spilled as cellular debris into the hematomas. The cores of the affected villi were populated with neutrophils and mononuclear cells. Between the chorion and the endometrium of the periplacentome were numerous leukocytes, erythrocytes, defoliated trophoblastic cells, and clusters of chlamydiae. Arterioles in the affected parts of the chorion were infiltrated by neutrophils; perivascular cuffing with mononuclear cells was common. The endometrial lining was focally denuded. Inflammatory cells infiltrated the endometrium and glands in the periplacentomes without noticeable involvement of the myometrium.

D. Fetal Responses to Chlamydial Infection

Changes in aborted ovine and bovine fetuses depend on the stage of gestation and duration of infection. Bovine fetuses aborted before the 6th month of gestation have blood-tinged, subcutaneous edema and increased amounts of clear, reddish pleural and peritoneal fluids. Fetuses of the 3rd trimester of gestation with infection persisting more than 20 days, have petechiae in the thymus, mucous membranes, and the subcutis. Ascites extensive enough to cause distention of the abdomen is observed in experimental and natural cases. The livers may be swollen, reddish-yellow, with mottled appearance and a coarsely nodular surface.[37,42-47]

Lambs that are aborted near term usually have a rather clean appearance and are in a well-preserved, fresh state. Numerous petechiae can be detected in the subcutis of many aborted lambs. Petechiae are seen most consistently in the skin of the legs, hip, head, and neck but also in the thymus, the salivary glands, and occasionally in lymph nodes. Livers of lambs from chlamydia-induced abortions may be congested and slightly swollen. Pinpoint, whitish foci are seen on the visceral and cut surfaces of livers from some lambs delivered near term. The lymph nodes are enlarged and edematous in chlamydia-infected, aborted calves and lambs.[43-47]

Individual histologic lesions of naturally infected ovine or bovine fetuses ranged from focal eosinophilic necrosis to pyogranulomatous inflammatory foci to chronic reticulo-endothelial hyperplasia. The reaction in the brain involved pleocellular accumulations within the adventitia of meningeal and parenchymal vessels. The portal areas of the fetal liver were affected by an inflammatory process that varied from moderate neutrophil and macrophage infiltration to abundance of monocytic cells. The portal vessels were the centers of most reactions, but the adventitia of central and hepatic venules was also involved.[44-47]

Microscopic lesions were not found in bovine fetuses examined sequentially earlier than 12 days following maternal inoculation. At this time hepatic and splenic lesions were detected, and most organs were involved after 20 days. Necrotic foci, sometimes with inflammatory reactions, were in the liver, spleen, kidneys, heart muscle, lungs, adrenal glands, and central nervous system.[46] These foci were midzonal in the liver and consisted of necrotic hepatocytes with small accumulations of monocytes and neutrophils. Phlebitis of central, sublobar, and portal veins occurred adjacent to inflammatory foci. There was vasculitis in liver, brain, heart and skeletal muscle, and in the gastrointestinal tract. The central nervous system had inflammatory foci composed of glial elements, histiocytes, and a few neutrophils in the white matter. Vasculitis with perivascular cuffing was associated with these foci, and with focal, nonsuppurative meningitis. The microscopic lesions in chlamydia-infected ovine and bovine fetuses appear remarkably compatible. Significantly, chlamydial agents were isolated from organs with the lesions described.[37,42,46,47]

The bovine fetus responded with chlamydial antibodies that were reactive by immunodiffusion but did not fix complement. Fetuses of the last trimester had elevated levels of IgM and IgG. The complexity of fetal responses to chlamydial infection was related to the progression of immunologic maturity of the fetus.[17]

VIII. UROGENITAL CHLAMYDIAL INFECTIONS

A. Seminal and Genital Transmission

Chlamydial agents were isolated from the semen or epididymis of six of ten bulls representing two different herds.[48] These young bulls had a seminal vesiculitis syndrome, and their semen was of unsatisfactory quality. A large number of leukocytes, low concentrations of sperm cells with poor mobility, and a high number of primary and secondary sperm cell abnormalities were present in ejaculates of these bulls. Chlamydiae have since been isolated from semen or genital organs of bulls and rams in different countries. Chlamydiae were also cultured from rams with epididymitis in sheep flocks with ewes experiencing chlamydial abortions.[17,49,50]

When bulls and rams were inoculated parenterally with chlamydial strains isolated from the epididymis, or from aborted lambs, or from synovial fluid of polyarthritic calves, they developed chlamydemia that lasted for 2 to 6 days.[51] Chlamydiae were first isolated from semen 3 to 8 days after inoculation. Of four inoculated bulls, three excreted chlamydiae in the semen, and this continued for 29 days. The rams excreted chlamydiae 3 to 22 days after inoculation, and their semen had unsatisfactory quality and became pus-like. Chlamydiae were isolated from the testes, epididymis, accessory sex glands, and from urine and kidneys of both rams and bulls 8 to 22 days after inoculation.

Orchitis, periorchitis, and epididymitis were observed after intraperitoneal inoculation of male guinea pigs with chlamydial strains isolated from this species.[52] When male guinea pigs were inoculated into the urethra with the guinea pig inclusion conjunctivitis agent, cytoplasmic inclusions were found in juxtaluminal superficial epithelial cells of the urethra and the bladder. The infection was transmitted to the vagina of sows that were bred to infected males.[53]

B. Lesions in Kidney Infections

Calves that were orally inoculated with chlamydial strains from polyarthritis harbored this infection in the kidneys and shed chlamydiae in high titers in the urine.[19,20] The kidneys of calves examined a week and later after inoculation had multiple, greyish-white foci visible through the capsule. Focal renal tubular dilation became severe and cystic, involving the distal convoluted tubules. The epithelial cells of such tubules were flattened. The lumen contained epithelial cell casts. Focal and interstitial inflammatory reactions of lymphocytes and plasma cells were dispersed throughout the cortex.[54]

The lesions were most pronounced 6 weeks post inoculation. Sagittal sections displayed these lesions as wedges of various depths in the cortex. The tissue changes advanced to a chronic focal nonsuppurative nephritis. The cellular reactions involved the medula but more often the cortex. Glomeruli surrounded by inflammatory reactions became atrophic, and their Bowman capsules had various degrees of fibrosis. The glomerular tufts had proliferative changes involving the epithelial, endothelial, and mesangial cells, with little encroachment on the patency of the blood capillaries but leading to obliteration of the urinary space of glomeruli.[54]

C. Infertility and Early Embryonic Death

Infertility and repeated breeding were the major problems in the cow herds bred by bulls naturally excreting chlamydiae in the semen.[48] The effect of chlamydiae present in the semen on fertilization and early embryonic development was investigated.[55] A known infectious dose of chlamydiae was added to high-quality semen, and cows were artificially inseminated. Incubation of normal semen with chlamydia-infected yolk sac homogenate for 4 hr at 38°C was not detrimental to progressive motility of the spermatozoa compared with the motility of control semen. None of ten cows inseminated with chlamydia-containing semen, but five of ten control cows inseminated with semen mixed with normal yolk sac homogenate, were pregnant 40 days after insemination. Biopsy samples from the endometrium contained cells with chlamydial inclusions when examined by immunofluorescence.[55]

The effect of chlamydiae on the process of fertilization was investigated in another experiment. One ovum was recovered from each of eight cows about 3 days after they were inseminated with chlamydia-contaminated semen. Six of the eight ova were fertilized since they had reached the blastomere stage of eight-cell embryos. Cells of these embryos did not contain chlamydial inclusions, and chlamydiae did not penetrate the zona pellucida. Accordingly, fertilization failure was not the cause of infertility. A probable hypothesis for the pathogenesis of infertility in seminal transmission of chlamydiae is: the agent multiplies in cells of the endometrium and thereby alters the environment of the embryo, resulting in early embryonic death. Endometritis was observed in uterine chlamydial infections of cows.[17]

The electron microscopic analysis of the genital infection of guinea pigs provides an example of the potential role of specific strains of *C. psittaci* as a cause of infertility in animals. Cells infected in the exocervix were squamous, columnar, and mucous epithelial cells of the mucosal surface. The ciliated epithelial cells of the Fallopian tubes were also productively infected. Lysis and disruption of infected cells or extrusion of entire infected cells from the mucous membranes occurred. Numerous leukocytes were attracted to the infected cells.[56-58]

Chlamydiae were also isolated from the uterus of Koala bears of a population that had a low reproductive rate. Female animals of this population had severe cloacal soiling and signs of conjunctivitis. The uterine walls were thickened and the lumen contained casts or pus. Endometritis, fibroplasia of the uterine wall, cyotic dilation of uterine glands with infiltration by lymphocytes and plasma cells characterized the histological changes. Salpingitis with submucosal invasion of plasma cells, lymphocytes, and macrophages was also observed.[59]

IX. MASTITIS AND TRANSMISSION THROUGH MILK

A. Naturally Occurring and Experimentally Induced Mastitis

Instances of mastitis of cows that were apparently caused by naturally occurring chlamydial infections were reported. Blanco-Loizelier[60] described mastitis caused by chlamydiae in a herd of milking cows in the Spanish province of Gerona. Within 2 months, 17 of 25 cows developed acute mastitis with severe reduction in, or transitory cessation of, milk production. Chlamydiae were also isolated from milk samples and udder tissues of a large herd of milking cows in East Germany.[61] These cows suffered from mastitis, and *Streptococcus agalactiae* was cultured from some samples. Chlamydiae were isolated by guinea pig inoculation and subsequent chicken embryo propagation from 5 pools of 34 milk samples and from mammary lymph nodes. Of 30 samples from udders and lymph nodes tested with the direct fluorescent antibody technique 28 were positive. Similar isolations were made from cows with mastitis in three other herds. Mycoplasma, yeasts, nocardia, and atypical mycobacteria were not isolated.

The nonlactating mammary glands of the pregnant cows in chlamydial abortion investigations did not appear to be affected.[37] Borovik and co-workers[62] observed mastitis in lactating pregnant cows inoculated with a bovine chlamydial strain. These cows aborted 32 to 42 days later. Chlamydiae were isolated from milk samples, which also had IgG$_1$ antibodies against chlamydiae.

Experimentally, mastitis was induced by intracysternal inoculation of different chlamydiae from sheep or cattle.[63-65,67] A severe mastitis, with marked alterations in the milk for a period of 8 days, was induced. The pregnant cows aborted 6 to 75 days after inoculation. Papadopoulos and Leontides[66] inoculated lactating ewes via the teat canal with an ovine chlamydial abortion strain and induced acute mastitis with signs of systemic infection. The milk contained large numbers of leukocytes, and milk secretion almost ceased during the febrile period 24 to 90 hr after infection. Chlamydiae were recovered from the milk for 13 days after inoculation.

The clinical signs observed by different investigators after intracysternal inoculation were similar. The cows developed fever as high as 41°C 1 to 3 days after inoculation. The temperature returned to normal levels after 6 days. The milk flow decreased sharply. The mammary secretion changed from milk to a yellow or amber-colored fluid with numerous white fibrin clots, which occasionally plugged the streak canal. The inoculated quarters became markedly swollen and hot. Milk production by infected cows almost ceased but recovered somewhat after 10 to 12 days, although it never returned to preinoculation levels.[64,65,67]

B. Pathogenesis and Pathological Changes

Chlamydial infectivity as high as 10^7 chicken embryo infectious doses per milliliter of secretion was present 2 to 6 days after inoculation, indicating abundant multiplication of the agents in the cells of the mammary gland.[64,65,67] The cellular contents of the mammary secretions began to increase 1 day after inoculation. Increasing numbers of leukocytes, monocytes, and epithelial cells were shed. Abundant elementary bodies were present, and many cells in the secretion had cytoplasmic chlamydial inclusions.

Under natural conditions the chlamydial infection may reach the mammary gland parenchyma through the teat canal, possibly originating from intestinal infections. The intracysternal experimental inoculation would be an exaggerated version of this mode of infection. Another mode of infection would involve localization of the infection in the mammary gland during a chlamydemic phase following oral, respiratory, or conjunctival infection, or by activation of the clinically inapparent intestinal infection, with penetration of the intestinal mucosa. The infection did not spread from the directly inoculated quarter to other quarters, but a systemic infection with chlamydemia ensued.

Chlamydial cytoplasmic inclusions were found in cells throughout the parenchyma 3 days after a quarter was inoculated. Necrosis of alveolar and duct epithelial cells was abundant and related to chlamydial inclusions in the cells. Swelling of endothelial cells and mononuclear infiltration into interstitial tissues occurred. The necrotic epithelial cells sloughed from the basement membrane, and fibrin, as well as eosinophilic exudates, poured from the interstitium into the lumen of mammary alveoli. The ducts were plugged by necrotic cells and fibrinous exudate. Edema with leukocyte infiltration and erythrocyte diapedesis were evident in the lamina propria of the gland cistern and of ducts. Perivascular infiltration of blood and lymph vessels was prominent 6 to 10 days after inoculation. Interstitial infiltration with lymphocytes and plasma cells caused separation of alveoli. Squamous metaplasia of duct epithelial cells developed.[64,67]

X. POLYARTHRITIS-POLYSEROSITIS

A. Arthropathogenic Potential

Chlamydial polyarthritis of lambs is a disease that occurs in epizootic proportions in the major sheep-raising regions of the U.S. The cause of the disease, which also is referred to as *stiff lamb disease,* was first identified in Wisconsin by Mendlowski and Segre[68] as a chlamydial infection of the synovial tissues, with inflammation of most diarthrodial joints of the limbs, leading to stiffness and lameness. A similar disease of lambs and calves — characterized by soreness of the joints, with lameness resulting from polyarthritis — was also recognized in the western U.S.[69,70] Chlamydiae were isolated from the joints of calves and lambs affected with this disease. Kölbl and associates[71] cultured chlamydiae from synovial specimens of 24 of 39 pigs affected with a chronic nonpurulent synovitis.

Chlamydial polyarthritis was observed in lambs on ranges, as well as in lambs from farm flocks and feedlots, and morbidity ranged from 2 to 75%. Rectal temperatures of affected lambs ranged from 39 to 42°C. The lambs had varying degrees of stiffness, lameness, anorexia, and conjunctivitis. Affected lambs were gaunt, depressed, reluctant to move or stand or bear weight on one or more limbs, and lingered behind the rest of the band. The highest incidence among sheep on ranges was observed in July, August, and September, while in feedlot lambs the disease is most prevalent in October, November, and December.[69,72]

The youngest calf with naturally occurring chlamydial polyarthritis was 4 days old when the first signs were detected, but more commonly the calves were several weeks old. As the disease progresses the calves become depressed, favor their legs, and assume a hunched position while standing. When recumbent they extend their legs passively, and the joints and tendons of the limbs are painful when palpated.[70]

B. Pathogenetic Events and Synovial Changes

Chlamydiae were isolated from the sites of the major lesions, the affected joints of naturally occurring cases of chlamydial polyarthritis of lambs and calves. These agents were also isolated from different organs and body excretions that included synovial tissues and fluids; lungs, liver, spleen, kidneys, brain, and lymph nodes; conjunctival scrapings; composite samples of mucosa and content from abomasum, duodenum, jejunum, ileum, and caecum; and samples of blood, cerebrospinal fluid, urine, and feces.[19,68,70,73,74]

A chlamydemic phase evidently precedes infection of the joint tissues in the pathogenesis of chlamydial polyarthritis. Polyarthritis can readily be reproduced experimentally by inoculating lambs or calves via the oral, subcutaneous, intramuscular, intravenous, or intra-articular routes. Three days after intra-articular inoculation of lambs, the directly inoculated joint was severely inflamed and had high chlamydial infectivity.[73] Lesions were not seen in other joints, but chlamydiae were recovered from the blood and fluid of uninoculated joints. Three days after intramuscular or intravenous inoculation of lambs, chlamydial infectivity was present in several joints, although gross lesions were minimal. Irrespective of the routes of inoculation, hip, knee, shoulder, and elbow joints of lambs examined on the 7th day of the experiment had moderate to severe inflammatory changes, and chlamydiae were isolated from the affected joints. Most joints of the experimental lambs remained affected for 21 days after inoculation. Maximum chlamydial infectivity was present in the joints 7 and 14 days after inoculation of the lambs and decreased progressively thereafter to become negative after the 28th day.

Calves inoculated orally or parenterally with chlamydial isolates from joints developed polyarthritis and polyserositis; chlamydiae were recovered from the affected joints, internal organs, and intestinal tract.[19] Because the intestinal tract was found to be infected in sheep and calves with naturally occurring chlamydial polyarthritis, oral infection occurs under field conditions. Following oral inoculation of calves with the chlamydial strain of polyarthritis, it multiplied in the mucosa of the small and large intestines, leading to primary diarrhea. Chlamydemia followed intestinal chlamydial multiplication and determined the further course of the disease.[19,20]

In naturally occurring and experimentally induced chlamydial polyarthritis of lambs and calves, the most striking tissue changes were articular and periarticular.[19,72,74] Radiologically, there was no discernible malalignment of the limbs or joints. Periarticular, subcutaneous edema, and fluid-filled, fluctuating synovial sacs contributed to joint enlargement of affected calves. The larger, freely movable, weight-bearing joints — such as hip, stiffle, tarsal, atlanto-occipital, shoulder, elbow, and carpal joints of calves — contained excessive, grayish-yellow, turbid synovial fluid (Figure 5). Grayish-yellow fibrin flakes and plaques of different sizes and shapes were in joints with advanced lesions. The fibrin plaques in the recesses of the affected joints adhered to the synovial membranes, and the joint capsules were thickened. The tendon sheaths of severely affected lambs and calves were distended

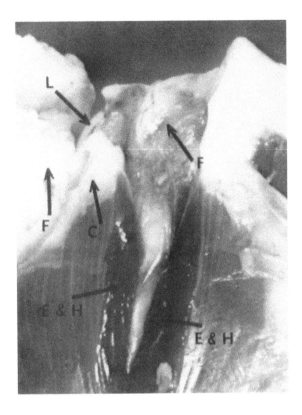

FIGURE 5. Lateral aspect of a stifle joint. Fibrin adhering to
inside lining of thickened joint capsule (C); intraarticular fibrin
(F); periarticular edema and hemorrhage (E & H); and intraarticular ligament (L). Courtesy of Dr. L. J. Shupe.

and contained creamy, grayish-yellow exudate. Surrounding muscles were hyperemic and
edematous having petechiae in their associated fascial planes.

The most consistent and striking histopathological changes involved the synovial membranes, joint capsule, tendon sheaths, ligaments, periarticular connective tissue, and muscles.
The inflammatory reaction in the synovium, tendon sheaths, and subsynovial tissues pro-
gressed from initial serous to fibrinopurulent inflammation to, finally, accumulations of
lymphoid cells and fibroplasia. Involvement of subsynovial connective tissue and muscle
was limited to accumulation of mononuclear cells in perivascular areas and increasing
fibrosis. The joint capsule and tendon sheat of advanced lesions had marked fibrotic thick-
ening, associated with an extensive inflammatory response that consisted primarily of infiltration by plasma cells, monocytes, lymphocytes, macrophages, and neutrophils. The
viable synoviocytes became swollen and hyperplastic and assumed a pseudostratified appearance (Figure 6). The tendon sheaths had marked fibroblastic proliferation characterized
by large, plump, round fibroblasts.[72] Chlamydial inclusions were found in fibroblasts, monocytes, and endothelial and synovial cells in thin sections of affected joint tissues.[74]

Changes in muscle were periarticular and adjacent to tendinous insertions. These changes
were best characterized as myositis because they consisted of infiltration with inflammatory
cells and accumulations of edema fluid between individual muscle fibers and between muscle
bundles. The reactive processes in muscle tissues appeared to be dominated by fibroblastic
proliferation suggestive of a musculotendofasciitis. Affected tendons and tendon sheaths
contained edema fluid and were infiltrated by inflammatory cells.[72]

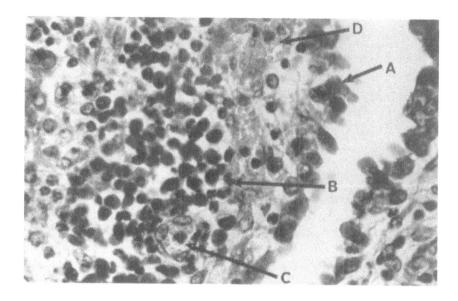

FIGURE 6.   Section of synovial tissue of a polyarthritic lamb illustrating pseudostratified appearance of synovial layers (A); infiltration with inflammatory cells (B); capillary (C); hyperemia and hemorrhage (D) in the underlying synovial layers, × 480.

XI. CHLAMYDIAL MENINGO-ENCEPHALOMYELITIS

Cattle may sporadically develop signs of nervous disease and encephalomyelitis following chlamydial infection. The causal agent of the disease called sporadic bovine encephalomyelitis was isolated in 1942[75,76] and later identified as a chlamydial strain with characteristics of bovine immunotype 2.[77] Tissue reactions in the brain and associated adverse neurological signs may be induced in the course of systemic chlamydial infections of other animal species, as well as man. Opossums with naturally occurring chlamydial infections developed paralysis of the hind legs and convulsions.[78] Meningeal and encephalitic lesions were found histologically in chickens, pigeons, and turkeys with chlamydial infections.[11]

The encephalitic clinical manifestations of chlamydial infections were investigated mainly in cattle. Depression and insufficient locomotor coordination are early clinical signs. The affected cattle become anorectic and develop a high fever approximately 2 weeks after exposure. Excessive salivation, dyspnea, and mild diarrhea are signs in the early stage. Recovery may follow this stage, but in most instances the cattle develop neural involvement manifested by walking difficulties, a stiff gait, and a tendency to stagger and circle, often falling over small obstacles. The limbs become progressively weaker, paralysis develops, and the animals become recumbent and may exhibit opisthotonus. The course of the disease usually lasts from 10 to 14 days, but in some instances affected cattle may survive for a month. This disease is seen primarily in cattle younger than 2 years, although older animals may be affected. The morbidity is low, but the mortality of stricken cattle is 50 to 60 percent. The disease may remain enzootic in one herd, but, not infrequently, outbreaks follow introduction of susceptible cattle into established herds.[79]

The dominant feature in the development of this disease is a generalized chlamydial infection. The chlamydial agent was isolated from the blood, brain, and cerebrospinal fluid and from the liver, spleen, kidneys, lungs, and peritoneal fluid.[76] Lesions in various organs are associated with vascular damage. Macroscopic lesions, when visible in the central nervous system, are primarily hyperemia and edema. After an acute course of the sickness, few gross lesions are found, but the serious cavities may contain an increased amount of yellowish fluid. Some chronic cases have serofibrinous exudate in serous cavities with peritonitis, pleuritis, or pericarditis. Microscopically, all parts of the brain and spinal cord have vasculitis with perivascular cuffing and parenchymal foci of inflammation composed of lymphoid and mononuclear cells, with occasional leukocytes. Similar inflammatory reactions may be seen in the leptomeninges. Small foci of granulomatous inflammation are found in kidneys, lungs, and livers. The degree of brain involvement may well be the critical factor that determines recovery or death. The types of cells infected in the brain have not been clearly identified by modern methods. The factors that lead to the severe reaction in the brain are not well understood and most likely involve immune-mediated injuries.[3,17]

XII. CONJUNCTIVITIS AND OCULAR DISEASE

While trachoma and inclusion conjunctivitis of man have a long history as specific and distinct clinical entities caused by *C. trachomatis*, the study of ocular chlamydial infections of animals is of rather recent history. Ocular diseases are now known to be caused by *C. psittaci* in several animal species.[80-86] Conjunctivitis and ocular involvement have long been recognized as clinical signs of chlamydiosis in pigeons, ducks, and geese.[10,11]

Yerasimides[81] was the first to isolate and identify a chlamydial agent from the conjunctival exudate of a cat suffering from acute catarrhal conjunctivitis. The significance of this infection of cats was later recognized as a distinct clinical entity related with feline pneumonitis.[31] An infectious agent with chlamydial properties was isolated from conjunctival samples of piglets affected with keratoconjunctivitis in Bulgaria.[82] In 1964 Murray identified inclusion conjunctivitis of guinea pigs as a prevalent chlamydial infection.[83]

Dickinson and Cooper[80] proposed that chlamydial agents were the cause of contagious keratoconjunctivitis of sheep in England. They based their conclusion on cytologic examinations of conjunctival samples and on chemotherapeutic treatment of the infection. Chlamydial agents were subsequently cultured in developing chicken embryos from conjunctival samples taken from sheep suffering from keratoconjunctivitis in the U.S.[84] Chlamydial agents were also isolated from conjunctival scrapings of cattle afflicted with keratoconjunctivitis in Czechoslovakia and Germany.[17] Interestingly, Rowe and co-workers isolated chlamydiae in cultured cells from conjunctival samples of a 1½-year-old calf of a free-living herd of African buffalos.[86] The studies of Cockram and Jackson[85] are of unique interest. Chlamydiae were isolated from eye samples of free-living koalas (*Phascolarctos cinereus*) severely affected with keratoconjunctivitis in Australia. Previous records indicated that severe eye conditions associated with blindness and mortality of koalas leading to major die-offs were observed and recorded as long ago as 1887. This infection of koalas is also considered to cause reproductive problems.[59]

The clinical signs in conjunctival infections of different animal species are probably similar. They have been studied in infections in cats, sheep, and guinea pigs. The infection may become endemic in catteries, and cats may remain affected with conjunctivitis for as long as one year. Kittens born to affected or clinically recovered queens may have severe conjunctivitis at the time their eyelids normally open. Cats developed a severe conjunctivitis 3 to 5 days after conjunctivial inoculation. On the 4th day, these cats invariably had a fever, which lasted for 24 hr. Chlamydial inclusion bodies were found in large numbers from the 3rd day after inoculation and persisted in decreasing numbers for 2 weeks. Experimentally inoculated cats had clinical evidence of conjunctivitis for as long as 90 days. Corneal involvement was not observed, but rhinitis and other signs of respiratory involvement were seen.[31] Infections of the gastric mucosa were also recently detected. This site of infection appears to be an extension of the feline pneumonitis syndrome caused by chlamydiae. Mild gastritis was observed, but most cats with persistent gastric infection appeared clinically normal.[87,88]

Early signs of chlamydial conjunctivitis of sheep consisted of chemosis and dilation of conjunctival vessels at the lid margins. The development of lymphoid follicles signaled intermediate stages of conjunctivitis. Lymphoid follicles began as small, discrete, pale, and elevated changes in the conjunctiva, enlarged to diameters of 3 mm, and became confluent, forming delicate pink-to-red folds in the lower fornix and the 3rd eyelid. Lymphoid follicles in the upper fornices developed into folds 2 to 4 mm high. Conjunctival hyperemia, edema, and follicle development caused swelling of the periorbital tissues. Lacrimal drainage was compromised, epiphora was a common sign, and abundant seropurulent eye discharge was observed to seal the eyelids with a crust. Perilimbal edema of the cornea was associated with signs of severe inflammatory involvement of the bulbar conjunctiva and its extension into the cornea as keratitis. Loops of blood vessels grew into the cornea at a rate of 1 mm per day. Neovascularization of the cornea occurred in the upper limbus, often extended into the cornea for distances of 5 mm, and became completely perilimbal. Both eyes were found to be equally affected in 80% of the lambs. The disease was self-limiting in uncomplicated cases because full resolution of lesions was observed. Complications consisted of opacity of the entire cornea and corneal ulceration.[89]

Of the lambs with conjunctivitis studied in the western U.S. 10 to 85% had polyarthritis, and virtually all lambs with polyarthritis had conjunctivitis. The same type of chlamydiae were isolated from conjunctiva, blood, and joint samples.[90]

Chlamydial inclusion conjunctivitis affects many herds of guinea pigs and is predominantly found in 4- to 8-week-old animals. A special effort is necessary to detect the inflammatory conjunctival reaction with its light, yellowish-white discharge. An intense conjunctivitis arises 3 to 6 days after experimental infection of guinea pig eyes. Watery to mucopurulent discharge, chemosis, and follicular hypertrophy in the palpebral conjunctiva develop. Localized subepithelial cellular infiltrates are frequently seen in the peripheral cornea. Pannus or micropannus form in most inoculated guinea pigs as early as 5 days after exposure. The keratoconjunctivitis is self-limiting and clears up within 3 to 4 weeks.[83,91]

The development of chlamydial inclusion bodies in epithelial and monocytic cells was associated with mononuclear and heterophilic cellular responses in cats, lambs, piglets, and guinea pigs. The epithelium and subepithelial tissues of the conjunctiva were infiltrated with polymorphonuclear leukocytes in the early stages, while subepithelial collections of macrophages and mononuclear cells were found in the peripheral limbus, and mononuclear cells dominated the histological changes in the advanced stages.[89,91]

Calves with systemic chlamydial infections leading to polyarthritis also developed conjunctivitis, and signs of blindness were observed in cattle with chlamydial meningo-encephalomyelitis. The intra-ocular structures involved were the retina and the optic nerve. Optic papillitis consisted of mononuclear inflammatory reactions extending from the lamina cribrosa to the head of the optic nerve. Histopathological alterations were present in the nerve fiber layer and the inner nuclear layer of the sensory retina. An inflammatory reaction extended towards the inner limiting membrane of the retina. Histiocytic cells, lymphocytes, and a few polymorphonuclear leukocytes formed inflammatory foci of granuloma-like appearance.[92] Bilateral anterior uveitis was observed after intravenous inoculation of albino rabbits with a chlamydial strain isolated from muskrats. This uveitis disappeared about 10 days later, but iritis persisted longer, and mononuclear cells invaded the iris and the ciliary body.[93]

XIII. INFECTIONS OF ECTOTHERMIC VERTEBRATES AND MOLLUSCS

A. Hepatitis and Chlamydial Infection of African Clawed Frogs (*Xenopus laevis*)

Widespread pyogranulomatous inflammation and large basophilic intracytoplasmic inclusions were detected in livers and other tissues of African clawed frogs of a commercial frog rearing operation.[94] The frogs were dying of a spontaneous disease of high morbidity and mortality. Chlamydiae were isolated in cell culture. The disease was reproduced by inoculating normal frogs with the cell culture isolate of *C. psittaci*.[94,95] This infection was also diagnosed in laboratory-bred frogs experiencing sudden death.[96]

The epizootic described by Newcomer and co-workers[94] who thus identified the first *C. psittaci* infection causing naturally occurring disease in ectothermic vertebrates, reduced a population of approximately 40,000 frogs to 20,000 frogs during an interval of 3 months from January to March. The frogs had been kept initially in a 5-acre pond. They were transferred indoors in October and housed in 4.5 m diameter aluminum and plastic swimming pools. The frogs were fed raw chopped beef liver which had been condemned for human consumption or a commercial frog ration saturated with fresh bovine blood.

The affected frogs exhibited signs of lethargy, disequilibrium, patchy cutaneous depigmentation and petechiation, and a bloated edematous appearance. They surfaced more frequently to breathe. The mean time from experimental inoculation until frogs became tarpid and died increased with higher dilutions of the inoculum. The 50% lethal dose was less than 10 inclusion-forming units when the frogs were inoculated by the dorsal lymph sac route.

Chlamydiosis of the African clawed frog is an experimentally reproducible disease with unique histopathological and ultrastructural features. These changes become manifest as diffuse pyogranulomatous inflammation in the liver, spleen, kidneys, heart, and lungs, and are associated with the presence of cytoplasmic basophilic inclusions. The incidence of diffuse granulomatous myocarditis and endocarditis was higher in naturally infected frogs than in experimentally infected frogs.[94-96]

The source of this chlamydial infection of frogs was not determined. The isolate induced diffuse, noniodine staining inclusions in McCoy cells. Sulfadiazine did not inhibit multiplication of this strain. Condemned beef liver or bovine blood was considered as a potential source of infection. The isolates from frogs were not compared antigenically or cytologically with bovine chlamydial strains. Immunotype 2 of bovine chlamydial strains (sporadic bovine encephalomyelitis) multiplied in cell cultures of fathead minnows (*Pimephales promelas*).[97] Alternatively, the infection of frogs may have been acquired from feral animals during the period of outdoor housing.

B. Chlamydia-Like Agents in Fish and Clams

Electron microscopic analysis of specimen from fish and clams revealed morphological evidence of infections with agents that have developmental stages similar to those of chlamydiae. Epithelial cells in the gills of Connecticut striped bass (*Morone saxalitis*) and of white perch (*Morone americanus*) affected with epitheliocystis, a chronic gill disease, had cytoplasmic inclusions compatible with those induced by chlamydiae in cells of warm blooded animals.[98] Digestive tubule cells of the hard clams (*Mercenaria mercenaria*) contained inclusions with pleomorphic stages of chlamydial development.[99] Interestingly, some of these inclusions resembled aberrant chlamydial forms similar to those found by Spears and Storz in cultured cells infected with strains of chlamydial polyarthritis under enhancing conditions.[100] Furthermore, crystal lattice arrays of small icosahedral particles were present in some of the reticulated forms in infections of hard clams. The hypothesis that these particles represented viruses was strengthened when Richmond and co-workers[101] succeeded in demonstrating 22 nm viruses in chlamydial strains isolated from respiratory and intestinal infections of ducks. These chlamydiaphages induced crystaline arrays within reticulate bodies of chlamydial development. These findings clearly encourage a search for chlamydiae in ectothermic animals. Attempts to cultivate these putative chlamydial agents from fish or clams are needed for further studies of their properties and significance.

XIV. ARTHROPODS AND THEIR ROLE IN PERPETUATION OR TRANSMISSION OF CHLAMYDIAE

Chlamydiae have been known to survive in arthropods over extended periods of time and may be isolated from naturally infected ectoparasites.[102] Weyer[103] artificially infected ticks (*Ornithodoros moubata*) and human body lice (*Pediculus humanus*). He could isolate chlamydiae from the ticks for 75 days and from lice for 21 days after infection. The avian chlamydial strain used was passaged five times in lice for a period of 83 days. *Dermacentor occidentalis* ticks were infected with chlamydiae of bovine abortion.[104] The chlamydiae apparently multiplied during nymphal engorgement, but isolations were not made from nymphal ticks after detachment or during subsequent metamorphosis. These authors did not establish that *D. occidentalis* is a biological vector. Chlamydiae were transmitted to pregnant cows with *Ornithodoros coriaceus* experimentally infected 25 and 48 days before. Also, chlamydiae were isolated from field-collected ticks 6 months after field collection.[105] An ovine chlamydial abortion strain was propagated in a cultured tick cell line.[106]

Electron microscopic studies on the hepatopancreas of the spider *Coelotes luctuosus* demonstrated that secretory cells contained cytoplasmic inclusions containing structures resembling the developmental forms of chlamydiae.[107] Of 17 individual spiders examined 2 had these inclusions. Similar chlamydia-like microorganisms were found to be associated with fatal disease of the spider *Pisaura mirabilis*.[108] These morphological observations are of interest and should encourage attempts to cultivate the infectious agents involved for further characterization.

XI. PERSPECTIVES AND VISTAS

The diseases resulting from infections by chlamydiae affect a wide range of different animal species, and they appear quite diverse. When analyzed comparatively, these diseases reflect in many respects a pattern of similarity despite the phylogenetic differences of the hosts. The disease mechanisms involve local infections of mucous membranes or pathogenetic progression involving penetration of mucous membranes with ensuing blood infectious phases. Other body compartments are invaded during these chlamydemic phases. Infections of specific target organs such as placenta and fetus, synovial tissue, liver or brain induce then unique clinical manifestations. What engenders the invasiveness is not known. Interestingly, some specific chlamydial strains associated with local infections also have invasive capacities. The species *C. psittaci* comprises chlamydiae with enormously diverse properties.

The host cell range in vivo of the few strains of *C. psittaci*, that were studied in depth, was surprisingly wide.[22,57] Direct and efficient cytocidal action resulting from the complex interaction of strains of *C. psittaci* with host cells probably represents one basis for disease induction.[109,110] The influence of chlamydiaphage on cytocidal functions of chlamydiae should be explored.[101] The latent or persistent and clinically inapparent states of chlamydial infections remain important epidemiologically. The in vitro models of persistence provide powerful tools to explore this state of host parasite interaction.[111,112]

The possible contribution of immune mediated reactions to specific disease manifestations has not been investigated to any significant extent. Dimorphic inflammatory responses with leukocytes along the mucosal surface, and mononuclear cell infiltration in the stroma are prominent in conjunctivitis, polyarthritis, placentitis, and enteritis.[25,40,41,72,89,91] Submucosal follicle formation is striking in conjunctivitis and polyarthritis caused by chlamydiae.[72,89] More information is needed on humoral and on cell-mediated immunological responses of different animal species to chlamydial antigens, in order to understand disease manifestations, recovery, and resistance in these infections.[113-115]

REFERENCES

1. **Moulder, J. W., Hatch, T. P., Kuo, C.-C., Schachter, J., and Storz, J.,** Genus *Chlamydia* Jones, Rake, and Stearns, in *Bergey's Manual of Systematic Bacteriology,* Vol. 1, Holt, J. G. and Krieg, N. R., Eds., Williams & Wilkins, Baltimore, Md., 1984, 729.
2. **Storz, J.,** in *Chlamydia and Chlamydia Induced Diseases,* Charles C. Thomas, Springfield, Ill., 1971, 358.
3. **Storz, J. and Krauss, H.,** Chlamydial infections and diseases of animals, in *Handbook on Bacterial Infections in Animals,* Vol. 5, Blobel, H. and Schliesser, T., Eds., Gustav Fischer Verlag, Jena, East Germany, 1985, 447.
4. **Meyer, K. F.,** The host spectrum of psittacosis-lymphogranuloma venereum (PL) agents, *Am. J. Ophthalmol.,* 63, 1225, 1967.
5. **Page, L. A.,** Interspecies transfer of psittacosis-LGV-trachoma agents: pathogenicity of two avian and two mammalian strains for eight species of birds and mammals, *Am. J. Vet. Res.,* 27, 397, 1966.
6. **Wang, S. P.,** A micro-immunofluorescence method. Study of antibody response to TRIC organisms in mice, in *Trachoma and Related Disorders Caused by Chlamydial Agents,* Nichols, R. L., Ed., Excerpta Medica, Amsterdam, 1971, 273.
7. **Perez-Martinez, J. A. and Storz, J.,** Antigenic diversity of *Chlamydia psittaci* of mammalian origin determined by microimmunofluorescence, *Infect. Immun.,* 50, 905, 1985.
8. **Eb, F. and Orfila, J.,** Serotyping of *Chlamydia psittaci* by the microimmunofluorescence test: isolates of ovine origin, *Infect. Immun.,* 37, 1289, 1982.
9. **McClenaghan, M., Herring, A. J., and Aitken, I. D.,** Comparison of *Chlamydia psittaci* isolates by DNA restriction endonuclease analysis, *Infect. Immun.,* 45, 384, 1984.
10. **Meyer, K. F.,** Ornithosis, in *Diseases of Poultry,* 5th ed., Biester, H. E. and Schwarte, L. H., Eds., Iowa State University Press, Ames, Iowa, 1965, 675.
11. **Page, L. A.,** Avian chlamydiosis (Ornithosis), in *Diseases of Poultry,* 7th ed., Hofstad, M. S., Ed., Iowa State University Press, Ames, Iowa, 1978, 337.
12. **Sadowski, J. M. and Minta, Z.,** Chlamydiosis of the air sacs in geese, *Bull. Vet. Inst. Pulawy.,* 23, 111, 1979.
13. **Grimes, J. E. and Clark, D.,** *Chlamydia psittaci* infections of pet and other birds in the United States in 1984, *J. Infect. Dis.,* 153, 374, 1986.
14. **Page, L. A.,** Observations on the involvement of wildlife in an epornitic of chlamydiosis in domestic turkeys, *J. Am. Vet. Med. Assoc.,* 169, 932, 1976.
15. **Page, L. A.,** Experimental ornithosis in turkeys, *Avian Dis.,* 3, 51, 1959.
16. **York, C. J. and Baker, J. A.,** A new member of the psittacosis-lymphogranuloma group of viruses that causes infection in calves, *J. Exp. Med.,* 93, 587, 1951.
17. **Perez-Martinez, J. A. and Storz, J.,** Chlamydial infections of cattle. I, II, *Mod. Vet. Pract.,* 66, 517, 1985.
18. **Kölbl, O.,** Untersuchungen über das Vorkommen von Miyagawanellen beim Schwein, *Wien. Tierärztl. Monatsschr.,* 56, 332, 1969.
19. **Storz, J., Shupe, J. L., Smart, R. A., and Thornley, R. W.,** Polyarthritis of calves: experimental induction by a psittacosis agent, *Am. J. Vet. Res.,* 27, 987, 1966.
20. **Eugster, A. K. and Storz, J.,** Pathogenic events in intestinal chlamydial infections leading to polyarthritis in calves, *J. Infect. Dis.,* 123, 41, 1971.
21. **Eugster, A. K., Joyce, B. K., and Storz, J.,** Immunofluorescence studies on the pathogenesis of intestinal chlamydial infections in calves, *Infect. Immun.,* 2, 351, 1970.
22. **Doughri, A. M., Altera, K. P., and Storz, J.,** Host cell range of chlamydial infection in the neonatal bovine gut, *J. Comp. Pathol.,* 83, 107, 1973.
23. **Doughri, A. M., Altera, K. P., Storz, J., and Eugster, A. K.,** Electron microscopic tracing of pathogenic events in intestinal chlamydial infections of newborn calves, *Exp. Mol. Pathol.,* 18, 10, 1973.
24. **Doughri, A. M., Altera, K. P., Storz, J., and Eugster, A. K.,** Ultrastructural pathologic changes in the chlamydia-infected ileal mucosa of newborn calves, *Vet. Pathol.,* 10, 114, 1973.
25. **Doughri, A. M., Young, S., and Storz, J.,** Pathologic changes in intestinal chlamydial infections of newborn calves, *Am. J. Vet. Res.,* 35, 939, 1974.
26. **Gönnert, R.,** Die Bronchopneumonie, eine neue Viruskrankheit der Maus, *Zentralbl. Bakt. I. Orig.,* 147, 161, 1941.
27. **Nigg, C.,** Unidentified virus which produces pneumonia and systemic infection in mice, *Science,* 95, 49, 1942.
28. **Baker, J. A.,** Virus causing pneumonia in cats and producing elementary bodies, *J. Exp. Med.,* 79, 159, 1944.

29. **Tolybekow, A. S., Wischnjakowa, L. A., and Dobin, M. A.,** Die ätiologische Bedeutung eines Erregers aus der Bedsoniengruppe für die enzootische Pneumonie der Schweine, *Monatschr. Vet. Med.,* 28, 339, 1973.

30. **Moorthy, A. R. S. and Spradbrow, P. B.,** *Chlamydia psittaci* infection of horses with respiratory disease, *Equine Vet. J.,* 10, 38, 1978.

31. **Cello, R. M.,** Ocular infections in animals with PLT (Bedsonia) group agents, *Am. J. Ophthalmol.,* 63, 1270, 1967.

32. **Gogolak, F. M.,** The histopathology of murine pneumonitis infection and the growth of the virus in the mouse lung, *J. Infect. Dis.,* 92, 254, 1953.

33. **Dungworth, D. L. and Cordy, D. R.,** The pathogenesis of ovine pneumonia. I. Isolation of a virus of the PL group, *J. Comp. Pathol. Ther.* 72, 49, 1962.

34. **Stamp, J. T., McEwen, A. D., Watt, J. A. A., and Nisbet, D. J.,** Enzootic abortion in ewes. I. Transmission of the disease, *Vet. Rec.,* 62, 251, 1950.

35. **Storz, J.,** Psittacosis infection of the fetus and its effect on the dam, *Zentbl. Vet. Med.,* 16, 1, 1969.

36. **Rodolakis, A., Boullet, C., and Souriau, A.,** *Chlamydia psittaci* experimental abortion in goats, *Am. J. Vet. Res.,* 45, 2086, 1984.

37. **Storz, J. and McKercher, D. G.,** Etiological studies on epizootic bovine abortion, *Zentralbl. Vet. Med.,* 9, 411, 1962.

38. **Schoop, G. and Kauker E.,** Infektion eines Rinderbestandes durch ein Virus der Psittakosis-Lymphogranuloma-Gruppe, Gehäufte Aborte im Verlauf der Erkrankungen, *Dtsch. Tierärztl. Wochenschr.,* 63, 233, 1956.

39. **Ata, F. A. and Storz, J.,** Chlamydial blood clearance in convalescent sheep, *Cornell Vet.,* 64, 25, 1974.

40. **Kwapien, R. P., Lincoln, S. D., Reed, D. E., Whiteman, C. E., and Chow, T. L.,** Pathologic changes of placentas from heifers with experimentally induced epizootic bovine abortion, *Am. J. Vet. Res.,* 31, 999, 1970.

41. **Novilla, M. N. and Jensen, R.,** Placental pathology of experimental enzootic abortion in ewes, *Am. J. Vet. Res.,* 31, 1983, 1970.

42. **Reed, D. E., Lincoln, S. D., Kwapien, R. P., Chow, T. L., and Whiteman, C. E.,** Comparison of antigenic structure and pathogenicity of bovine intestinal chlamydia isolate witn an agent of epizootic bovine abortion, *Am. J. Vet. Res.,* 36, 1141, 1975.

43. **Studdert, M. J.,** *Bedsonia* abortion of sheep. II. Pathology and pathogenesis with observations on the normal ovine placenta, *Res. Vet. Sci.,* 9, 57, 1968.

44. **Storz, J., Call, J. W., Jones, R. W., and Miner, M. L.,** Epizootic bovine abortion in the intermountain region: some recent clinical, epidemiologic and pathologic findings, *Cornell Vet.,* 57, 21, 1967.

45. **Storz, J. and Whiteman, C. E.,** Bovine chlamydial abortions, *Bovine Pract.,* 16, 71, 1981.

46. **Lincoln, S., Kwapien, R. P., Reed, D. E., Whiteman, C. E., and Chow, T. L.,** Epizootic bovine abortion: clinical and serologic responses and pathologic changes in extra-genital organs of pregnant heifers, *Am. J. Vet. Res.,* 30, 2105, 1969.

47. **Studdert, M. J. and Kennedy, P. C.,** Enzootic abortion of ewes, *Nature,* 203, 1088, 1964.

48. **Storz, J., Carroll, E. J., Ball, L., and Faulkner, L. C.,** Isolation of a psittacosis agent (Chlamydia) from semen and epididymis of bulls with seminal vesiculitis syndrome, *Am. J. Vet. Res.,* 29, 549, 1968.

49. **Rodolakis, A. and Bernard, K.,** Isolation of *Chlamydia ovis* from genital organs of rams with epididymitis, *Bull. Acad. Vet. Fr.,* 50, 65, 1977.

50. **Pienaar, J. G. and Schutte, A. P.,** The occurrence and pathology of chlamydiosis in domestic and laboratory animals: a review, *Onderstepoort J. Vet. Res.,* 42, 77, 1975.

51. **Storz, J., Carroll, E. J., Stephenson, E. H., Ball, L., and Eugster, A. K.,** Urogenital infection and seminal excretion after inoculation of bulls and rams with chlamydiae, *Am. J. Vet. Res.,* 37, 517, 1976.

52. **Storz, J.,** Über eine natürliche Infektion eines Meerschweinchenbestandes mit einem Erreger aus der Psittakose-Lymphogranuloma Gruppe, *Zentralbl. Bakteriol. I Orig.,* 193, 432, 1964.

53. **Mount, D. T., Bigazzi, P. E., and Barron, A. L.,** Experimental genital infection of male guinea pigs with the agent of guinea pig inclusion conjunctivitis and transmission to females, *Infect. Immun.,* 8, 925, 1973.

54. **Doughri, A. M., Eugster, A. K., and Altera, K. P.,** Pathological studies on chlamydial kidney infections in newborn calves, *Arab Develop. J. Sci. Technol.,* 1, 10, 1978.

55. **Bowen, R. A., Spears, P., Storz, J., and Seidel, G. E., Jr.,** Mechanisms of infertility in genital tract infections due to *Chlamydia psittaci* transmitted through contaminated semen, *J. Infect. Dis.,* 138, 95, 1978.

56. **Barron, A. L., White, H. J., and Rank, R. G., and Soloff, B. L.,** Target tissues associated with genital infection of female guinea pigs by the chlamydial agent of guinea pig inclusion conjunctivitis, *J. Infec. Dis.,* 139, 60, 1979.

57. **Soloff, B. L., Rank, R. G., and Barron, A. L.,** Ultrastructural studies of chlamydial infection in guinea-pig urogenital tract, *J. Comp. Pathol.,* 92, 1982.

58. **Soloff, B. L., Rank, R. G., and Barron, A. L.,** Electron microscopic observations concerning the in vivo uptake and release of the agent of guinea-pig inclusion conjunctivitis *(Chlamydia psittaci)* in Guinea-pig exocervix, *J. Comp. Pathol.,* 95, 1985.

59. **McColl, K. A., Martin, R. W., Gleeson, L. J., Handasyde, K. A., and Lee, A. K.,** Chlamydia infection and infertility in the female koala *(Phascolarctos cinereus), Vet. Res.,* 115, 655, 1984.

60. **Blanco-Loizelier, A.,** Aislamiento de un agente del grupo psittacosis linfogranuloma venereo (PLV) en mastitis bovinas, *Rev. Patronato Biol. Anim.,* 13, 179, 1969.

61. **Wehnert, C., Wehr, J., Teichmann, G., Mecklinger, S., and Zimmerhackel, W.,** Untersuchungen über die Beteiligung von Chlamydien an einem Mastitisgeschehen, *Wiss. Z. Humboldt Univ. Berlin Math. Naturwiss. Reihe,* 29, 71, 1980.

62. **Borovik, R. V., Kurbanov, I. A., and Terskikh, I. I.,** Specific activity of cattle immunoglobulins in experimental chlamydia infection, *Vopr. Virusol.,* 4, 485, 1978.

63. **Boulanger, P. and Bannister, G. L.,** Abortion produced experimentally in cattle with an agent of the psittacosis lymphogranuloma venereum group of virus, *Can. J. Comp. Med. Vet. Sci.,* 23, 259, 1959.

64. **Corner, A. H., Bannister, G. L., and Hill, D. P.,** A histological study of the effects of enzootic abortion of ewes virus in the lactating bovine mammary gland, *Can. J. Comp. Med. Vet. Sci.,* 32, 372, 1968.

65. **Ronsholt, L. and Basse, A.,** Bovine mastitis induced by a common intestinal *Chlamydia psittaci* strain, *Acta Vet. Scand.,* 22, 9, 1981.

66. **Papadopoulos, O. and Leontides, S.,** Mastitis produced experimentally in sheep with an ovine abortion Chlamydia, *Zentralbl. Vet. Med. [B],* 19, 655, 1972.

67. **Mecklinger, S., Wehr, J., Horsch, F., and Seffner, W.,** Zur experimentellen Chlamydienmastitis. *Wiss. Z. Humboldt Univ. Berlin Math. Naturwiss. Reihe,* 29, 75, 1980.

68. **Mendlowski, B. and Segre, D.,** Polyarthritis in sheep. I. Description of the disease and experimental transmission, *Am. J. Vet. Res.,* 21, 68, 1960.

69. **Storz, J., Shupe, J. L., James, L. F., and Smart, R. A.,** Polyarthritis of sheep in the intermountain region caused by a psittacosis-lymphogranuloma agent, *Am. J. Vet. Res.,* 24, 1201, 1963.

70. **Storz, J., Smart, R. A., Marriott, M. E., and Davis, R. V.,** Polyarthritis of calves: isolation of psittacosis agents from affected joints, *Am. J. Vet. Res.,* 27, 633, 1966.

71. **Kölbl, O., Burtscher, H., and Hebenstreit, J.,** Polyarthritis bei Schlachtschweinen: microbiologische, histologische und fleischhygienische Untersuchungen und Aspekte, *Wien. Tierärztl. Monatsschr.,* 57, 355, 1970.

72. **Shupe, J. L. and Storz, J.,** Pathologic study of psittacosis lymphogranuloma polyarthritis of lambs, *Am. J. Vet. Res.,* 25, 943, 1964.

73. **Storz, J., Shupe, J. L., Marriott, M. E., and Thornley, W. R.,** Polyarthritis of lambs induced experimentally by a psittacosis agent, *J. Infect. Dis.,* 115, 9, 1965.

74. **Cutlip, R. C. and Ramsey, F. K.,** Ovine chlamydial polyarthritis: sequential development of articular lesions in lambs after intraarticular exposure, *Am. J. Vet. Res.,* 34, 71, 1973.

75. **McNutt, S. H.,** A preliminary report on an infectious encephalomyelitis of cattle, *Vet. Med.,* 35, 228, 1940.

76. **Wenner, H. A., Harshfield, G. S., Chang, T. W., and Menges, R. W.,** Sporadic bovine encephalomyelitis. II. Studies on the etiology of the disease, isolation of nine strains of an infectious agent from naturally infected cattle, *Am. J. Hyg.,* 57, 15, 1953.

77. **Schachter, J., Banks, J., Sugg, N., Sung, M., Storz, J., and Meyer, K. F.,** Serotyping of chlamydia: isolates of bovine origin, *Infect. Immun.,* 11, 904, 1975.

78. **Rocca-Garcia, M.,** Viruses of the lymphogranuloma-psittacosis group isolated from opossums in Columbia, Opposum virus A, *J. Infect. Dis.,* 85, 275, 1949.

79. **Harshfield, G. S.,** Sporadic bovine encephalomyelitis, *J. Am. Vet. Med. Assoc.,* 156, 466, 1970.

80. **Dickinson, L. and Cooper, B. S.,** Contagious conjunctivo-keratitis of sheep, *J. Pathol. Bacteriol.,* 78, 257, 1959.

81. **Yerasimides, T. G.,** Isolation of a new strain of feline pneumonitis virus from a domestic cat, *J. Infect. Dis.,* 106, 290, 1960.

82. **Pavlov, P., Milanov, M., and Tschilev, D.,** Recherches sur la rickettsiose kerato-conjunctivale du porc en Bulgarie, *Ann. Inst. Pasteur,* 105, 450, 1963.

83. **Murray, E. S.,** Guinea pig inclusion conjunctivitis virus. I. Isolation and identification as a member of the psittacosis-lymphogranuloma-trachoma group, *J. Infect. Dis.,* 114, 1, 1964.

84. **Storz, J., Pierson, R. E., Marriott, M. E., and Chow, T. L.,** Isolation of psittacosis agents from follicular conjunctivitis of sheep, *Proc. Soc. Exp. Biol. Med.,* 125, 857, 1967.

85. **Cockram, F. A. and Jackson, A. R. B.,** Keratoconjunctivitis of the koala bear caused by chlamydial agents, *J. Wildlife Dis.,* 17, 497, 1981.

86. **Rowe, L. W. R., Hedger, R. S., and Smale, C.,** The isolation of a *Chlamydia psittaci*-like agent from a free-living African buffalo *(Syncerus caffer), Vet. Rec.,* 103, 13, 1978.

87. **Hargis, A. M., Prieur, D. J., and Gaillard, E. T.,** Chlamydial infection of the gastric mucosa in twelve cats, *Vet. Pathol.,* 20, 170, 1983.

88. **Gaillard, E. T., Hargis, A. M., Prieur, D. J., Evermann, J. F., and Dhillon, A. S.,** Pathogenesis of feline gastric chlamydial infection, *Am. J. Vet. Res.,* 45, 2314, 1984.

89. **Hopkins, J. B., Stephenson, E. H., Storz, J., and Pierson, R. E.,** Clinical characteristics of eye lesions of lambs affected with chlamydial conjunctivitis and polyarthritis, *J. Am. Vet. Med. Assoc.,* 163, 510, 1973.

90. **Stephenson, E. H., Storz, J., and Hopkins, J. B.,** Properties and frequency of isolation of chlamydiae from eyes of lambs with conjunctivitis and polyarthritis, *Am. J. Vet. Res.,* 35, 177, 1974.

91. **Kazdan, J. J., Schachter, J., and Okumoto, M.,** Inclusion conjunctivitis, *Am. J. Ophthalmol.,* 64, 116, 1967.

92. **Doughri, A. M. and Storz, J.,** Ocular histopathologic lesions in chlamydial infection of newborn calves, *Arab Develop. J. Sci. Technol.,* 1, 5, 1978.

93. **Iversen, J., Spalatin, J., Fraser, C., and Hanson, R.,** Ocular involvement with *Chlamydia psittaci* (Strain M56) in rabbits inoculated intravenously, *Can. J. Comp. Med.,* 38, 298, 1974.

94. **Newcomer, C. E., Anver, M. R., Simmons, J. L., Wilcke, B. W., Jr., and Nace, G. W.,** Spontaneous and experimental infections of *Xenopus laevis* with *Chlamydia psittaci, Lab. Anim. Sci.,* 32, 680, 1982.

95. **Wilcke, B. W., Newcomer, C. E., Anver, M. R., Simmons, J. L., and Nace, G. W.,** Isolation of *Chlamydia psittaci* from naturally infected African clawed frogs *(Xenopus laevis), Infect. Immun.,* 41, 789, 1983.

96. **Howerth, E. W.,** Pathology of naturally occurring chlamydiosis in African clawed frogs *(Xenopus laevis), Vet. Pathol.,* 21, 28, 1984.

97. **Solis, J. and Mora, E. C.,** Development cycle and cytochemistry of the sporadic bovine encephalomyclitis agent in fathead minnow cell culture, *Am. J. Vet. Res.,* 30, 1985, 1969.

98. **Wolke, R. E., Wyand, D. S., and Khairallah, L. H.,** A light and electron microscope study of epitheliocystis disease in the gills of Connecticut striped bass *(Morone saxatilis)* and white perch *(Morone americanus), J. Comp. Pathol.,* 80, 559, 1970.

99. **Harshbarger, J. C., Chang, S. C., and Otto, S. V.,** Chlamydiae (with phages), mycoplasmas, and rickettsiae in Chesapeake Bay bivalves, *Science,* 196, 666, 1976.

100. **Spears, P. and Storz, J.,** Changes in the ultrastructure of *Chlamydia psittaci* produced by treatment of the host cell with DEAE-dextran and cycloheximide, *J. Ultrastruct. Res.,* 67, 152, 1979.

101. **Richmond, S. J., Stirling, P., and Ashley, C. R.,** Virus infecting the reticulate bodies of an avian strain of *Chlamydia psittaci, FEMS Microbiol. Lett.,* 14, 31, 1982.

102. **Eddie, B., Radovsky, F. J., Stiller, D., and Kumada, N.,** Psittacosis lymphogranuloma venereum (PL) agents in ticks, fleas and native mammals in California, *Am. J. Epidemiol.,* 90, 449, 1969.

103. **Weyer, F.,** Zur Frage der Rolle von Arthopoden als Reservoir des Psittakoseerregers., *Z. Tropenmed. Parasitol.,* 21, 146, 1970.

104. **Caldwell, H. D. and Belden, E. L.,** Studies of the role of *Dermacentor occidentalis* in the transmission of bovine chlamydial abortion, *Infect. Immun.,* 7, 147, 1973.

105. **McKercher, D. G., Wada, E. M., Ault, S. K., and Theis, J. H.,** Preliminary studies on transmission of chlamydia to cattle by ticks *(Ornithodoros coriaceus), Am. J. Vet. Res.,* 41, 922, 1980.

106. **Shatkin, A. A., Beskina, S. R., Medvedeva, G. I., and Grokhovasaya, I. M.,** Cultivation of the agent of enzootic abortion of sheep in a continuous line of *Hyalomma* tick embryonal cells, *Med. Parazitol. Parazit. Bolezni,* 46, 420, 1977.

107. **Osaki, H.,** Electron microscopic observations of chlamydia-like microorganism in hepatopancreas cells of the spider, *Coelotes luctuosus, Acta Arachnol.,* 25, 23, 1973.

108. **Morel, G.,** Isolément de deux chlamydiales (rickettsies) chez un arachnide: L'araignée *Pisaura mirabilis* Cl., *Experientia,* 34, 344, 1978.

109. **Todd, W. J. and Storz, J.,** Ultrastructural cytochemical evidence for activation of lysosomes in the cytocidal effect of *Chlamydia psittaci, Infect. Immun.,* 12, 638, 1975.

110. **Todd, W. J., Doughri, A. M., and Storz, J.,** Ultrastructural changes in host cellular organelles in the course of the chlamydial developmental cycle, *Zentralb. Bakteriol. I. Orig.,* 236, 359, 1976.

111. **Moulder, J. W., Levy, N. J., and Schulman, L. P.,** Persistent infection of mouse fibroblasts (L cells) with *Chlamydia psittaci:* evidence for a cryptic chlamydial form, *Infect. Immun.,* 30, 874, 1980.

112. **Perez-Martinez, J. A. and Storz, J.,** Persistent infection of L Cells with an ovine abortion strain of *Chlamydia psittaci, Infect. Immun.,* 50, 453, 1985.

113. **Senyk, G., Sharp, M., Stites, D. P., Hanna, L., Keshishyan, H., and Jawetz, E.,** Cell-mediated immune response to chlamydial antigens in guinea pigs injected with inactivated chlamydiae, *Med. Microbiol. Immunol.,* 168, 91, 1980.

114. **Page, L. A.,** Stimulation of cell-mediated immunity to chlamydiosis in turkeys by inoculation of chlamydial bacterin, *Am. J. Vet. Res.,* 39, 473, 1978.

115. **Schmeer, N., Perez-Martinez, J. A., Schnorr, K., and Storz, J.,** Dominance of *Chlamydia psittaci*-specific IgG_2 isotypes in naturally and experimentally infected cattle, *Vet. Immunol. Immunopathol.,* in press.

Chapter 10

HOST RESPONSE

Cho-chou Kuo

TABLE OF CONTENTS

I. INTRODUCTION

Chlamydial infections induce prominent systemic immune responses in the host, including both humoral and cell mediated responses. These are accompanied by severe inflammatory reactions at the site of infection. The host's immune response may succeed in clearing the infection. However, when such is not the case, the infection often results in chronicity and latency and severe tissue damage, especially after relapse or repeated infection. While chronicity and latency are general characteristics of intracellular parasitism, chlamydia's unique developmental life cycle within host cells is an important contributing factor. Chlamydiae display tissue tropism in the natural host which is pathogen specific. However, the underlying immunohistopathological changes in the different anatomic sites are remarkably similar. Although the immune mechanisms of the host's resistance to chlamydia have yet to be elucidated, we have begun to discern the role of inflammatory and immunologic responses in the pathogenesis of some chlamydial infections, i.e., ocular trachoma, tubal infertility, and pneumonitis. These diseases will be used as examples for discussion. The subjects of this chapter on host response to chlamydial infection will contain tissue tropism, inflammatory response, role of inflammatory reaction in pathogenesis and resistance, and factors affecting these reactions. Other subjects on host immune responses can be found in Chapter 11, Stimulation of Immune Response, and Chapter 12, Role of Immune Response.

II. TISSUE TROPISM AND TARGET CELLS

A. *Chlamydia trachomatis*

The species *C. trachomatis* is differentiated into three biovars based on host specificity and virulence properties. The three biovars are trachoma, lymphogranuloma venereum (LGV), and mouse. The organisms of trachoma and LGV biovars are the primary human pathogens. Although lower mammals, including monkeys and mice, can be experimentally infected with trachoma and LGV organisms, diseases similar to those described in humans can only be induced in monkeys. Mice are the natural host for the mouse biovar. The organism of the mouse biovar is not pathogenic to man and the susceptibility of other animals has not been tested.

The diseases caused by the trachoma biovar fall into three main categories: ocular, respiratory, and urogenital infections. This differentiation is not due to differences in the inherent pathogenesis of the organisms, but is determined by the initial sites of colonization with subsequent spread of the infection to the contiguous organs. The initial colonization occurs in the mucous membranes of the conjunctiva, nasopharynx, rectum, urethra, and cervix. The infection can spread from the nasopharynx to the trachea, bronchus, and lung parenchyma; from the urethra to the prostate and epididymis in males; and from the cervix to the endometrium, salpinx, and peritoneum in females. The target cells are squamocolumnar epithelial cells of the mucous membrane of the conjunctiva, urethra, and cervix;[1-4] ciliated columnar epithelial cells of the trachea and bronchus;[5,6] secretory cells of salpinx;[7] and possibly mononuclear phagocytes in the lung parenchyma.[6]

The LGV biovar gains entrance to the host by initially infecting the epithelial cells of the mucous membrane of the genitourinary tract and the adjacent skin.[8] The primary lesion is transient and usually not detected. The organisms are then carried to the draining lymph nodes (inguinal in male and pelvic in female), and subsequently multiply within the mononuclear phagocytes to establish infection.[1,9,10] Thus, the main manifestation of the disease caused by the LGV biovar is lymphadenitis, which is distinctly different from infections due to the trachoma biovar.

In the natural host, i.e., mice, the organism of the mouse biovar preferentially infects the lung to cause pneumonitis.[1] Chlamydial inclusions have been located in the bronchial epithelial cells and in the macrophages of the lung parenchyma.[11] In addition, the genital organs are also susceptible to experimental infection.[12,13]

B. *Chlamydia psittaci*

C. psittaci has a much broader host range than *C. trachomatis*. The degree of host specificity is usually low and tissue tropism is not prominent. Therefore, a single strain may produce more than one disease entity in a single host species.[1] (See Chapter 9 for host range and disease diversity). *C. psittaci* is comprised of two groups, avian and mammalian. The natural hosts for avian psittacosis agents include many species of wild and domestic birds such as parrots, parakeets, pigeons, turkeys, and ducks. Mammalian psittacosis agents also have a wide host range including wild, domesticated, and laboratory animals such as cattle, sheep, cats, mice, guinea pigs, and koala bears. The anatomic sites infected by avian *C. psittaci* strains are mainly the intestinal and respiratory tracts. The target cells are the epithelium of these organs and mononuclear phagocytes in the intestinal subepithelium and lung interstitium. Mammalian *C. psittaci* strains infect many organs and tissues in different animals including intestinal, genital, and respiratory tracts, conjuctiva, placenta and fetus, and synovial tissue. The target cells are the epithelial cells and probably also infiltrating mononculear phagocytes. Thus *C. psittaci* has both greater virulence and broader tissue tropism than *C. trachomatis*.

III. INFLAMMATORY RESPONSE

A. General Features

The local inflammatory reaction to chlamydial infection is characterized by heavy infiltration of the tissue by many types of inflammatory cells. During the acute stage, neutrophilic polymorphonuclear leukocytes (PMNs) are the predominant cells in the infiltrate. At the subacute and chronic stages, mononuclear leukocytes consisting mainly of lymphocytes characterize the infiltrate. Also present, are plamsa cells, eosinophils, and mononuclear phagocytes (macrophages). Plasma cells are typically found in large numbers in *C. trachomatis* conjunctivitis,[14,15] trachoma,[16,17] endometritis,[18,19] and in LGV lymphadenitis.[20] Eosinophils are seen in increased numbers in *C. trachomatis* infant pneumonitis.[5,21] During an acute exacerbation or reinfection, PMNs and lymphocytes reappear in large quantities. The same histological picture of cell infiltration is repeated but in a heightened response. The inflammatory signs at the site of infection include redness, swelling and edema, and a mucopurulent discharge.

B. Characteristic Histopathological Findings

1. Lymphoid Follicle Formation

A characteristic histopathological finding in ocular[3,14-17] and genital[18,22,23] infections with the trachoma biovar of *C. trachomatis* is lymphoid follicle formation (Figure 1). This is observed in natural infections in humans[3,14,16,18,22] and also in experimental infections in monkeys.[15,17,23] In *C. trachomatis* conjunctivitis and trachoma, a heavy infiltration of PMNs and lymphocytes in the submucosa is seen at the acute stage.[14,16,17] Later as the number of PMNs decrease, lymphocytes become relatively more numerous. These lymphocytes form clusters which later develop into distinct lymphoid follicles as the disease progresses. The epithelium over the follicle is reduced to a one to two cell layer thickness or totally denuded. Thus, the follicles become visible by clinical observation. This sign is used as a diagnostic criterion for *C. trachomatis* conjunctivitis and trachoma. Follicles are not observed in neonatal conjunctivitis unless the disease remains active for longer than 3 months because the conjunctiva is deficient in lymphoid tissue at birth.[14] Follicles are also observed by colposcopy in *C. trachomatis* cervicitis.[22] Histologically, the center of the follicle is composed of large

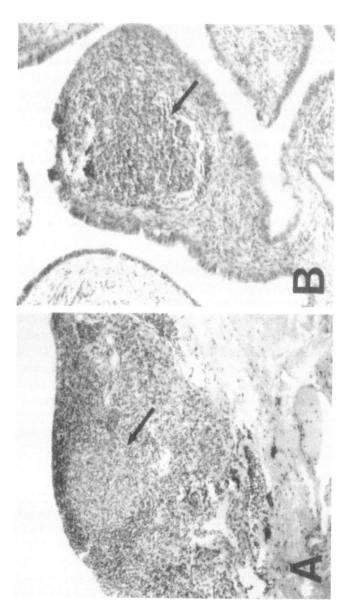

FIGURE 1. Lymphoid follicles (arrows) in experimental monkey infections with the trachoma biovar of *Chlamydia trachomatis*. (A) Conjunctiva; (B) salpinx. Magnification × 25. (Courtesy of Dorothy L. Patton, University of Washington, Seattle.)

mononuclear cells, mostly lymphoblasts and macrophages. The middle zone consists of medium sized lymphocytes.[16] The periphery are small lymphocytes. Immunocytological identification with leukocyte-specific monoclonal antibodies reveal that the center of the follicle is comprised of B cells and macrophages, and the peripheral region or ''cap'' is made up of T cells.[17] Follicle formation is also a prominent histopathological feature in *C. trachomatis* cervicitis,[22] endometritis,[18,19] and salpingitis.[23] The process of follicle formation is accelerated upon relapse or reinfection.

2. Granuloma Formation

LGV lesions are characterized by granulomatous inflammation.[8,20,24,25] The histopathological features vary from a loose infiltrate of monocytes to an organized aggregation of mature macrophages and epithelioid formations.[20] The typical findings of a progressive lesion in the lymph node are (1) accumulation of neutrophilic PMNs in minute necrotic foci; (2) rapid hyperplasia of lymphocytes associated with a profuse infiltration of plasma cells; (3) enlargement and coalescence of small suppurative foci to form stellate abscesses; (4) further maturation and transformation of macrophages into epithelioid cells; (5) proliferation of fibroblasts at the marginal zone to form a suppurative granuloma as the lesion ages; and (6) healing by cicatrization. The prominent cellular components in LGV lesions are monocytes/macrophages and plasmacytes. This histopathological feature is distinct from that of trachoma lesions, which are characterized by the accumulation of lymphocytes and the development of lymphocyte aggregates into lymphoid follicles. The granulomatous inflammation is also the characteristic feature of the lesions in other sites such as in LGV-induced proctitis, and is in contrast to the nongranulomatous inflammatory changes of proctitis induced by the tranchoma biovar.[24] A granuloma-like reaction has also been observed in intestinal infection of calves with *C. psittaci*.[26]

The granulomatous inflammatory reaction is manifested in the host's other immune responses which are unique among chlamydial infections. These include induction of strong delayed hypersensitivity (Frei skin test),[27] high antibody titers (specific to LGV),[28] and hypergammaglobulinemia (nonspecific).[8] Lesions such as sinus and fistula formation, lymphatic obstruction, and urethral and intestinal stricture result from extensive scarring.[8,20]

C. Contribution to Pathogenesis

The primary purpose of the inflammatory reaction is to rid the host of invading organisms. However, the inflammatory reaction may secondarily inflict injury to the host, thus contributing to the development of a unique pathology. Such complications are often observed in chlamydial infections. This is due to the intracellular nature of the infection and the ability of chlamydia to induce strong inflammatory reactions. An example of this response, the granuloma formation observed in LGV, has been presented above. Two more examples will be presented in this section: one on ocular trachoma and the other on tubal infertility.

1. Ocular Trachoma

Two forms of *C. trachomatis* conjunctivitis are observed, inclusion conjunctivitis and trachoma. These two forms have different epidemiological backgrounds. Inclusion conjunctivitis is seen in newborns infected at the time of birth during parturition through an infected birth canal[14,29,30] or in sexually active adults who inoculate their eyes with secretions from infected genitalia.[29,30] The disease is self-limited and rarely leaves sequelae, although it often runs a chronic course. Conversely, trachoma is a severe form of conjunctivitis causing damage to the conjunctiva (scarring) and cornea (pannus formation or neovascularization), and may result in blindness.[29] The disease is seen in endemic areas of the world such as in northern and sub-Sahara Africa and in the Middle East (see Chapter 8).

In the past, inclusion conjunctivitis and trachoma were considered separate diseases because of their distinctive clinical manifestations and different epidemiological backgrounds. However, it is now generally accepted that inclusion conjunctivitis and trachoma are different stages of the same disease, with the inclusion conjunctivitis syndrome occurring at the first infection and trachoma appearing after reinfection. The differences in clinical manifestations are due more to the host's immunological response than to differences in the agents. The evolvement of this concept has been demonstrated elegantly by Grayston and Wang in their epidemiological studies[31] and experimental monkey studies,[32,33] and is summarized in their recent publications.[34,35] The concept that trachoma is a disease of reinfection and/or relapse is further supported by an experimental monkey model of reinfection devised by Taylor and colleagues to produce pannus and scar formation.[36]

As described above, the histological findings of the inflammatory reaction in the conjunctiva are characterized by intensive infiltration of PMNs and lymphocytes in the acute stage and lymphoid follicle formation in the subacute and chronic stages.[3,14-17] The inflammation causes tissue damage including both exfoliation and thinning of the epithelium. This is followed by a proliferation of epithelial cells and formation of papillae. Newborns are not able to form follicles due to the lack of lymphoid tissue in their conjunctiva at birth. Thus, papillary hypertrophy is the primary manifestation of conjunctivitis in the newborns.[14] The epithelium over the follicle becomes thin or even denuded. As the follicles age, central degeneration and necrosis ensue. The healing process involves proliferation of fibroblasts, formation of fibrotic tissue, and subsequent scarring. This inflammatory process involves the palpebral, bulbar, and limbal (margin of cornea) conjunctiva. Therefore, the cornea may show epithelial or subepithelial keratitis and infiltration which is most marked in the upper third portion. In reinfection, pannus develops first in the upper part from a swollen limbus to marginal infiltration of the cornea and then progresses downward in the cornea over a period of months and years.[37] Further progression of pannus and scarring of the cornea will eventually lead to blindness. In relapse or reinfection, this inflammatory process, particularly lymphoid follicle formation, is repeated in an accelerated manner, and will result in more extensive tissue damage. These histopathological changes have been illustrated by Patton and Taylor in the monkey model.[17] In their study, lymphoid follicles did not develop until 18 weeks after the primary inoculation, while definite lymphoid follicles could be seen as early as 2 weeks after the secondary inoculation (Figure 1A). Thus, the characteristic pathological changes seen in chronic ocular *C. trachomatis* infection are the result of an intensive inflammatory reaction with infiltration of PMNs, lymphocytes, plasmacytes, and mononuclear phagocytes in association with the host's humoral and cellular immunological responses. Grayston et al. have called trachoma an immunopathologic disease whose severity increases with reinfection.[34,35]

2. Tubal Infertility

Tubal infertility is a serious complication of *C. trachomatis* salpingitis.[38] The incidence of tubal infertility has continued to rise during the past two decades as a consequence of increasing numbers of women suffering from pelvic inflammatory disease due to *C. trachomatis*. Similar to ocular trachoma, tubal infertility is a disease of immunopathology and occurs more frequently after reinfection. The incidence of infertility is 15% following a single infection, doubles after each episode of salpingitis, and reaches over 50% after the third episode.[39]

The histopathological findings of acute salpingitis in women have been characterized by profuse infiltration of PMNs and lymphocytes in all layers of the salpinx and by desquamation of epithelial cells.[40] Similar histological findings were observed with experimental salpingitis in monkeys.[7,23,41] The sequelae that lead to infertility are the damage to the endothelium, deciliation of the epithelial cells, and the distal tubal obstruction especially at the fimbriated end of the fallopian tube, due to scarring. These factors result in either an impairment or interruption in ovum transport.[23,41]

The pathogenesis of tubal infertility has been studied by Patton and colleagues in a monkey model.[7,23] The characteristic inflammatory response observed after a primary infection was a heavy and extensive lymphocyte infiltration in the submucosa and the muscular layer with extension into the epithelium. Immunohistological staining with monoclonal antibodies specific to lymphocyte markers revealed that most of the infiltrating lymphocytes were T cells. A small number of T cell subpopulations (helper and cytotoxic/suppressor) and B cells were identified. Extensive epithelial damage due to deciliation and increased plasmalemmal alterations of nonciliated cells occurred in patchy areas. Epithelial cell degeneration was seen with the damaged cells in close proximity to lymphocytes, which suggests an immunological basis for the tissue destruction. A more severe tubal inflammation was induced following repeated infection. The pathology included peritubal adhesions, perihepatitis, epithelial cell damage, and fibrous adhesions of mucosal folds. The inflammatory infiltration was characterized by a heavy collection of lymphocytes and the development of lymphoid follicles with active germinal centers in the submucosa (Figure 1B). There were also increased numbers of eosinophils and plasmacytes. Lymphocytes were often seen infiltrating the epithelium in close proximity to degenerating epithelial cells (Figure 2). These observations illustrate that the pathogenesis of tubal infertility is strikingly similar to that of ocular trachoma.

D. Contribution to Host Resistance

1. Clearance of Organisms

The chlamydicidal ability of professional phagocytes, PMNs, and monocytes/macrophages, has been studied in vitro. Studies by Yong et al. showed that the organisms of both the trachoma and LGV biovars of *C. trachomatis* are rapidly inactivated by exposure to human PMNs.[42] A decrease of 3.0 to 3.5 logs in viable titer was observed after 60 min of interaction in vitro. Both normal PMNs and PMNs from myeloperoxidase (MPO) deficient patients were equally chlamydicidal. Similarly, PMNs from patients with chronic granulomatous disease (PMNs that lack an oxidative bactericidal system) were also strongly chlamydicidal, especially with LGV organisms. However, these PMNs were less effective than normal PMNs in the inactivation of trachoma organisms. These findings suggest that oxygen dependent systems, including the MPO system, are not essential antichlamydicidal systems within PMNs and that nonoxidative antimicrobial systems, most of which reside within PMN lysosomes, are important in chlamydicidal activity. However, both trachoma and LGV organisms were rapidly inactivated by the cell-free MPO-H_2O_2-halide system. Therefore, the involvement of the MPO system cannot be excluded. An ultrastructural study using a cytochemical marker revealed that chlamydiae are degraded within phagosomes which have fused with lysosomes.[43] These in vitro studies clearly indicate the importance of PMNs in the eradication of chlamydiae. PMNs are the predominant cells in the infiltrate during the acute phase when the bacterial load is the greatest.

The toxicity of macrophages for chlamydia has been studied using freshly isolated mouse peritoneal macrophages.[44,45] It was shown that macrophages are toxic to chlamydiae. However, a small number of chlamydiae may escape killing by macrophages by evading the fusion of lysosomes with phagosomes. They then multiply within the macrophages. The ability of chlamydiae to grow within macrophages can be correlated with the observed difference in virulence among chlamydial strains. Chlamydial strains causing psittacosis and LGV grow better within macrophages than trachoma organisms. Yong et al. demonstrated that human monocytes are also bactericidal to both the trachoma and LGV organisms, although they are less efficient than PMNs.[46] Nevertheless, almost complete inactivation of organisms occurred 48 hr after monocytes were inoculated. However, the LGV organism could infect and survive in almost 100% of monocytes that had been cultured for 8 days or more before inoculation. Conversely, the trachoma organism continued to be killed by such

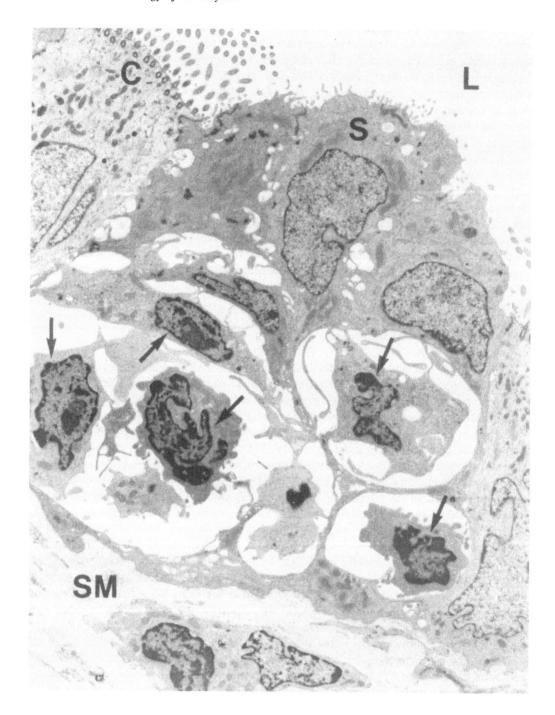

FIGURE 2.   Transmission electron micrograph of the endosalpinx in the experimental monkey infection with the trachoma biovar of *Chlamydia trachomatis*, 14-days post inoculation. Note infiltrating lymphocytes (arrows) in close proximity to degenerating epithelial cells and edematous tissue. (C) Ciliated cells; (S) secretory cells; (L) lumen; and (SM) submucosa. Magnification × 3,000. (Courtesy of Dorothy L. Patton, University of Washington, Seattle.)

cells, or cells that had been cultured for as long as 21 days. Monocytes also contain oxygen-dependent bactericidal enzymes, one of which is MPO, but in much lower concentration than in PMNs. These enzymes are gradually depleted when monocytes are cultured in vitro and are totally absent after 8 days of culture, a time which corresponds to that required for the monocytes to transform into macrophages.[47] The capability of LGV to survive in macrophages reflects the difference in virulence and tissue tropism of the LGV biovar from that of the trachoma biovar. As mentioned above, LGV stimulates granuloma formation in which monocytes undergo transformation to macrophages and then to epithelioid cells.

Chlamydia-infected cells may be killed by cytotoxic T cells, especially if the chlamydial antigens are expressed on the host cell surface. By the immunofluorescence antibody test, Richmond and Stirling demonstrated chlamydial antigens on the cell surface of McCoy cells infected with *C. trachomatis*.[48] However, this finding has not been confirmed. Studies on cell-mediated cytotoxicity for chlamydia-infected target cells have been controversial. Lammert reported that cytotoxic cells induced in mice after infection with *C. psittaci* meningopneumonitis agent killed meningopneumonitis-infected mouse fibroblasts in vitro.[49] However, her results could not be repeated in *C. trachomatis* by Pavia and Schachter[50] or Qvigstad and Hirshberg.[51] Other cell-mediated cytotoxicity systems such as antibody-dependent cell cytotoxicity and natural killer cells may also participate in host defenses. However, these mechanisms have not been fully evaluated. A study by Onsrud and Qvigstad showed no significant difference in the natural killer cell activity of healthy women, women who had *C. trachomatis* salpingitis 1 to 2 years earlier, and women who had cervicitis 1 to 2 months earlier.[52]

The importance of inflammatory reactions in host resistance against chlamydial infection has been studied in an animal model by Stephens et al.[53] Mice were treated with cyclophosphamide or cortisone acetate to suppress the inflammatory response. Daily administration of the drugs was begun 2 days before intranasal inoculation with serovar B trachoma organisms and continued for 6 days after inoculation with cyclophosphamide, or for 4 days with cortisone acetate. Titration of organisms recovered from the lungs showed that in cyclophosphamide-treated mice the infection was resolved after 12 days, whereas in cortisone-treated mice a significant titer remained after 17 days and until the end of the observation period. In contrast, no organisms were recovered from control mice after day six. Histologically, the ability of these drugs to inhibit the inflammatory reaction correlated with the respective course of chlamydial pneumonitis in treated mice (Figure 3). Compared with the profuse cellular response and the short duration of infection in the lungs of control mice, cortisone treatment prevented cellular infiltration and revealed the highest and most enduring titers of chlamydiae. Cyclophosphamide treatment resulted in mild cellular infiltration with both intermediate titers and duration of infection.

2. Chronicity and Latency

In the natural hosts, chlamydiae are successful parasites since chronic or latent infections occur frequently. Perinatally infected infants may continue shedding trachoma organisms from the nasopharynx and rectum for as long as 2.5 years if untreated or inadequately treated.[54] Latent infections also occur in birds.[55] The infection may be reactivated when many birds are caged together and the disease spreads among the colony. The mechanisms that parasites use to establish persistent infections in natural hosts are complicated. They probably involve multiple mechanisms from both the host and parasite and are partly related to the host's immune status in general as well as the amount of inflammatory reaction induced at the site of infection. The immunity produced by chlamydial infection in the host is usually incomplete and of short duration. Although an antibody response in the host is observed,[28] and the antibody neutralization of cell culture infection can be demonstrated in vitro,[56,57] there is no evidence that humoral antibodies are able to eliminate the agent from the host.

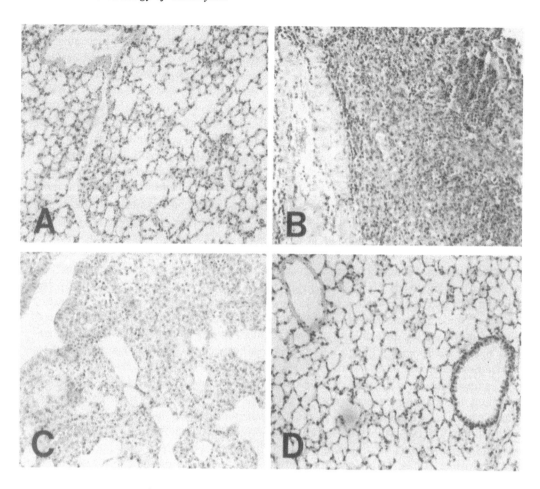

FIGURE 3.   Lung histology of mouse pneumonitis due to serovar B *Chlamydia trachomatis*, 2-days post inoculation. (A) Control mouse inoculated with HeLa cell materials; no inflammatory reaction except a mild congestion. (B) Inoculated mouse; extensive infiltration of polymorphonuclear leukocytes in the interstitium and the terminal bronchioles. (C) Inoculated mouse treated with cyclophosphamide; moderate infiltration in the lung parenchyma, sparing the bronchial lumens. (D) Inoculated mouse treated with cortisone acetate; paucity of cellular infiltrate. Magnification × 150. (Courtesy of Richard S. Stephens, University of California at Berkeley.)

The evidence presented above seems to suggest that cellular immunity is the host's main defense against chlamydia.

From in vitro studies, we have learned some facts that might contribute to our understanding of persistent infection. Horoschak and Moulder,[58] and Bose and Liebhaber[59] demonstrated that when chlamydiae-infected cells divided, one or both daughter cells might receive inclusions. This suggests that chlamydiae may be passed vertically and may avoid extracellular destruction by the host's microbicidal systems and immune mechanisms. Normally, unrestricted chlamydial growth will result in the death of host cells. Therefore, factors that slow down the chlamydial growth cycle, such as deprivation of metabolites needed for reproduction of chlamydia,[60,61] incomplete inhibition of chlamydia by antibiotic treatment, and lymphokine inhibition of the maturation of reticulate bodies into elementary bodies (See Section IV.C. below), may favor the maintenance of persistent infection. These are all reversible inhibitions; therefore, chlamydiae may resume growth when these factors are removed. Another factor that might contribute to persistent infection is the ability of chlamydiae to grow in macrophages and thus evade the host's cellular immunity. One evasive mechanism proposed by Moulder is ''cryptic infection''. Moulder and colleagues established persistent infections of both *C. psittaci*[62] and *C. trachomatis*[63] in mammalian cell cultures. When mouse fibroblasts were infected with psittacosis agent 6BC at sufficient titers, the chlamydiae destroyed most of the cells. However, one in every 10^5 to 10^6 cells became persistently infected. If the cultures were maintained indefinitely by replenishment with fresh medium, the cultures would undergo cycles of host cell proliferation followed by multiplication of the parasite. This was demonstrated by an increased number of cells carrying inclusions, which then resulted in the massive destruction of host cells. Since the persistently infected cells showed no inclusions by Giemsa stain and were resistant to superinfection, Moulder called it cryptic infection.[64] He postulated that "the fluctuations in host and parasite density occur because some factor periodically sets off the conversion of cryptic chlamydial forms into reticulate bodies that multiply and differentiate into infectious elementary bodies in a conventional chlamydial developmental cycle".[62] The application of this hypothesis to natural infection needs further investigation.

IV. MODULATION OF THE INFLAMMATORY REACTION

If the inflammatory reaction is a factor in pathogenesis and resistance, we should be able to alter the pathogenesis and resistance by modulating the inflammatory reaction. The effects of many modulators of the inflammatory response have been studied in vivo in natural and experimental infections, and in in vitro experiments. The modulators investigated are gonadal hormones and glucocorticosteroids, immunosuppressive drugs, and mediators of leukocytes such as lymphokines. These factors will be discussed briefly in this section.

A. Hormonal Factors

The effect of estrogen and progesterone on chlamydial genital infection has been documented in clinical observations and in animal experiments. Epidemiological observational studies have shown that the prevalence of *C. trachomatis* infection is two to three times higher in oral contraceptive users than in nonusers.[65-67] Oral contraceptives contain both estrogen and progesterone which have been shown to enhance chlamydial genital infection in animal models. Barron and colleagues, who have extensively studied oculogenital infection in the model of guinea pig inclusion conjunctivitis, showed that estrogen prolongs cervical infection[68] and enhances ascending genital tract infection.[69] Tuffrey and Taylor-Robinson showed that in mice progesterone is a critical factor in the development of genital tract infection by the L2 serovar of *C. trachomatis*.[70] The effects of gonadal hormones on host susceptibility are probably due to physiological change rather than an influence on the inflammatory response. Oral contraceptives cause ectropion of endocervical columnar epithelium.[71] Therefore, it is possible that more susceptible cells are made accessible to infection. Alternatively, the higher prevalence may be due to improved detection rather than increased infection because ectropion makes specimen collection easier. In animal experiments, the blood levels of estrogen in the guinea pigs treated with effective doses of estrogen were within the peak levels observed during the natural cycle in the control animals.[69] The authors concluded that the altered pathogenesis effected by estrogen treatment is not due to excessive high levels, but is due to the persistence of physiological concentrations, as seen in women taking contraceptives. Progesterone is also thought to act on the target cells of the epithelium of the genital tract because progesterone treatment aborts the estrous cycle and therefore possibly prevents the shedding of target epithelial cells.[70]

It has been reported that oral contraceptive use reduces the risk of pelvic inflammatory disease by half.[72,73] However, this claim has been disputed.[67] Contraceptives affect the physiology of the reproductive tract,[67] causing thickened cervical mucus, reduced menstrual flow, and decreased myometric activity. All of these effects might retard the upward spread of infection. A report by Svensson et al. on laparoscopic observations of women with the first episode of salpingitis showed that women using oral contraceptives had a significantly less inflammation of the fallopian tubes than did women using other contraceptive methods.[73] This suggests that oral contraceptives may have an anti-inflammatory effect. It is possible that only the inflammation is suppressed, not the infection. If this is the case, oral contraceptives may mask the infection. Such a situation may prevent the patient from seeking treatment and consequently lead to a more serious tubal pathology such as infertility and ectopic pregnancy. Both diseases are now recognized consequences of *C. trachomatis* genital infection.

B. Immunosuppression

Several immunosuppressants, including glucocorticosteroids,[53,74] cyclophosphamide,[53,75,76] and antilymphocyte antisera,[77] have been studied for their effects on chlamydial infection in animal models. Generally, treatments with these immunosuppressants produce a more severe and prolonged infection in experimental animals.[53,75-77] It may reactivate the infection.[32,74,78] These reagents have often been used as tools for the analysis of immune mechanisms of chlamydial infection. Chlamydial infection of immunosuppressed hosts is described in more detail in Chapter 12. As mentioned, these immunosuppressants are also anti-inflammatory. Thus, they may suppress the inflammatory cell infiltration and affect pathogenesis and resistance to infection.

C. Mediators of Leukocytes

The role of lymphokines in cellular resistance to chlamydia has been studied intensively in cell culture systems in recent years. Byrne and Faubion found that Concanavalin A-stimulated lymphokines from mouse immune spleen cells suppressed the intracellular development of *C. psittaci* 6BC in mouse peritoneal macrophages.[79] The inhibition was bacteriostatic rather than bactericidal because replication resumed after the lymphokines were removed from the culture medium. Byrne and Krueger later showed that mouse lymphokines also exerted antichlamydial activity in mouse fibroblasts.[80] The effective mediator was identified as gamma interferon because the antichlamydial activity of the lymphokines was neutralized specifically with anti-gamma- but not anti-beta-interferon.

The antichlamydial activity of human interferons has also been evaluated. Rothermel et al. obtained human interferons from human peripheral mononuclear cells by stimulation with Concanavalin A and studied their inhibitory activity against *C. psittaci* 6BC in human macrophages derived from monocytes.[81] The effective factor was identified as gamma-interferon because the antichlamydial activity was neutralized by monoclonal antibodies to human gamma- but not alpha- or beta-interferon. Human interferons also demonstrated antichlamydial activity in nonprofessional phagocytic cells.[82] De La Maza and colleagues[82] showed that recombinant human interferons inhibited the yield of infectious *C. trachomatis* (an LGV) in both human (HeLa) and mouse (McCoy) cells. This suggests that the interferon activity may cross species boundaries.[82] Ultrastructural observations of the growth cycle of chlamydiae in cells treated with interferon showed an inhibition of the transformation of elementary bodies to reticulate bodies with a consequent decrease in both the number of inclusions and the yield of infectious particles.[83,84] In summary, these studies revealed that the gamma-interferon has antichlamydial activity for both phagocytic and nonphagocytic cells; that the inhibitory effect is bacteriostatic rather than bactericidal and the suppression is reversible; and that intracellular development, rather than the uptake of chlamydiae is inhibited.

The mediators of inflammation in chlamydial infection have been less studied than the mediators of cellular immunity. Megran et al. showed that chlamydia may activate complement to generate chemotaxin C5a, a potent chemotactant for PMNs.[85] This mechanism may be partially responsible for the intensive and predominant PMN infiltration seen during the acute stage of chlamydial infection. Kuo and Grayston demonstrated that the lymphokines produced by antigen stimulation of lymph node lymphocytes from immunized guinea pigs induced skin erythema in the normal guinea pig skin.[86] The same lymphokines also exerted an antiproliferative effect on mouse fibroblasts.

These studies illustrate that biological mediators of leukocytes may participate in modulating both the inflammatory reaction and the cellular immunity, and may contribute to the immunopathology and persistence of chlamydial infection.

V. CONCLUSIONS

The inflammatory reaction is a double edged sword. It functions to eliminate the invading microorganisms and heal the tissue injury inflicted by the infection. However, in achieving this goal, the inflammatory reaction also produces irreversible damage to the host. This is exemplified by chlamydial infection. Chlamydial infection induces an intensive inflammatory tissue reaction. This inflammatory response is essential for controlling chlamydial multiplication within the host. However, if the host fails to clear the infection, the infection may lapse into latency or persist as a low grade chronic infection. Upon reactivation or reinfection, this cellular response is accelerated and intensified. Although the secondary response is more effective in clearing the infection, it is more detrimental to the host. This scenario is exemplified by trachoma and tubal infertility caused by *C. trachomatis*. Both diseases are the result of chronic or repeated reinfection. At the final stage, chlamydiae are not easily isolated and the disease manifestations are characterized by the immunopathology of the host's response to the infection. This concept of chlamydial infection as a disease of immunopathology may influence our strategies in its treatment and prevention.

REFERENCES

1. **Moulder, J. W., Hatch, T. P., Kuo, C.-C., Schachter, J., and Storz, J.,** Genus *Chlamydia* Jones, Rakes, and Stearns 1945,55, in *Bergey's Manual of Systematic Bacteriology,* Vol. 1, Krieg, N. R., Ed., Williams & Wilkins, Baltimore, Md., 1984, 729.
2. **Braley, A. E.,** Inclusion blenorrhea, *Am. J. Ophthalmol.,* 21, 1203, 1938.
3. **Mitsui, Y. and Suzuki, A.,** Electron microscopy of trachoma virus in section, *A. M. A. Arch. Ophthalmol.,* 56, 429, 1956.
4. **Swanson, J., Eschenbach, D. A., Alexander, E. R., and Holmes, K. K.,** Light and electron microscopic study of *Chlamydia trachomatis* infection of the uterine cervix, *J. Infect. Dis.,* 131, 678, 1975.
5. **Harrison, H. R., Alexander, E. R., Chiang, W.-T., Giddens, W. E., Jr., Boyce, J. T., Benjamin, D., and Gale, J. L.,** Experimental nasopharyngitis and pneumonia caused by *Chlamydia trachomatis* in infant baboons: histologic comparison with a case in human infant, *J. Infect. Dis.,* 139, 141, 1979.
6. **Chen, W.-J. and Kuo, C.-C.,** A mouse model of pneumonitis induced by *Chlamydia trachomatis:* morphological, microbiological, and immunological studies, *Am. J. Pathol.,* 100, 365, 1980.
7. **Patton, D. L., Halbert, S. A., Kuo, C.-C., Wang, S.-P., and Holmes, K. K.,** Host response to primary *Chlamydia trachomatis* infection of the fallopian tube in pig-tailed monkeys, *Fertil. Steril.,* 40, 829, 1983.
8. **Koteen, H.,** Lymphogranuloma venereum, *Medicine,* 24, 1, 1945.
9. **Miyagawa, Y., Mitamura, T., Yaoi, H., Ishii, N., Nakajima, H., Okanishi, J., Watanabe, S., and Sato, K.,** Studies on the virus of lymphogranuloma inguinale *Nicolas, Favre* and *Durand, Jpn. J. Exp. Med.,* 13, 1, 1935.
10. **Klotz, S. A., Drutz, D. J., Tam, M. R., and Reed, K. H.,** Hemorrhagic proctitis due to lymphogranuloma venereum serogroup L2. Diagnosis by fluorescent monoclonal antibody, *N. Engl. J. Med.,* 308, 1563, 1983.
11. **Gogolak, F. M.,** The histopathology of murine pneumonitis infection and the growth of the virus in the mouse lung, *J. Infect. Dis.,* 92, 254, 1953.

12. **Barron, A. L., White, H. J., Rank, R. G., Soloff, B. L., and Moses, E. B.,** A new animal model for the study of *Chlamydia trachomatis* genital infection: infection of mice with the agent of mouse pneumonitis, *J. Infect. Dis.,* 143, 63, 1981.

13. **Swenson, C. E., Donegan, E., and Schachter, J.,** *Chlamydia trachomatis*-induced salpingitis in mice, *J. Infect. Dis.,* 148, 1101, 1983.

14. **Thygeson, P.,** The etiology of inclusion blenorrhea, *Am. J. Ophthalmol.,* 17, 1019, 1934.

15. **Dawson, C. R., Mordhorst, C. H., and Thygeson, P.,** Infection of rhesus and cynomologus monkeys with egg-grown viruses of trachoma and inclusion conjunctivitis, *Am. J. Ophthalmol.,* 98, 167, 1962.

16. **Jin, X., Zhang, X., Zhang, W., Wang, X., Sun, X., Zhao, W., and Luo, S.,** Pathogenesis of trachoma, *Chin. Med. J.,* 93, 827, 1980.

17. **Patton, D. L. and Taylor, H. R.,** The histopathology of experimental trachoma: ultrastructural changes in the conjunctival epithelium, *J. Infect. Dis.,* 153, 870, 1986.

18. **Winkler, B., Gallo, L., Reumann, W., Richart, R. M., Mitao, M., and Crum, C. P.,** Chlamydial endometritis: a histological and immunohistochemical analysis, *Am. J. Surg. Pathol.,* 8, 771, 1984.

19. **Paavonen, J., Kiviat, N., Brunham, R. C., Stevens, C. E., Kuo, C.-C., Stamm, W. E., Miettinen, A., Soules, M., Eschenbach, D. A., and Holmes, K. K.,** Prevalence and manifestations of endometritis among women with cervicitis, *Am. J. Obstet. Gynecol.,* 152, 280, 1985.

20. **Smith, E. B. and Custer, R. P.,** The histopathology of lymphogranuloma venereum, *J. Urol.,* 63, 546, 1950.

21. **Arth, C., von Schmidt, B., Grossman, M., and Schachter, J.,** Chlamydia pneumonitis, *J. Pediatr.,* 93, 447, 1978.

22. **Paavonen, J., Vesterinen, E., Meyer, B., and Saksela, E.,** Colposcopic and histologic findings in cervical chlamydial infection, *Obstet. Gynecol.,* 59, 712, 1982.

23. **Patton, D. L., Kuo, C.-C., Wang, S.-P., and Halbert, S. A.,** Distal tubal obstruction induced by repeated *Chamydia trachomatis* salpingeal infections in pig-tailed macaques, *J. Infect. Dis.,* 155, 1292, 1987.

24. **Quinn, T. C., Goodell, S. E., Mkrtichian, E., Schuffer, M. D., Wang, S.-P., Stamm, W. E., and Holmes, K. K.,** *Chlamydia trachomatis* proctitis, *N. Engl. J. Med.,* 305, 195, 1981.

25. **Brunham, R. C., Kuo, C.-C., and Chen, W.-J.,** Systemic *Chlamydia trachomatis* infection in mice: a comparison of lymphogranuloma venereum and trachoma biovars, *Infect. Immun.* 48, 78, 1985.

26. **Doughri, A. M., Young, S., and Storz, J.,** Pathologic changes in intestinal chlamydial infection of newborn calves, *Am. J. Vet. Res.,* 35, 939, 1974.

27. **Frei, W.,** Eine neue Hautreaktion bei 'Lymphogranuloma inguinale', *Klin. Wchnschr.,* 4, 2148, 1925.

28. **Wang, S.-P., Grayston, J. T., Kuo, C.-C., Alexander, E. R., and Holmes, K. K.,** Serodiagnosis of *Chlamydia trachomatis* infection with the micro-immunofluorescence test, in *Nongonococcal Urethritis and Related Infections*, Hobson, D. and Holmes, K. K., Eds., American Society for Microbiology, Washington, D.C., 1977, 237.

29. **Jones, B. R.,** Ocular syndromes of TRIC virus infection and their possible genital significance, *Br. J. Vener. Dis.,* 40, 3, 1964.

30. **Schachter, J., Rose, L., and Meyer, K. F.,** The venereal nature of inclusion conjunctivitis, *Am. J. Epidemiol.,* 85, 445, 1967.

31. **Woolridge, R. C., Grayston, J. T., Perrin, E. B., Yang, C.-Y., Cheng, K.-H., Chang, I.-H.,** Natural history of trachoma in Taiwan school children, *Am. J. Ophthalmol.,* 63, 1313, 1967.

32. **Wang, S.-P. and Grayston, J. T.,** Trachoma in the Taiwan monkey, *Macaca cyclopis*, *Ann. N. Y. Acad. Sci.,* 98, 177, 1962.

33. **Wang, S.-P. and Grayston, J. T.,** Pannus with experimental trachoma and inclusion conjunctivitis agent infection of Taiwan monkeys, *Am. J. Ophthalmol.,* 63, 1133, 1967.

34. **Grayston, J. T., Yeh, L.-J., Wang, S.-P., Kuo, C.-C., Beasley, R. P., and Gale, J.,** Pathogenesis of ocular *Chlamydia trachomatis* infections in humans, in *Nongonococcal Urethritis and Related Infections*, Hobson, D. and Holmes, K. K., Eds., American Society for Microbiology, Washington, D.C., 1977, 113.

35. **Grayston, J. T., Wang, S.-P., Yeh, J.-L., and Kuo, C.-C.,** Importance of reinfection in the pathogenesis of trachoma, *Rev. Infect. Dis.,* 7, 717, 1985.

36. **Taylor, H. R., Johnson, S. L., Prendergast, R. A., Schachter, J., Dawson, C. R., and Silverstein, A. M.,** An animal model of trachoma. II. The importance of repeated reinfection, *Invest. Ophthalmol. Vis. Sci.,* 23, 506, 1982.

37. **Thygeson, P.,** The limbus and cornea in experimental and natural human trachoma and inclusion conjunctivitis, *Ann. N.Y. Acad. Sci.,* 98, 201, 1962.

38. **Center for Disease Control,** *Chlamydia trachomatis* infections, policy guidelines for prevention and control, Morbidity and Mortality Weekly Report, 34(suppl.), 35, 1985.

39. **Svensson, L., Mardh, P.-A., and Westrom, L.,** Infertility after acute salpingitis with special reference to *Chlamydia trachomatis, Fertil. Steril.,* 40, 322, 1983.

40. **Moller, B. R., Westrom, L., Ahrons, S., Ripa, K. T., Svensson, L., von Mecklenburgh, C., Henrikson, H., and Mardh, P.-A.,** *Chlamydia trachomatis* infection of the fallopian tubes: histological findings in two patients, *Br. J. Vener. Dis.,* 55, 422, 1979.

41. **Ripa, K. T., Moller, B. R., Mardh, P.-A., Freundt, E. A., and Melsen, F.,** Experimental acute salpingitis in grivet monkeys provoked by *Chlamydia trachomatis, Acta Pathol. Microbiol. Scand. Sect. B,* 87, 65, 1979.

42. **Yong, E. C., Klebanoff, S. J., and Kuo, C.-C.,** Toxic effect of human polymorphonuclear leukocytes on *Chlamydia trachomatis, Infect. Immun.,* 37, 422, 1982.

43. **Yong, E. C., Chi, E. Y., Chen, W.-J., and Kuo, C.-C.,** Degradation of *Chlamydia trachomatis* in human polymorphonuclear leukocytes: an ultrastructural study of the peroxidase positive phagolysosomes, *Infect. Immun.,* 53, 427, 1986.

44. **Wyrick, P. B. and Brownridge, E. A.,** Growth of *Chlamydia psittaci* in macrophages, *Infect. Immun.,* 19, 1054, 1978.

45. **Kuo, C.-C.,** Cultures of *Chlamydia trachomatis* in mouse peritoneal macrophages: factors affecting organism growth, *Infect. Immun.,* 20, 439, 1978.

46. **Yong, E. C., Chi, E. Y., and Kuo, C.-C.,** Differential antimicrobial activity of human mononuclear phagocytes against the human biovars of *Chlamydia trachomatis, J. Immunol.,* 139, 1297, 1987.

47. **Murray, H. W. and Cartelli, D. M.,** Killing of intracellular *Leishmania donovani* by human mononuclear phagocytes: evidence for oxygen-dependent and -independent leishmanicidal activity, *J. Clin. Invest.,* 72, 32, 1983.

48. **Richmond, S. J. and Stirling, P.,** Localization of chlamydial group antigen in McCoy cell monolayers infected with *Chlamydia trachomatis* or *Chlamydia psittaci, Infect. Immun.,* 34, 561, 1981.

49. **Lammert, J. K.,** Cytotoxic cells induced after *Chlamydia psittaci* infection in mice, *Infect. Immun.,* 35, 1011, 1982.

50. **Pavia, C. S. and Schachter, J.,** Failure to detect cell-mediated cytotoxicity against *Chlamydia trachomatis*-infected cells, *Infect. Immun.,* 39, 1271, 1983.

51. **Qvigstad, E. and Hirschberg, H.,** Lack of cell-mediated cytotoxicity towards *Chlamydia trachomatis* infected target cells in humans, *Acta Pathol. Microbiol. Immunol. Scand., Sect. C,* 92, 153, 1984.

52. **Onsrud, M. and Qvigstad, E.,** Natural killer cell activity after gynecologic infections with chlamydia, *Acta Obstet. Gynecol. Scand.,* 63, 613, 1984.

53. **Stephens, R. S., Chen, W.-J., and Kuo, C.-C.,** Effects of corticosteroids and cyclophosphamide on a mouse model of *Chlamydia trachomatis* pneumonitis, *Infect. Immun.,* 35, 680, 1982.

54. **Bell, T. A., Stamm, W. E., Kuo, C.-C., Holmes, K. K., and Grayston, J. T.,** Chronic *Chlamydia trachomatis* infection in infants, in *Chlamydial Infections: Proceedings of the Six International Symposium on Human Chlamydial Infections,* Oriel, D., Ridgway, G., Schachter, J., Taylor-Robinson, D., and Ward, M., Eds., Cambridge University Press, Cambridge, England, 1986, 305.

55. **Meyer, K. F. and Eddie, B.,** Latent psittacosis infections in shell parakeets, *Proc. Soc. Exp. Biol. Med.,* 30, 484, 1933.

56. **Caldwell, H. D. and Perry, L. J.,** Neutralization of *Chlamydia trachomatis* infectivity with antibodies to the major outer membrane protein, *Infect. Immun.,* 38, 745, 1982.

57. **Lucero, M. E. and Kuo, C.-C.,** Neutralization of *Chlamydia trachomatis* cell culture infection by serovar-specific monoclonal antibodies, *Infect. Immun.,* 50, 595, 1985.

58. **Horoschak, K. D. and Moulder, J. W.,** Division of single host cells after infection with chlamydia, *Infect. Immun.,* 19, 281, 1978.

59. **Bose, S. K. and Liebhaber, H.,** Deoxyribonucleic acid synthesis, cell cycle progression, and division of *Chlamydia*-infected HeLa 229 cells, *Infect. Immun.,* 24, 953, 1979.

60. **Morgan, H. R.,** Latent viral infection of cells in tissue culture. I. Studies on latent infection of chick embryo tissues with psittacosis virus, *J. Exp. Med.,* 103, 37, 1956.

61. **Hatch, T. P.,** Competition between *Chlamydia psittaci* and L cells for host isoleucine pools: a limiting factor in chlamydial multiplication, *Infect. Immun.,* 12, 211, 1975.

62. **Moulder, J. W., Levy, N. J., and Schulman, L. P.,** Persistent infection of mouse fibroblasts (L cells) with *Chlamydia psittaci*: evidence for a cryptic chlamydial form, *Infect. Immun.,* 30, 874, 1980.

63. **Lee, C. K. and Moulder, J. W.,** Persistent infection of mouse fibroblasts (McCoy cells) with a trachoma strain of *Chlamydia trachomatis, Infect. Immun.,* 32, 822, 1981.

64. **Moulder, J. W., Levy, N. J., Zeichner, S. L., and Lee, C. K.,** Attachment defect in mouse fibroblasts (L cells) persistently infected with *Chlamydia psittaci, Infect. Immun.,* 34, 285, 1981.

65. **Ripa, K. T., Svensson, L., Mardh, P.-A., and Westrom, L.,** *Chlamydia trachomatis* cervicitis in gynecologic outpatients, *Obstet. Gynecol.,* 52, 698, 1978.

66. **Kinghorn, G. R. and Waugh, M. A.,** Oral contraceptive use and prevalence of infection with *Chlamydia trachomatis* in women, *Br. J. Vener. Dis.,* 57, 187, 1981.

67. **Washington, A. E., Gove, S., Schachter, J., and Sweet, R. L.,** Oral contraceptives, *Chlamydia trachomatis* infection, and pelvic inflammatory disease. A word of caution about protection, *JAMA,* 253, 2246, 1985.

68. **Rank, R. G., White, H. J., Hough, A. J., Jr., Pasley, J. N., and Barron, A. L.,** Effects of estradiol on chlamydial genital infection of female guinea pigs, *Infect. Immun.,* 38, 699, 1982.

69. **Pasley, J. N., Rank, R. G., Hough, A. J., Cohen, C., and Barron, A. L.,** Effects of various doses of estradiol on chlamydial genital infection in ovariectomized guinea pigs, *Sex. Transm. Dis.,* 12, 8, 1985.

70. **Tuffrey, M. and Taylor-Robinson, D.,** Progesterone as a key factor in the development of a mouse model for genital-tract infection with *Chlamydia trachomatis, FEMS Microbiol. Lett.,* 12, 111, 1981.

71. **Goldacre, M. J., London, N., Watt, B., Grant, G., London, J. D. O., McPherson, K., and Vessey, M. P.,** Epidemiology and clinical significance of cervical erosion in women attending a family planning clinic, *Br. Med. J.,* 1, 748, 1978.

72. **Rubin, G., Ory, H. W., and Layde, P. M.,** Oral contraceptives and pelvic inflammatory disease, *Am. J. Obstet. Gynecol.,* 144, 630, 1982.

73. **Svensson, L., Westrom, L., and Mardh, P.-A.,** Contraceptives and acute salpingitis, *JAMA,* 251, 2553, 1984.

74. **Yang, Y.-S., Kuo, C.-C., and Chen, W.-J.,** Reactivation of *Chlamydia trachomatis* lung infection in mice by cortisone, *Infect. Immun.,* 39, 655, 1983.

75. **Modabber, F., Bear, S. E., and Cerny, J.,** The effect of cyclophosphamide on the recovery from a local chlamydial infection. Guinea-pig inclusion conjunctivitis (GPIC), *Immunology,* 30, 929, 1976.

76. **White, H. J., Rank, R. G., Soloff, B. L., and Barron, A. L.,** Experimental chlamydial salpingitis in immunosuppressed guinea pigs infected in the genital tract with the agent of guinea pig inclusion conjunctivitis, *Infect. Immun.,* 26, 728, 1979.

77. **Rank, R. G. and Barron, A. L.,** Effect of antilymphocyte serum on the course of chlamydial genital infection in female guinea pigs, *Infect. Immun.,* 41, 876, 1983.

78. **Ormsby, H. L., Thompson, G. A., Cousineau, G. G., Lloyd, L. A., and Hassard, J.,** Topical therapy in inclusion conjunctivitis, *Am. J. Ophthalmol.,* 35, 1811, 1952.

79. **Byrne, G. I. and Faubion, C. L.,** Lymphokine-mediated microbistatic mechanisms restrict *Chlamydia psittaci* growth in macrophages, *J. Immunol.,* 128, 469, 1982.

80. **Byrne, G. I. and Krueger, D. A.,** Lymphokine-mediated inhibition of *Chlamydia* replication in mouse fibroblasts is neutralized by anti-gamma interferon immunoglobulin, *Infect. Immun.,* 42, 1152, 1983.

81. **Rothermel, C. D., Rubin, B. Y., and Murray, H. W.,** Gamma-interferon is the factor in lymphokine that activates human macrophages to inhibit intracellular *Chlamydia psittaci* replication, *J. Immunol.,* 131, 2542, 1983.

82. **De La Maza, L. M., Peterson, E. M., Goebel, J. M., Fennie, C. W., and Czarniecki, C. W.,** Interferon-induced inhibition of *Chlamydia trachomatis*: dissociation from antiviral and antiproliferative effects, *Infect. Immun.,* 47, 719, 1985.

83. **Rothermel, C. D., Bryne, G. I., and Havell, E. A.,** Effect of interferon on the growth of *Chlamydia trachomatis* in mouse fibroblasts (L cells), *Infect. Immun.,* 39, 362, 1983.

84. **De La Maza, L. M., Goebel, J. M., Czarniecki, C. W., and Peterson, E. M.,** Ultrastructural analysis of the growth cycle of *Chlamydia trachomatis* in mouse cells treated with recombinant human alpha-interferons, *Exp. Mol. Pathol.,* 41, 227, 1984.

85. **Megran, D. W., Stiver, H. G., and Bowie, W. R.,** Complement activation and stimulation of chemotaxis by *Chlamydia trachomatis, Infect. Immun.,* 49, 670, 1985.

86. **Kuo, C.-C. and Grayston, J. T.,** Studies on delayed hypersensitivity with trachoma organisms. III. Lymphokines, *J. Immunol.,* 112, 540, 1974.

Chapter 11

STIMULATION OF IMMUNE RESPONSE*

Dwight M. Williams

TABLE OF CONTENTS

*   This work was supported by grants AI-22566-02 and AI-22380-01 from the National Institutes of Health.

I. STIMULATION OF IMMUNE RESPONSE

A variety of host immune responses are generated by infection with *Chlamydia*. These involve both cellular and humoral responses. Further, some of these responses have a suppressor rather than stimulator function. The relative role of these responses in host defenses against infection remains unclear and controversial. This review will address the types of immune responses described in humans and animals infected with *Chlamydia*.

A. Polymorphonuclear Response

Polymorphonuclear leukocytes (PMN) are part of the early inflammatory response to chlamydial infection in both man and animals. For example, PMNs are seen in human chlamydial urethritis,[1] cervicitis,[2] proctitis,[3] and penumonia[4] — often in a mixed cellularity infiltrate. In a murine model of chlamydial penumonia, heterophils are the predominant early inflammatory response followed later by mononuclear cells.[5,6] In vitro, Yong et al. demonstrated that human PMNs can inactivate both trachoma and LGV biovars of *Chlamydia*.[7] Both biovars were also inactivated by a cell-free myeloperoxidase-H_2O_2-halide system. However, PMNs from myeloperoxidase deficient patients inactivated *Chlamydia* normally, while those from chronic granulomatous disease patients were less effective. It was not possible, therefore, to define the exact mechanism of the chlamydicidal activity of the PMNs. More recently, Hammerschlag et al., in an in vitro system, have shown by electron microscopy, ingestion of *C. trachomatis* by human PMNs, in the presence of serum, with degranulation and many intracellular elementary bodies (EB) in phagolysosomes.[8] Unopsonized elementary bodies (EB) did not induce a significant chemiluminescence response while opsonized organisms did. Immune serum was not needed, but *Chlamydia* opsonized with sera containing < 100 mg of IgG were ineffective in stimulating chemiluminescence. Complement depletion did not diminish the effect of normal serum. Soderlung et al. found in their system, however, that type-specific but not group-specific antisera increased the chemiluminescence response induced in human PMNs by *C. trachomatis* in vitro.[9] In contrast, association of *Chlamydia* with PMNs was not increased by type-specific antiserum. Zvillich and Sarov[10] have recently described the interaction between $L_2/434/Bu$ EB and human PMNs. Two different types of phagocytic vacuoles were observed. In type one, a single EB was tightly surrounded by the vacuolar membrane. In the other type, one or more intact or degenerated EB could be seen. They postulate that the two types of vacuoles might indicate two mechanisms of entry of EB into PMNs. In the presence of specific antibody, a fuzzy coat was observed on EB within vacuoles. This might serve as a marker for entry of opsonized organisms into PMNs. A marked chemiluminescent response was observed in these studies as well. Thus, ample data suggest an interaction between *Chlamydia* and human PMN with chemiluminescence and probably at least partial chlamydial inactivation in vitro. The role of antibody in the interaction is controversial. In the mouse model of pneumonia,[6] EB are also present within heterophils by electron microscopy early in infection, implicating this cell in host response. However, the susceptible nude athymic mouse continues to have a prominent heterophil response during infection. The more resistant nu/+ mouse (heterozygous littermate) converts to a mononuclear response later in infection, casting some doubt as to the ultimate efficacy of the heterophil host response at least in the absence of active cell-mediated immunity (CMI) and antibody function.[6] However, no studies have compared the relative antichlamydial activity of human PMNs and murine heterophils. Therefore, data from mice in this area must be extrapolated with caution to humans.

B. Antibody

Infection with *Chlamydia* both in animal models and human infection elicits a specific antibody response. For example, Brunham et al., studying 95 women with *C. trachomatis* cervical infection, showed both a specific IgM and IgG response.[11] In the serum IgM, antibody increased with the number of organisms found in the cervix. In contrast to serum antibody, local antibody in cervical secretions showed an inverse correlation with the number of organisms isolated. The relationship was most striking with secretory IgA.[11] In animal models, specific local and systemic antibody has been stimulated by genital infection in the mouse, guinea pig, and monkey,[12-14] eye infection in the guinea pig,[15] and lung infection in the mouse.[6,16,17] Thymic dependence of antibody has to date been more clearly shown in the mouse than in the guinea pig.[18] By either microimmunofluorescence or enzyme-lined immunosorbent assay (ELISA) methodology, athymic nude mice have been shown to have a markedly impaired ability to make antibody in response to chlamydial infection while their heterozygous littermates do.[13,19-21] This inability extends both to IgM, IgG, and IgA serum antibody[20] and secretory antibody.[13,55] The ability to make antibody is restored to nude mice by thymic transplantation[19,20] from a neonatal heterozygous or homogyzous littermate, although serum antibody production is somewhat delayed in these animals compared to normal animals. Thus, antibody production in the murine model is strongly dependent on a T lymphocyte response. It is of interest that while the nonantibody producing athymic mice were more susceptible to infection with mouse pneumonitis agent (MoPn, murine *C. trachomatis*[13,19]), although this was much less striking in a genital model employing a human strain,[21] C5 complement-deficient animals and animals treated with cobra venom factor were not more susceptible to primary infection.[17] Susceptibility to reinfection, where antibody present at the time of infection might play a more important role in host defense, was not tested in complement-deficient animals. In vitro, the need for complement in neutralization of *Chlamydia* by specific antibody has been variable. In some studies, complement has been an absolute requirement for neutralization.[22,23] In others, it has had an enhancing effect.[24,25] In at least two studies, significant neutralization has been possible without it.[25,26] EB of *C. trachomatis* in vitro activate the complement cascade with generation of C5a and subsequent chemotaxigenic effects on normal human plasma for PMN.[27] Serovarspecific antibody is effective in neutralization in vitro.[25] Antibodies which bind to the outer membrane surface in a homogenous pattern by immune electron microscopy may be effective in neutralizing, while those which bind to either EB or reticulate bodies, a patchy distribution, may not neutralize.[23] Antibody directed at the major outer membrane protein (MOMP) will neutralize infection in vitro at a stage after attachment of *Chlamydia* to the host cell.[24,26] Opsonization of organisms with antibody to MOMP is also protective in vivo.[17] In vitro, normal human sera have variable abilities to inactivate *C. trachomatis* with a range of 10 to 100% inactivation.[28] Normal mouse sera also have some inactivating ability against MoPn compared to normal saline in in vivo opsonization studies, but immune serum was much more effective.[55]

The role of antibody in vivo is controversial and will be explored in Chapter 12. Briefly, in the murine model of pneumonia, adoptively transferred antibody provided protection to both nude and heterozygous mice, by a criterion of reduced mortality, in mice with a background flora which included Gram-negative rods.[20] These mice presumably had active background CMI. In contrast, intravenous antibody was not effective when the mice were converted to germfree and given a defined flora nonpathogenic to mice.[17] However, opsonization of organisms with specific antibody prior to infection, or immune serum given intranasally prior to intranasal chlamydial infection, was protective, with death prevented in nu/+ mice (which have intact CMI and antibody production) but only delayed death in nude athymic mice.[17] In contrast, in the guinea pig model of mucosal genital infection (as opposed to the invasive model of mouse pneumonia above), local antibody appeared to be the dominant modality of immunity.[12,13,18] This will be discussed in greater detail in the next chapter.

An interesting interaction of T and B cell function has recently been explored by Bard and Levitt[29] in vitro. Here L$_2$ serovar organisms were found to bind to and activate B cells in a polyclonal manner. Optimal immunoglobulin production occurred only when T cell factors were present. Perhaps in analogous fashion, a prominent plasma cell infiltrate is seen only in the lungs of nu/+ thymus-containing mice infected with *Chlamydia*.[6] In the nude mouse, despite large amounts of *Chlamydia* antigen, there is no plasma cell response consistent with the idea that a potential polyclonal activation of B cells in vivo may similarly depend on T cell factors.

C. Cell-Mediated Immunity
1. Macrophages

Infection with *Chlamydia* stimulates a variety of cell-mediated immune responses. Of these, the macrophage is one of the best studied, although many questions remain even here. It must be remembered that *C. psittaci* and *C. trachomatis* may not have identical interaction with macrophages. Macrophages from the guinea pig, for example, can be infected with meningopneumonitis agent, with cells from immunized animals being more resistant.[30] *C. psittaci* survives and grows in macrophages in part by inhibiting phagolysosomal fusion.[31,32] In the case of *C. trachomatis*, the epithelial cell is generally the preferred cell for replication. LGV biovar organisms multiply best in macrophages[33] while trachoma biovar organisms achieve limited growth.[33,34] The ability to multiply may correlate with differences in the ability of the biovars to inhibit the cells phagolysosomal fusion.[35] Infection with *C. psittaci* leads to increased levels of Ia antigen on macrophages[36] from the murine peritoneum, suggesting activation. In a mouse pneumonia model employing MoPn, immunologically intact mouse lungs had frequent lipid laden macrophages with prominent secondary lysosomes.[6] Functional activation in this model is suggested by increased resistance to secondary infection with *Histoplasma capsulatum*. However, mice were not more resistant to *Listeria* superinfection in a *C. psittaci* peritoneal model.[36] Similarly, in chronic LGV biovar infection in humans, macrophages had numerous mitochondria with many secondary lysosomes[37] and large granulations, again consistent with activation. Dividing organisms were rare. Thus both animal and human data are consistent with the concept that infection by *Chlamydia* induces activation of macrophages. The role of this cell in host defense, however, remains relatively controversial. It may be more important in infection with *C. psittaci* and/or LGV biovar *C. trachomatis* than with the usual ocular/genital serovars of *C. trachomatis*.

2. Lymphocytes

In our murine model of invasive infection with murine *C. trachomatis* (MoPn), athymic mice are significantly more susceptible than their heterozygous littermates, both by mortality and duration of infection.[6,17,19,20] Resistance can be transferred by neonatal thymus transplantation, thus demonstrating a T cell dependence of infection. Rank et al.[13] have similarly found in a genital model of infection with the same biovar that nude mice developed chronic nonresolving infection, whereas heterozygous mice resolved their infection in 20 days.

Transfer of fractionated T cell-enriched whole spleen cells (T+WSC) from nu/+ donors to nude recipients has indicated that more than one T cell population is important in host defense against MoPn.[6] Specifically, cells with both the L3T4+ phenotype and the Lyt 2+ phenotype are involved.[56] Studies transferring T+WSC from nu/+ donors to nude recipients at various durations of infection in the donor mice suggest that L3T4+ cells are dominant in transferring protection, particularly if the donor has been infected for only a short period of time. At later time periods, *both* L3T4+ and Lyt 2+ cell populations are involved. Later T cell populations are consistently protective while the protection offered by early T cells may be variable. In addition, mixing experiments employing late WSC populations from which either L3T4+ or Lyt 2+ cells have been removed by panning, show that a mixture of L3T4+ or Lyt 2+ cells provides better protection than either alone.[56] Thus, apparently more than one T cell subset is involved in maximal host resistance to MoPn. The role of the L3T4+ cell may well be to produce lymphokines (see below). The function of the Lyt 2+ cell remains unclear.

During infection, both antigen-specific lymphocyte transformation can be shown and ear or footpad swelling can be demonstrated in animal models of infection;[6,38-41] and antigen-specific lymphocyte transformation in human infections is also demonstrated.[42,43] In the mouse mucosal infection (genital tract) model, delayed-type hypersensitivity (DTH) reactions were usually not observed until late in infection (day 25),[41] whereas they appeared at day 5 to 7[38] or day 15[6] in the pneumonia model. In the pneumonia model, both DTH and antigen-specific lymphocyte transformation responses were T cell-dependent, occurring in nu/+ but not nude mice.[6] For example, on day 15 of infection, Chlamydia-infected nu/+ mice had 13.2 ± 2.7 μm in specific ear swelling in response to antigen. Infected nude mice had 4.3 ± 5.1 μm and control nu/+ and nude mice 4.0 ± 4.1 and 4.8 ± 4.4 μm, respectively. Mean lymphocyte transformation data on day 15 followed the same pattern with (cpm × 10^{-3}) 46.55 ± 20.28, 2.54 ± 1.24, 3.34 ± 1.39, and 1.79 ± 0.52 μm in the same groups, respectively. At day 10 of infection, splenocytes stimulated with MoPn produced interferon. This production was abolished by treatment of the splenocytes with anti-L3T4+ antibody plus complement, but not anti-Lyt 2+ plus complement.[57]

Thus, evidence exists for stimulation of T lymphocyte responses by Chlamydia infection. Some of these responses are beneficial in helping to protect the host.

In this regard, however, it should be noted that evidence for induction of suppressor T cells during chlamydial eye infection in monkeys[44] exists as well. This was most evident in multiply immunized or live-organism-challenged animals late in infections.[44] In a model of chlamydial salpingitis in the pig-tailed monkey, the inflammatory response was predominantly one of T cells of the suppressor/cytotoxic group.[45] Histopathology showed epithelial degeneration, but it was not possible to directly correlate this degeneration with the T cell infiltrate; thus, it was not clear if this cell was playing a suppressor or cytotoxic role. Suppression of immune response has also been observed during primary infection of C. psittaci in an intraperitoneal model in mice,[40] although it was not clear that the suppressor cell was a T cell. In other nonchlamydial models, cells of the Lyt 2+ (suppressor/cytotoxic) phenotype have been shown to secrete a lymphokine that is directly bactericidal.[46] No similar data exists for Chlamydia, but this should be further evaluated.

Induction of cytotoxic activity against Chlamydia-infected L cells has also been described during murine infection with C. psittaci.[48,49] In the studies of Lammert,[48] cytotoxicity was at first nonspecific, but became specific for infected target cells during infection and was effected by spleen cells. In related studies, Byrne and Kreuger[49] employed supernatant fluid prepared by incubation of C. psittaci immunized spleen cells with mitogen (Con A). This supernatant fluid induced cytotoxicity in mouse fibroblasts infected with C. psittaci. No cytotoxicity was observed against uninfected L cells or by supernatant fluid generated from nonimmunized spleen cells. Both studies of cytotoxicity required prolonged incubation (20 hrs) of effector cells or supernatant with targets for demonstration of cytotoxicity. Traditional rapid cytotoxicity (4 hr) probably more characteristic of cytotoxic T cell activity was not seen by Pavia and Schachter[50] in a murine model of C. trachomatis infection using syngeneic fibroblasts infected with an LGV biovar as a target and spleen cells as effectors. Similarly, we have not been able to demonstrate this in the MoPn model employing spleen cells as effectors and MoPn-infected 3T3 cells as targets.[55] Recently Quigstad and Hirschberg,[51] using C. trachomatis-specific cloned human T cells, were also unable to demonstrate T-cell-dependent cytotoxicity towards C. trachomatis-infected target cells, although most clones generated apparently were OKT4+,OKT8−. A reasonable postulate at this point may be that chlamydial induced cytotoxicity is due to stimulation of a lymphotoxin releasing populations of lymphocytes that are not classic cytotoxic T cells or release of tumor necrosis factor. More study in this area is needed.

Recent studies in our laboratory[52] have shown that natural killer (NK) activity is increased in both the spleen and lung early in infection with MoPn in both nu/+ and nu/nu mice. Peak activity is at approximately day five post infection, with a return toward baseline by day ten. Ablation of the rise in NK activity by giving nu/+ nice antiasialo GM-1 antibody (which we showed blocked the normal rise in NK activity in lung at day 5) or stimulation of NK activity by giving poly IC or interferon α and β did not affect quantitative tissue counts of MoPn or significantly influence survival.[52] Although further in vitro and in vivo studies are needed to assess the role of the increased NK activity observed, it is likely to be one of immunomodulation rather than host protection. The relationship of the NK target directed cytotoxicity to the early nonspecific cytotoxicity described by Lammert[48] (see above) is unclear, but NK cells could play a role in immunopathology in this model as well as immunoregulation.

II. CONCLUSIONS

The role of PMNs, antibody, and parameters of CMI in host resistance remains poorly defined and will be discussed in detail in the following chapter. I will comment only briefly here. Experiments have indicated that protection afforded by giving mice specific antibody intranasally prior to intranasal infection with *Chlamydia* is better in the nu/+ mouse (which has intact CMI and intact antibody function) than in the nude mouse (which lacks both). Further, studies of protection afforded by stimulation of cellular immunity to *Chlamydia* by prior infection with *H. capsulatum* infection suggest that intranasal antibody plus prior stimulation of cellular immunity can be additive in effect.[17] Thus, several aspects of the immune response stimulated by chlamydial infection may be of importance in host resistance. In particular, response controlled by L3T4+ lymphocytes (and their equivalents in humans), including lymphokines and local antibody as well as those of Lyt 2+ T lymphocytes, need further evaluation for their role in host defenses. Further, cytokines needed to be evaluated for their role in potentially detrimental reactions in the host which might lead to clinical trachoma or tubal scarring. Specifically, the role of cytotoxins generated during infection (see above), and monokines (such as interleukin 1 or tumor necrosis factor which may be released from macrophages during infection and which may have a role in stimulating fibroblast proliferation[53,54]) need to be investigated in terms of their possible deleterious effects in vivo. Finally, the role of suppressor cells, both T cells and macrophages potentially activated during chlamydial infection, need further examination for their beneficial or possibly deleterious effects in modulating host repsonse.

REFERENCES

1. **Kraus, S. J.,** Semiquantitation of urethral polymorphonuclear leukocytes as objective evidence of non-gonococcal urethritis, *Sex. Trans. Dis.,* 9, 52, 1982.
2. **Kiviat, N. B., Paavonen, J. A., Brockway, J., Critchlow, C. W., Brunham, R. C., Stevens, C. E., Stamm, W. E., Kuo, C-C., DeRoven, T., and Holmes, K. K.,** Cytologic manifestations of cervical and vaginal infections, *J. Am. Med. Assoc.,* 253, 989, 1985.
3. **Quinn, T. C., Goodell, S. E., Mkrtichian, P. A. C., Shuffler, M. D., Wang, S-P., Stamm, W. E., and Holmes, K. K.,** *Chlamydia tranchomatis* proctitis, *N. Engl. J. Med.,* 305, 195, 1981.
4. **Frommell, G. T., Bruhn, F. W., and Schwartzman, J. D.,** Isolation of *Chlamydia tranchomatis* from infant lung tissue, *N. Engl. J. Med.,* 296, 1150, 1977.
5. **Kuo, C-C., Chen, W-J., Brunham, R. C., and Stephens, R. S.,** A mouse model of *Chlamydia trachomatis* infection, in *Chlamydial Infections,* Mardh, P-A., Holmes, K. K., Oriel, J. D., Piot, P., and Schachter, J., Eds., Elsevier, Amsterdam, 1982, 379.
6. **Williams, D. M., Schachter, J., Coalson, J. E., and Grubbs, B.,** Cellular immunity to the mouse pneumonitis agent, *J. Infec. Dis.,* 149, 630, 1984.
7. **Yong, E. C., Klebanoff, S. J., and Kuo, C-C.,** Toxic effect of human polymorphonuclear leukocytes on *Chlamydia trachomatis, Infect. Immun.,* 37, 422, 1982.

8. **Hammerschlag, M. R., Suntharalingam, K., and Fikrig, S.,** The effect of *Chlamydia trachomatis* on luminol-dependent chemiluminescence of human polymorphonuclear leukocytes: requirements for opsonization, *J. Infect. Dis.,* 151, 1045, 1985.

9. **Soderlung, G., Dahlgren, C., and Kihlstrom, E.,** Interaction between human polymorphonuclear leukocytes and *Chlamydia trachomatis, FEMS Microbiol. Lett.,* 22, 21, 1984.

10. **Zvillich, M. and Sarov, I.,** Interaction between human polymorphonuclear leukocytes and *Chlamydia trachomatis* elementary bodies: electron microscopy and chemiluminescent response, *J. Gen. Microbiol.,* 131, 2627, 1985.

11. **Brunham, R. C., Kuo, C-C., Cles, L., and Holmes, K. K.,** Correlation of host immune response with quantitative recovery of *Chlamydia trachomatis* from the human endocervix, *Infect. Immun.,* 39, 1491, 1983.

12. **Rank, R. G., White, H. J., and Barron, A. L.,** Humoral immunity in the resolution of genital infection in female guinea pigs infected with the agent of guinea pig inclusion conjunctivitis, *Infect. Immun.,* 26, 573, 1979.

13. **Rank, R. G., Soderberg, L. S. F., and Barron, A. L.,** Chronic chlamydial genital infection in congenitally athymic nude mice, *Infect. Immun.,* 48, 847, 1985.

14. **Patton, D. L., Halbert, S. A., Kuo, C-C., Wang, S-P., and Holmes, K. K.,** Host response to primary *Chlamydia trachomatis* infection of the fallopian tube in pig-tailed monkeys, *Fertil. Steril.,* 40, 829, 1983.

15. **Murray, E. S., Charbonnet, L. T., and MacDonald, A. B.,** Immunity to chlamydial infections of the eye, *J. Immunol.,* 110, 1518, 1973.

16. **Kuo, C-C. and Chen, W-J.,** A mouse model of *Chlamydia trachomatis* pneumonitis, *J. Infect. Dis.,* 141, 198, 1980.

17. **Williams, D. M., Schachter, J., Weiner, M. H., and Grubbs, B.,** Antibody in host defense against mouse pneumonitis agent (murine *Chlamydia trachomatis), Infect. Immun.,* 45, 674, 1984.

18. **Rank, R. G. and Barron, A. L.,** Effect of antithymocyte serum on the course of chlamydial genital infection in female guinea pigs, *Infect. Immun.,* 41, 876, 1983.

19. **Williams, D. M., Schachter, J., Drutz, D. J., and Sumaya, C. V.,** Pneumonia due to *Chlamydia trachomatis* in the immune compromised (nude) mouse, *J. Infect. Dis.,* 143, 238, 1981.

20. **Williams, D. M., Schachter, J., Grubbs, B., and Sumaya, C. V.,** The role of antibody in host defense against the agent of mouse pneumonitis, *J. Infect. Dis.,* 145, 200, 1982.

21. **Taylor-Robinson, D., Tuffrey, M., and Folder, P.,** Some aspects of animal models for *Chlamydia trachomatis* genital infections, in *Chlamydia Infections,* Mardh, P-A., Ed., Elsevier, Amsterdam, 1982, 375.

22. **Lucero, M. E. and Kuo, C-C.,** Neutralization of *Chlamydia trachomatis* in cell culture infection by serovar-specific monoclonal antibodies, *Infect. Immun.,* 50, 595, 1985.

23. **Clark, R. B., Nachamkin, I., Schatzki, P. F., and Dalton, H. P.,** Localization of distinct surface antigens on *Chlamydia trachomatis* HAR-13 by immune electron microscopy with monoclonal antibodies, *Infect. Immun.,* 38, 1273, 1982.

24. **Howard, L. V.,** Neutralization of *Chlamydia trachomatis* in cell culture, *Infect. Immun.,* 11 698, 1975.

25. **Caldwell, H. D. and Perry, L. J.,** Neutralization of *Chlamydia trachomatis* infectivity with antibodies to the major outer membrane protein, *Infect. Immun.,* 38, 745, 1982.

26. **Peeling, R., Maclean, I. W., and Brunham, R. C.,** In vitro neutralization of *Chlamydia trachomatis* with monoclonal antibody to an epitope on the major outer membrane protein, *Infect. Immun.,* 46, 484, 1984.

27. **Mergran, D. W., Stiver, H. G., and Bowie, W.,** Complement activation and stimulation of chemotaxis by *Chlamydia trachomatis, Infect. Immun.,* 49, 670, 1985.

28. **Osborn, M. F., Johnson, A. P., and Taylor-Robinson, D.,** Susceptibility of different serovars of *Chlamydia trachomatis* to inactivation by normal human serum, *Genitourin. Med.,* 61, 244, 1985.

29. **Bard, J. and Levitt, D.,** *Chlamydia trachomatis* (I_2 serovar) binds to distinct subpopulations of human peripheral blood leukocytes, *Clin. Immunol. Immunopathol.,* 38, 150, 1986.

30. **Benedict, A. A. and McFarland, C.,** Growth of meningopneumonitis virus in normal and immune guinea pig monocytes, *Nature,* 181, 1742, 1958.

31. **Wyrick, P. B. and Brownridge, E. A.,** Growth of *Chlamydia psittaci* in macrophages, *Infect. Immun.,* 19, 1054, 1978.

32. **Wyrick, P. B., Brownridge, E. A., and Ivins, B. E.,** Interaction of *Chlamydia psittaci* with mouse peritoneal macrophages, *Infect. Immun.,* 19, 1061, 1978.

33. **Kuo, C-C.,** Culture of *Chlamydia trachomatis* in mouse peritoneal macrophages: factors affecting organism growth, *Infect. Immun.,* 20, 439, 1978.

34. **Yong, E. C., Chi, E. Y., and Kuo, C-C.,** Differential antimicrobial activity of human mononuclear phagocytes against human biovars of *Chlamydia trachomatis, J. Immunol.,* 139, 1297, 1987.

35. **Yong, E. C., Chi, E. Y., and Kuo, C-C.,** Differences in the survival mechanisms of the human biovars of *Chlamydia trachomatis* are responsible for their differential virulence in human mononuclear phagocytes, in Abstr. 86th Annu. Mtg. Am. Soc. Microbiol., #D161, Washington, D.C., March, 1986.

36. **Paulnock, D. M., Huebner, R. E., Guagliardi, L., Lietzke, R. M., Albrecht, R. M., and Byrne, G. I.**, Acquired resistance to *Chlamydia:* induction and characterization of activated macrophages from immunized mice, in Abstr. 86th Annu. Mtg. Am. Soc. Microbiol., #618, Washington, D.C., March, 1986.

37. **Alacoque, B., Cloppet, H., Dumontel, C., and Moulin, G.**, Histological, immunofluorescent, and ultrastructural features of lymphogranuloma venereum, *Br. J. Vener. Dis.*, 60, 390, 1984.

38. **Kuo, C-C. and Chen, W-J.**, A mouse model of *Chlamydia trachomatis* penumonitis, *J. Infect. Dis.*, 141, 198, 1980.

39. **Harrison, H. R., Lee, S. M., and Lucas, D. O.**, *Chlamydia trachomatis* pneumonitis in the C57 BL/K⁵J mouse: pathologic and immunologic features, *J. Lab. Clin. Med.*, 100, 953, 1982.

40. **Lammert, J. K. and Wyrick, P. B.**, Modulation of the immune response as a result of *Chlamydia psittaci* infection, *Infect. Immun.*, 35, 537, 1982.

41. **Barron, A. L., Rank, R. G., and Moses, E. B.**, Immune responses in mice infected in the genital tract with mouse pneumonitis agent (*Chlamydia trachomatis* biovar), *Infect. Immun.*, 44, 82, 1984.

42. **Brunham, R. C., Martin, D. H., Kuo, C-C., Wang, S. P., Stevens, C. E., Hubbard, T., and Holmes, K. K.**, Cellular immune response during uncomplicated genital infection with *Chlamydia trachomatis* in humans, *Infect. Immun.*, 34, 98, 1981.

43. **Hanna, L., Schmidt, L., Sharp, M., Stites, D. P., and Jawetz, E.**, Human cell-mediated immune responses to chlamydial antigens, *Infect. Immun.*, 23, 412, 1979.

44. **Young, E. and Taylor, H. R.**, Immune mechanisms in chlamydial eye infection: cellular immune response in chronic and acute disease, *J. Infect. Dis.*, 150, 745, 1984.

45. **Patton, D. L.**, Immunopathology and histopathology of experimental chlamydial salpingitis, *Rev. Infect. Dis.*, 7(6), 746, 1985.

46. **Markham, R. B., Pier, G. B., Goellner, J., and Mizel, S. B.**, *In vitro* T cell-mediated killing of *Pseudomonas aeruginosa*: the role of macrophages and T cell subsets in T cell killing, *J. Immunol.*, 134, 4112, 1985.

47. **Young, E. and Taylor, H. R.**, Immune mechanisms in chlamydial eye infection: development of T suppressor cells, *Invest. Ophthalmol. Vis. Sci.*, 27(4), 615, 1986.

48. **Lammert, J. K.**, Cytotoxic cells induced after *Chlamydia psittaci* infection in mice, *Infect. Immun.*, 35, 1011, 1982.

49. **Byrne, G. I. and Krueger, D. A.**, In vitro expression of factor-mediated cytotoxicity activity generated during the immune response to *Chlamydia* in the mouse, *J. Immunol.*, 134, 4189, 1985.

50. **Pavia, C. S. and Schachter, J.**, Failure to detect cell-mediated cytotoxicity against *Chlamydia trachomatis*-infected cells, *Infect. Immun.*, 39, 1271, 1983.

51. **Quigstad, E. and Hirschberg, H.**, Lack of cell-mediated cytotoxicity towards *Chlamydia trachomatis* infected target cells in humans, *Acta Pathol. Microbiol. Immunol. Scand. Sect. C*, 92, 153, 1984.

52. **Williams, D. M., Schachter, J., and Grubbs, B.**, A study of NK cells in infection with the mouse pneumonitis agent, *Infect. Immun.*, 55, 223, 1987.

53. **Schmidt, J. A., Mizel, S. B., Cohen, D., and Greer, I.**, Interleukin I, a potential regulator of fibroblast proliferation, *J. Immunol.*, 128, 2177, 1982.

54. **Bendtzen, K.**, Biologic properties of interleukins, *Allergy*, 38, 219, 1983.

55. **Williams, D. M. and Schachter, J.**, unpublished observations.

56. **Williams, D. M., Kung, J., and Schachter, J.**, Immunity to the Mouse Pneumonitis Agent in Chlamylial Infection, Oriel, D., Ed., Cambridge University Press, 1986, 465 and unpublished data.

57. **Byrne, G. I., Grubbs, B., Dickey, T. J., Schachter, J., and Williams, D. M.**, Interferon in recovery from pneumonia due to *Chlamydia trachomatis* in the mouse, *J. Infect. Dis.*, 156, 993, 1987.

Chapter 12

ROLE OF THE IMMUNE RESPONSE

Roger G. Rank

TABLE OF CONTENTS

I. INTRODUCTION

The interaction between chlamydiae and their hosts, as in all host-parasite relationships, is a dynamic affair which is influenced by a multitude of biochemical and physiological factors, not the least of which is the immunologic response of the host. Chlamydiae, as all microorganisms, are a complex composite of antigens, which are able to elicit a variety of humoral and cellular immune responses. Some of these responses may be superfluous while others may be important in controlling or eliminating the offending organism. Still other immune responses may be coconspirators in the production of disease. In this chapter, we will concentrate our discussion on the role of specific humoral and cellular immune mechanisms in the resolution of and resistance to chlamydial infections, as well as those immunopathological mechanisms contributing to chlamydial disease. Since the effectiveness and participation of various immune components may be governed by the location of the parasite in the body, it will be important that we consider the role of the immune response as it pertains to the disease process in a particular organ system, i.e., the eye, respiratory tract, and the genital tract.

II. EFFECT OF IMMUNE COMPONENTS AGAINST CHLAMYDIAE IN VITRO

A variety of immune mechanisms have been tested for their effect on chlamydiae in vitro. While the relevance of these studies to specific disease entities is difficult to assess, the information they provide may establish important paths for investigators to follow in in vivo studies.

A. Antibody-Mediated Mechanisms

Antibody has been found to have both neutralizing and opsonizing functions against chlamydiae in vitro. Several investigators have demonstrated that sera from trachoma patients or rabbits immunized with whole organisms were able to neutralize the infectivity of *C. trachomatis* in tissue culture.[1-4] Byrne and Moulder[5] reported that *C. psittaci* (6BC) infection of mouse L cells could also be neutralized by specific antiserum. Using ^{14}C-labeled organisms, they found that the antibody apparently prevented ingestion of the elementary bodies (EBs). This mechanism was supported by Ainsworth et al.[6] and Allan et al.[7] who observed that specific antiserum would inhibit attachment of either *C. trachomatis* or GPIC to McCoy cells. More recently, Caldwell and Perry[8] described neutralization of infectivity of LGV-434 EBs for HeLa cells with IgG antibodies raised against the major outer membrane protein (MOMP). However, in contrast to previous studies, they observed that the antibody-treated organisms were internalized by the cells at the same rate as untreated EBs, but that further growth was apparently inhibited. When organisms were incubated with anti-MOMP Fab fragments, the infectivity of the chlamydiae was not affected. Thus, intact IgG molecules are required for the internalization and subsequent destruction of the organisms. Peeling et al.[9] also reported that a monoclonal antibody directed against a species-specific epitope on the MOMP of *C. trachomatis* was able to neutralize the organism, but likewise did not impede attachment of the EBs to the host cells. In these two studies,[8,9] complement was not necessary for neutralization to occur while other studies have required complement.[1,3,10] Whether complement aids in the inhibition of infectivity by direct cytotoxicity against chlamydiae or by some other mechanism cannot be gleaned from the available data, although it is quite possible that different pathways of neutralization may occur depending on the isotype of the antibody, the molecular target of the antibody, and the host cell involved.

Opsonization of chlamydiae has also been observed in vitro. Wyrick et al.[11,12] found that incubation of *C. psittaci* (Cal 10) EBs with antibody resulted in maximum uptake by stimulated or unstimulated murine peritoneal macrophages. Furthermore, the antibody permitted the fusion of the phagosomes containing chlamydiae with lysosomes so that degradation of the organisms occurred within 6 hr. Normally, in the absence of antibody, phagosome-lysosome fusion does not occur and the chlamydiae complete their developmental cycle.

B. Cell-Mediated Mechanisms

Cell-mediated immune mechanisms have also been examined in vitro. Two basic defense mechanisms are usually considered to be encompassed by cell-mediated immunity (CMI), cell-mediated cytotoxity and lymphokine-mediated activation of macrophages. With regard to the former, Lammert[13] was able to demonstrate that spleen cells from mice previously infected with *C. psittaci* (Cal 10) were cytotoxic for macrophages or L cells infected with the homologous organism, but were also found to be cytotoxic for uninfected target cells, albeit to a lesser degree. In contrast, cell-mediated cytotoxicity could not be demonstrated by cells derived from spleens, mesenteric lymph nodes, or the peritoneal cavities of mice infected intravenously 1 to 3 weeks earlier with LGV.[14] A concern in this study was that cell-cytotoxicity may have been prevented by the apparent absence of detectable chlamydial antigen on the surface of the target cells. Similarly, Qvigstad and Hirschberg,[15] using human peripheral blood mononuclear cells from previously infected patients and or human T cell clones, did not detect any *Chlamydia*-specific cytotoxicity. However, both the peripheral blood mononuclear cells and the T cell clones were antigen-reactive since they were able to produce a proliferative response when stimulated by chlamydial antigen. Moreover, they were able to show that the target cells did indeed possess chlamydial surface antigen and that their system could measure cytotoxicity when effector cells primed with allogeneic antigen were exposed to the same chlamydial-infected target cells.

Lymphokine-mediated macrophage activation has also been described in vitro by Byrne and colleagues.[16] In their system, spleen cells were obtained from mice which has recovered from infection with the 6BC strain of *C. psittaci* and cultured in the presence of concanavalin A for 24 hr. The supernatants were removed and added to cultures of macrophages. This resulted in activation of the macrophages; and subsequent infection of these cells revealed a marked suppression of chlamydial growth. This same cell-free supernatant was also cytotoxic for chlamydiae growth in mouse L cells.[17] The identity of the lymphokine was confirmed by inhibition of the cytoxicity by anti-gamma interferon but not anti-alpha/beta interferon. That gamma interferon can activate macrophages is not unexpected, since it has now been equated with macrophage activation factor.[18] Rothermel et al.,[19] in an analogous system, stimulated human peripheral blood mononuclear cells with concanavalin A and treated macrophages with the lymphokine-containing supernatant. They also noted inhibition of *C. psittaci* in the treated-macrophages and attributed this activity to gamma interferon. However, it should be noted that both classes of interferon have been shown to have anti-chlamydial activity, although gamma interferon is generally more potent.[20,21] Interestingly, the mechanism by which interferon inhibits chlamydial growth appears to be different than that employed against viruses.[22] Thus, with regard to control of chlamydial infections, these studies implicate a potentially important role for gamma interferon as an effector arm of CMI, functioning by direct inhibition of chlamydial growth or by activation of macrophages.

III. INFECTIONS OF THE EYE

Chlamydial infections of the eye can be divided into two basic categories, trachoma and inclusion conjunctivitis. While the two entities differ basically with regard to immunotypes involved and chronicity of infection, they are similar in many ways. In fact, trachoma probably results from repeated infections while inclusion conjunctivitis represents the early or primary infection(s).[23,24] In the following discussion, in an attempt to present an overall view of ocular immune mechanisms operative in chlamydial infections, both disease entities in humans as well as animal models will be considered together, since it is unlikely that radically different mechanisms are involved in the different systems.

Furthermore, when analyzing the basic immune mechanisms which the host employs in its response to an infectious agent, it is important to differentiate between those mechanisms which are used to resolve a primary infection and those mechanisms which are used to prevent reinfection, since it is conceivable that they may be different. Thus, care will be taken to discuss resolution of a primary infection and resistance to reinfection separately.

A. Resolution of Infection

It is well documented that individuals with inclusion conjunctivitis infections do indeed resolve their infections with or without drug intervention. Similarly, primates infected with *C. trachomatis* and guinea pigs infected with GPIC also resolve their infections naturally.[25] It is thus reasonable to assume, that at least in these infections, immune mechanisms are capable of successfully halting the course of the infection and probably eliminating the organism. As one would expect, both antibody and CMI responses can be measured in humans and animals infected ocularly with chlamydiae, but the exact roles of each in the resolution of infection have been difficult to identify.

When the temporal relationship between resolution of infection and the appearance of both antibody and CMI responses is examined, one generally finds a positive correlation with both responses. In guinea pigs infected in the conjunctiva with GPIC, the infection usually resolves within 15 to 20 days. The demise of the infection corresponds with the appearance of IgG antibodies in serum and secretory IgA antibodies in tears by day 11 with a maximum in both isotypes being reached by day 20.[26,27] The development of CMI responses as measured by delayed-type hypersensitivity (DTH) skin tests and macrophage migration inhibition assays was also temporally associated with the resolution of the infection, although positive responses were seen as early as 5 days after infection.[26] Sacks et al.[28] reported similar findings in owl monkeys infected with immunotype B of *C. trachomatis*. Leukocyte inhibition responses were positive 9 days after infection, while antibodies in serum and eye secretions did not appear until day 14, at which time the infection had already diminished markedly in intensity. While it is tempting to extrapolate from these experiments a stronger role for CMI because of its earlier debut, consideration must be given to the sensitivity of the assays and the frequency of measurement, which in many cases was several days.

More assertive data implicating an important role for antibody in the resolution of GPIC ocular infection was presented by Modabber et al.[29] in which they utilized the immunosuppressive drug, cyclophosphamide, to selectively depress the humoral immune response of guinea pigs. Guinea pigs were infected in the conjunctiva and were injected with 300 mg/kg of cyclophosphamide 1 day later. This treatment resulted in a significant delay in the appearance of antibodies to GPIC in serum and secretions, even though delayed-type hypersensitivity reactions to chlamydial antigen were not affected to any great degree. Resolution of the infection was delayed from 4 to 18 days and roughly corresponded with the appearance of high titers of antibody. These data suggest an important role for the humoral immune response in the resolution of chlamydial ocular infection. While antibody seems to be quite important, the participation of CMI mechanisms cannot be ruled out. Mull and Thompson[30] reported that suppression of CMI by treatment of guinea pigs with anti-lymphocyte globulin produced infections of greater length than the controls even though the antibody response in serum and eye secretions was similar to controls.

The exact mechanisms by which antibody and CMI participate in the resolution of ocular infection remain undefined; although based on data to be discussed in the following section, it is reasonable to assume that secretory antibodies function by neutralizing the infectivity of chlamydiae in the eye. Whether this occurs by preventing attachment, internalization, or replication after ingestion, is unknown. Further, opsonization of the organisms with subsequent phagocytosis by polymorphonuclear leucocytes or macrophages may also be a possible mechanism. There have been no studies regarding mechanisms of CMI in ocular chlamydial infections to date.

B. Resistance to Reinfection

Most studies on resistance to reinfection in the eye have focused on the role of antibody. Indeed, there is no firm evidence that CMI plays a major role in the protective immune response, but then the classical systems for studying CMI such as adoptive transfer experiments are not feasible in primate systems and have not been attempted in the guinea pig model. Unfortunately, a model for ocular infection does not exist in the mouse, the most convenient and, potentially, the most productive model for the study of CMI.

As has been described earlier, sera from patients with trachoma were able to neutralize the infectivity of chlamydiae for cells in tissue culture.[1] Since high levels of antibody, particularly secretory IgA (sIgA) have been found in eye secretions, these too were examined for neutralizing activity.[31,32] As expected, eye secretions from children with trachoma, when incubated with C. trachomatis, were able to neutralize the infectivity of the organisms when inoculated into the eyes of owl monkeys. It is interesting that in these studies the onset of the infection was merely delayed, suggesting that neutralization was incomplete. The incompleteness of the neutralization probably represents a dose response phenomenon, since it was noted that the use of lower-titered secretions resulted in an even shorter delay in the onset of infection. Pearce et al.[33] reported similar findings when they inoculated guinea pig eyes with different concentrations of GPIC incubated with antibody. Only the lower concentrations of organisms were completely neutralized. This concept is of potential importance when considering immunity to infection, in that the ratio of antibody to organisms might be critical in the establishment of infection. Disease might still occur if at least some organisms are able to complete a life cycle and thereby release antigen which could elicit immunopathological changes.

Whether the mechanism of neutralization in vivo is by prevention of attachment or by inhibition of the developmental cycle after ingestion has not been confirmed. Pearce et al.[33] have accrued evidence supporting the former mechanism by adding GPIC incubated with immune serum or immune tears to fresh guinea pig conjunctival tissue in vitro. They observed a decreased association of the chlamydiae with the tissue and found the effect to be related to the antibody titer.

While the relative success of the above experimentation implies that neutralization may be an important immune mechanism in resistance to infection, it is certainly important that such work be supported by in vivo studies. There have been numerous studies in which various immunization regimens have been given to humans, primates, or guinea pigs; and in many cases, some degree of protective immunity has been induced to ocular challenge.[34-36] In general, since both antibody and CMI responses are usually produced, it has been difficult to correlate protection with a particular response. Moreover, the presence of both serum and secretory antibodies resulting from natural infection as well as artificial immunization correlates with immunity to a challenge infection, so that it is difficult to assign a predominant role to either.[37-39]

Several studies have attempted to ascertain whether the local or the systemic antibody response bears the ultimate responsibility for providing protection to reinfection. Murray et al.[40] immunized guinea pigs with formalin-killed GPIC intraperitoneally and were able to elicit neutralizing antibodies in serum. When challenged in the eye, no protection was seen. Only those animals having had a previous ocular infection and having neutralizing antibodies in eye secretions were protected. In support of these data, Watson et al.[41] passively administered serum antibody to GPIC to naive guinea pigs prior to ocular challenge with infectious organisms. At the time of challenge, animals receiving immunoglobulin from immune animals had high titers in serum but no antibody in eye secretions. On the other hand, control guinea pigs which had been previously infected had significant titers in both serum and eye secretions. Upon challenge, the course of the infection in passively-immunized guinea pigs was no different than animals given normal serum, while all previously-infected animals were totally resistant. Of course, it is possible that the natural infection activated CMI which may have also participated in the protective immune response. Similar findings were reported by Orenstein et al.[42] when they passively transferred immune serum to owl monkeys and challenged them with *C. trachomatis*. In humans, the importance of a local humoral immune response is implied by a high rate of inclusion conjunctivitis in neonates despite the presence of maternal antibody in their serum.[43] In a different approach, Malaty et al.[27] treated guinea pigs in the external eye with either immune serum, immune tears, or normal serum 5 times daily for 2 days before and 5 days after challenge with infectious GPIC. The immune serum delayed infection and decreased the level of inclusions. Immune tears had a similar but less marked effect.

Nichols et al.[44] presented additional evidence that protective immunity is mediated via the mucosal-associated lymphoid system. They gave live GPIC to guinea pigs by the oral route and then challenged the animals in the eye and genital tract either 11 or 22 days later. Interestingly, in both sites, the pathological response and the level of infection were decreased compared to the untreated controls, suggesting that animals could be stimulated at one mucosal site and that reactive cells could home to other mucosal sites. Similarly, in our laboratory, we have found that both IgG and IgA antibodies could be detected in ocular secretions quite soon following a primary infection in the genital tract, in the absence of ocular infection.[45] That there is indeed preferential homing of lymphocytes associated with the mucosal-associated lymphoid system, which includes the Peyer's patches, mesenteric lymph nodes, respiratory tract, mammary glands, salivary glands, genital tract, and most probably, the eye, has been well documented.[46-48]

The above studies indicate an important role for local antibody. The implication is that sIgA is the immunoglobulin present in ocular secretions providing the protective function; however, IgG can also be measured in secretions. It remains possible that serum-derived IgG may appear in secretions as a result of pathological changes in the conjunctiva initiated by the chlamydial infection and may participate in the protective response. However, it should be emphasized that none of these studies rule out an active contribution of CMI.

An interesting and significant feature of trachoma and inclusion conjunctivitis is that individuals can be reinfected.[24] This also seems to be the case in the monkey and guinea pig models.[37,49-51] Long-term solid immunity simply does not seem to develop in the course of natural infection. In addition, immunization trials in humans, when successful, produced immunity which was at best short-lived.[34] In fact, most studies on resistance to challenge infection, whether following natural infection or immunization, have judged protective immunity in terms of an abbreviated infection or infection with a lessened pathological response and not the complete absence of organisms. For example, when guinea pigs were challenged 35 days or less after a primary infection, only a few inclusions could be detected; but when they were challenged more than 60 days after the primary infection, inclusions could be demonstrated consistantly for a week.[51] One other variable affecting the degree of immunity is the size of the inoculating dose. Ahmad et al.[50] found that at 1 month after the initial infection, guinea pigs were refractory to low challenge doses but became reinfected by higher inocula. Moreover, the length of the incubation period was shortened and the intensity of the infection increased as the challenge dose was raised.[50,52] These data are not unexpected because it is quite possible that antibody or immune cells, even in optimal conditions, might not be present in sufficient amounts to deal with overwhelming numbers of organisms.

The length of time that animals possess a high degree of immunity is relatively short and seems to correspond to a decrease in the titer of antibody in ocular secretions but not serum. Malaty et al.[27] observed in guinea pigs infected with GPIC that while serum IgG levels remained elevated for in excess of 210 days after infection, sIgA and IgG titers in tears reached peak levels at 20 to 30 days and decreased dramatically to low levels by day 60. When guinea pigs were challenged in the eye at 60, 90, or 120 days later, a significant correlation between the level of sIgA and IgG in secretions and the susceptibility to reinfection with low doses of chlamydiae was seen while there was no correlation with serum IgG. Pearce et al.[33] also found that immunity to GPIC ocular infection waned between 63 to 80 days after infection and was associated with a decrease in tear sIgA titer.

Similar observations have been reported by Taylor and colleagues in a primate model for trachoma using serovars A and E of *C. trachomatis*.[53] In this model trachoma could be produced by repeated inoculation of the organism at 1 or 15 week intervals. Serum IgG and IgM levels increased quickly and remained elevated for the duration of the observation period in animals with weekly inoculations. IgA developed more slowly than IgG in tears, but both persisted at high levels for the course of the experiment. In animals receiving inoculations at 15 week intervals, only serum IgG was elevated for long periods of time. Again, IgA developed more slowly than IgG in tears, but both diminished rather markedly within several weeks. Although the disease process was relatively unaltered following successive challenges, the ability to isolate viable chlamydiae from the conjunctiva was modified. In the weekly inoculation group, organisms could only be detected in the time period before the production of high titers of IgA in tears. In the 15 week interval group, chlamydiae were isolated in the primary infection until IgA reached high levels, and then again only for short periods very early following reinfection. The disappearance of the organisms was associated with the increase in IgA and IgG titers in tears.

While the protective antigens on chlamydiae with regard to ocular infections have not been identified, the antibody response to specific chlamydial antigens has been described by immunoblot analysis on sera and tears from infected monkeys.[54] IgA antibody to MOMP appeared in tears about day 14 and was still strong at day 56. Antibodies to a 60Kd, 54Kd, and LPS developed between days 21 to 28. A good response to MOMP was noted in the sera of all animals, but there was great variability in presence and intensity of the response to the other antigens.

The reason for the decline in local antibody is not clear; however, one explanation would be the relatively short half-life of IgA.[55] Thus, once the infection is resolved, the immunogenic stimulus disappears and antibody production gradually diminishes. An alternative explanation would include the negative regulatory effects of suppressor cells. In fact, Young and Taylor[56] have found that depletion of suppressor T cells from peripheral blood lymphocytes of immunized and chronically-infected monkeys resulted in a substantial increase in the proliferative response to chlamydial antigen. They further noted that significant suppressor cell activity did not develop until after animals had received a second inoculation or had received multiple weekly inoculations.[57] The appearance of suppressor cells seemed to be associated with a decrease in circulating IgG. With these data in mind, it is feasible that certain populations of suppressor cells might be elicited that are able to turn off the local antibody response. Indeed, the response measured in the proliferative assay has been shown, at least with certain viral antigens, to be primarily due to the proliferation of T helper cells.[58] Suppression of the T helper cell could influence both antibody and CMI responses.

C. Contribution of Immune Response to the Disease Process in Trachoma

Pathologically, inclusion conjunctivitis and the early stages of trachoma are similar, and it appears simply that the development of trachoma is caused by repeated inoculation with the agent, while inclusion conjunctivitis is merely the result of a single chlamydial infection.[23,24] In studies with human volunteers, individuals inoculated with chlamydiae isolated from either trachoma or inclusion conjunctivitis cases developed infections characteristic of inclusion conjunctivitis; but when previously infected individuals were inoculated, more severe disease occurred.[59]

Experimentation in animal models has supported this concept. Wang and Grayston[60] observed that monkeys infected ocularly with different strains of *C. trachomatis* only developed pannus if they had had a previous exposure to the organism. Taylor et al.[49,53] demonstrated very convincingly in cynomolgus monkeys that reinfection was the significant factor in the development of trachoma, and that it did not seem to matter whether the organism was one classically associated with trachoma (immunotype A) or one associated with oculo-genital infections (immunotype E). Previously, Monnickendam and colleagues[61] had described the pathogenesis of GPIC infection in the eyes of guinea pigs. This disease resembles very closely that of inclusion conjunctivitis in humans. However, most interestingly, they also found that by multiply reinfecting the animals, they could elicit chronic conjunctivitis and pannus analogous to trachoma.[51] With these data in mind, the real significance of the short-lived immunity described above becomes evident. It is this inability to mount a long-term protective immune response which allows continual reinfection and is therefore the key factor permitting the development of trachoma.

The consequence of multiple ocular infections is the development of hypersensitivity to chlamydial antigens which results in the classical pathological changes characteristic of trachoma. The potential role of hypersensitivity was recognized during the development of vaccines in primate models. Collier and Blyth[62] observed that immunization with adjuvant vaccines produced more severe disease in baboons upon challenge when compared to unimmunized controls. Similar deleterious effects were also seen in monkeys given varying vaccine preparations.[63] In both cynomolgus monkey and guinea pig models, the predominant cell present in the inflammatory infiltrate is the mononuclear cell, including lymphocytes, monocytes, and macrophages, obviously reminiscent of a DTH reaction.[49,51] Interestingly, chlamydiae were only detected in small numbers or not at all following reinoculations, despite the presence of marked pathological changes.[51,53] These data would suggest that simply the introduction of chlamydial antigen is sufficient to induce the disease state, and that persistent infection is not required. Moreover, Monnickendam et al.[52] demonstrated that significant pathological changes occurred even when low levels of infectious GPIC were used for inoculation.

In a recent study, Watkins and Caldwell[64] were able to elicit conjunctivitis in guinea pigs, previously infected with GPIC, with a Triton X-100 soluble extract of GPIC elementary bodies. The clinical disease was identical, if not more severe, than that induced with live organisms. The extract was found to contain lipopolysaccharide and some protein, but conspicuously absent was the MOMP. Surprisingly, the same reaction in the guinea pig could be induced with similar extracts of *C. trachomatis* (immunotypes B and H) and *C. psittaci* (strain Mn), suggesting that at least some of the determinants eliciting clinical disease via DTH are genus-specific. These results are most significant in that it may now be possible to identify specific epitopes responsible for the induction of ocular disease.

IV. INFECTIONS OF THE RESPIRATORY TRACT

In this section, the role of the immune response in chlamydial infections of the respiratory tract will be limited to those caused by *C. trachomatis*. In addition, while specific mechanisms relating to resolution of infection and resistance to reinfection will be covered, the nature of the experimental data precludes division into resolution and resistance and lends itself more to a discussion of the humoral versus cellular components of the immune mechanisms which participate in chlamydial respiratory infections.

Perhaps the most significant aspect of chlamydial respiratory infections caused by *C. trachomatis* with regard to immunity is that they almost invariably occur in immunocompromised patients, primarily the newborn infant, but also adults whose immune systems have been suppressed for one reason or another. This immediately implies that in immunologically intact individuals, the immune response in the respiratory tract is capable of either resolving an infection before clinical disease becomes apparent, keeping the disease process at a subclinical level or at worst allowing a low grade infection which remains undiagnosed.

Information on immune mechanisms in human chlamydial respiratory disease remains at a premium. It is interesting to note that in contrast to inclusion conjunctivitis in the newborn, which develops very soon after birth (5 to 12 days), the onset of pneumonitis ranges from 4 to 11 weeks after birth.[65] While there may be microbiological considerations for this delay, it is tempting to speculate that maternal IgG may be initially protective, but as it diminishes in titer, the child becomes susceptible. A significant drop in circulating IgG in the neonate does indeed occur in this 4 to 12 week period.[66] Once the infection becomes established, the infant elaborates high levels of IgM and IgG, but these antibodies by themselves may be insufficient to resolve the infection as some of the data described below will suggest. If IgA is required for resolution, the child is at a major disadvantage, since IgA levels are at less than 20% of adult levels for the first year of life.[66]

The study of immune mechanisms in the neonate has been difficult to approach experimentally. Animal models for neonatal pneumonia have been described in the baboon and guinea pig, but immune mechanisms have not to date been investigated in these systems.[67,68] The majority of information on the role of the immune response in chlamydial pneumonia has been derived from the study of the infection in adult mice. Obviously, care must be taken in extrapolating these data to neonatal animals; however, the information gained in the adult, which can successfully manage the infection, may provide an understanding as to why the neonate cannot handle the infection as readily. Furthermore, the immunology of the lung is extremely complex. Depending on the system studied, one may find important roles for IgM, IgG, sIgA, cytotoxic T cells, lymphokines, and macrophages. Moreover, the reactions and participating elements may even vary with the location within the lung.[69]

Two murine models for respiratory infections with human strains of *C. trachomatis* have been described and basic immunological responses reported. Kuo and Chen[70] infected mice intranasally with five different immunotypes (B, C, D, G, and L_2) and noted that the L_2 immunotype appeared to be the most virulent. When serum antibody and CMI as assessed by DTH were measured in mice infected with immunotype B, resolution of the infection was more closely associated with the appearance of the DTH reaction, although antibody appeared shortly thereafter. In a different study in which diabetic mice were inoculated with immunotype H, lymphocyte proliferative responses also appeared earlier than antibody.[71]

Perhaps the most detailed study of immune mechanisms in chlamydial respiratory infections has been reported by Williams and colleagues[72] using the agent of mouse pneumonitis (MoPn), a *C. trachomatis* biovar and natural parasite of the mouse. They originally found that when MoPn was inoculated intranasally into congenitally athymic nude mice (nu/nu), the mice routinely died from a fulminant pneumonia in 11 to 30 days; whereas heterozygote (+/nu) controls would generally recover form the infection, depending on the dose.[72] Moreover, when nu/nu mice were grafted with a neonatal thymus, they, too, recovered. Restoration of the thymus resulted in the production of serum IgG and IgA antibodies to MoPn, although they were delayed in appearance when compared to +/nu mice.[73] These data indicated very strongly that T cells play an essential role in the recovery of mice from chlamydial respiratory infection.

The participation of antibody was assessed by first passively administering immune serum intraveneously to nu/nu mice on multiple days after chlamydial challenge.[73] This treatment prolonged the survival of the nude mice when compared to normal serum-treated controls, but did not prevent death from occurring. When immune serum was given intranasally to +/n and nu/nu mice shortly before infection with a lethal dose of MoPn, 70% of the +/nu mice survived the infection compared to none of the mice given normal serum.[74] While all of the nude mice eventually died, their survival was markedly prolonged.

To ascertain that contact between antibody and organisms was achieved and to gain some insight into the mechanism by which antibody protects the host, they opsonized MoPn with various serum and lung lavage fluid preparations.[74] When nude mice were inoculated intranasally with organisms opsonized with immune serum, survival was again prolonged. Both IgG-rich and IgA-rich fractions protected the animals equally as well. Similarly, whole lavage fluid and a fraction containing predominantly IgA were found to be protective. Sera and lavage fluids were determined to be free of interferon activity, and lavage fluids were also free of other lymphokines having deterimental effects on chlamydiae so that participation of CMI in these experiments could be ruled out. Finally, prolongation of the infection could also be produced by opsonization of MoPn with murine antibodies to MOMP, suggesting that MOMP may possess the key determinants against which a protective immune response must be directed. The significant aspect of these experiments is not only that antibody is capable of providing a degree of protection, but that it is apparently insufficient to single-handedly resolve the infection. This would indicate that CMI may also be intimately involved in the resolution of a chlamydial respiratory infection.

To investigate the role of CMI in this model, Williams et al.[75] first adoptively transferred spleen cells from immune mice (+/nu) to nu/nu mice on the same day as intranasal inoculation with MoPn. All mice receiving immune cells recovered completely from the infection while none of the mice given normal spleen cells survived. As expected, antibody (IgG and IgA) also developed to high levels in the sera of recipients of immune cells. Serum antibodies appeared in normal cell recipients as well, but not as quickly, nor to the same level. When spleen cells were fractionated into T cell-enriched and B cell-enriched populations and injected into nude mice, recipients of T cells had a higher survival rate than recipients of B cells, while macrophage-enriched spleen cells had no protective effect whatsoever. However, when T cells were further separated into Lyt 1.2-positive (T helper, T_{DTH} cells) and Lyt 2.2-positive (cytotoxic T cells and T suppressor cells) fractions and injected separately into nu/nu recipients, neither was able to enhance survival, thus indicating a requirement for both populations, but still making it difficult to assign a defined role for CMI.

They also explored possible mechanisms by which CMI may be participating in the resolution of the infection.[75] An obvious effector mechanism of CMI in the lung would be the production of activated macrophages. Pathological examination of +/nu mice revealed a more marked mononuclear infiltrate and morphological evidence for activated macrophages when compared to nude mice which had mainly an acute inflammatory response. To determine whether this mechanism could potentially be operative, +/nu mice which had recovered from an MoPn infection were challenged intranasally with *Histoplasma capsulatum* and were found to have an enhanced survival rate when compared to control +/nu mice. In contrast, no protection was seen in MoPn-infected nu/nu mice. Interestingly, nude mice given a sublethal infection of *H. capsulatum*, were more resistant to MoPn challenge than previously uninfected nude mice, suggesting that nonspecific immune mechanisms, possibly mediated by activated macrophages, could be effective in combatting a chlamydial infection.

The above data would suggest that at least in +/nu mice, lymphokines are released from T cells which have some antichlamydial effect either alone or via macrophages. Indeed, lavage fluid from infected +/nu mice was found to have an inhibitory effect on *C. psittaci* in vitro.[76] The presence of lymphokines might also include gamma interferon; however, when lavage fluids were assayed for interferon, none could be detected.[74] Nevertheless, a potential contribution of gamma interferon cannot be ruled out, since it is possible that it may not persist for long periods of time in tissue or may be produced in low levels which are effective only within the immediate microenvironment in which it is produced.

A possible role for natural killer cells in the resolution of chlamydial respiratory infection has been suggested by the enhanced susceptibility of beige mice to intranasal inoculation of moderate to low doses of MoPn.[76] Beige mice are characterized by having a selective defect in natural killer (NK) cell function while having normal T cell- and macrophage-mediated cytotoxic mechanisms.[77] The restoration of functional NK cells to beige mice by the adoptive transfer of spleen cells from mice heterozygous for the beige trait was able to protect the mice against MoPn challenge. These data imply that NK cells may participate in resolution of chlamydial infection possibly by direct cytotoxicity of MoPn-infected cells by NK cells, or by an antibody-dependent cell-mediated cytotoxicity mechanism in which NK cells are known to participate.[77]

V. INFECTIONS OF THE GENITAL TRACT

Chlamydial genital infections have become recognized as a major health problem, and there has accumulated a vast literature on the incidence and diagnosis of the disease. That antibody and CMI responses develop in infected individuals is also well documented. However, information on the role of the immune response in resolution of and resistance to chlamydial genital infections is somewhat lacking. The data are logistically and technically difficult to obtain in humans, in whom there is still not complete certainty that resolution without antibiotic intervention occurs or that immunity to reinfection develops. As in other disease entities associated with *Chlamydia*, animal models of genital tract infections have shed the most light on the immune mechanisms involved. It should also be noted that the bulk of research of immune mechanisms in chlamydial genital infections has been pursued in the female, and only minimal information is available in the male.

A. Resolution of Infection

A major question yet unanswered with regard to human genital infections with chlamydiae is how long does the infection last and is the immune system capable of resolving the infection? It is generally accepted that the infections may persist for long periods of time and have been documented to last for at least a year in men and women.[34] However, in many cases, clinical disease may be absent even though the organism is present; so it might be suggested that the infection, i.e., the clinically apparent infection, has resolved. It is indeed striking that in all of the animal models studied in which chlamydiae were inoculated either intraurethrally or intravaginally, including male,[78,79] and female guinea pigs,[80] female mice,[81,82] female marmosets,[83] and male and female cats,[84] the infection course was finite. It would be quite surprising if the same did not also occur in humans.

Both antibody and CMI responses have been measured in patients exposed to *Chlamydia* via the genital route; nevertheless, for obvious reasons, it has been impossible to determine the role of either in resolution of a genital infection. Using female guinea pigs infected with GPIC, we studied the role of the humoral immune response in resolution of the infection by selective immunosuppression of animals with cyclophosphamide.[85] While daily infections of 25 mg/kg suppressed both humoral immunity and CMI, a regimen of 150 mg/kg given at 9 day intervals, beginning 1 day after infection, was found to suppress the IgG antibody response in serum and IgA response in genital secretions, but to leave CMI intact.[85,86] As a result of daily injections, the infection was prolonged and increased in intensity until all the animals had died by day 36. Interestingly, when humoral immunity alone was absent even in the presence of functional CMI, the infection did not resolve but persisted until all animals finally died, possibly as a result of toxicity related to the infection. Animals receiving only cyclophosphamide at 9 day intervals showed no significant physical effects. When a lower dose of cyclophosphamide was administered at 9 day intervals, the antibody response was delayed but did eventually reach normal titers. Resolution of the infection was also delayed but occurred as antibody levels increased. Similar findings were obtained when humoral immunity alone was suppressed in male guinea pigs inoculated intraurethrally with GPIC.[79] These data indicate very strongly that the humoral immune response is essential for the resolution of a chlamydial genital infection in male and female animals.

The importance of the local antibody response in the genital tract was suggested by studies on the effect of estradiol on GPIC infection in female guinea pigs. Hormone treatment extended the course of infection by about 10 days,[87] but when serum antibody was measured in these animals, no difference between hormone-treated and untreated animals was noted. In contrast, the development of peak levels of IgG and IgA antibodies in secretions of hormone-treated guinea pigs was delayed by about 10 days when compared to control animals.[88,89] Thus, the resolution of the infection was associated with the production of antibodies in secretions but not with antibodies in serum. Nevertheless, this does not rule out the possibilty that serum antibodies may appear in secretions via transudation and participate in the local immune response. It is indeed interesting to note that in humans, an inverse correlation between the level of sIgA measured in cervical secretions and the recovery of organisms from the cervix was found, suggesting a possible role for sIgA in the control of the infection.[90]

In female mice infected intravaginally with MoPn, the development of high titers of serum and secretion antibody also occurred concurrently with the resolution of the infection while DTH developed somewhat later.[91] Of significance was the observation that nude mice, when infected via the genital route, were totally unable to clear their infections.[92] These mice produced only minimal titers of antibody and lacked CMI. However, when reconstituted with a neonatal thymus or with spleen cells from immune or normal +/nu mice, they were able to recover from the infection normally, although the recovery rate in immune cell recipients was accelerated.[93] Even though these data do not implicate specifically humoral or CMI immune responses in the resolution of the infection, they do indicate an important role for the T cell and provide a convenient model to assess the role of various cell populations. In contrast to these data, Tuffrey et al.[94] found that nu/nu and +/nu mice were able to recover from genital infection with LGV in a comparable fashion, and that transfer of immune serum or immune spleen cells had no influence on the course of the infection.[95] However, the relevance of this model is difficult to evaluate since LGV is not a natural parasite of the mouse, and progesterone treatment of the mice was required in order for the infection to be established at all.[96]

While intact CMI in the absence of humoral immunity in the guinea pig is insufficient to effect recovery, it may still have a vital function in the resolution of the infection. Indeed, when guinea pigs were deprived of CMI with antithymocyte serum, the infection was reduced to low levels but was not completely cleared until the treatment was discontinued.[97] The effect was apparently due to the CMI depression, since normal levels of antibodies developed in serum and genital secretions. Thus, it would appear that recovery from chlamydial genital infection requires the participation of both the humoral immune response and CMI.

B. Resistance to Reinfection

Just as with the study of a primary genital infection in humans, it has been difficult to formulate definitive statements regarding the development and nature of protective immunity. Epidemiological studies seem to favor the concept that some degree of immunity to reinfection does occur. Alani et al.[98] observed that chlamydiae were more frequently isolated from men experiencing their first bout of nongonococcal urethritis (NGU) than from men with a history of NGU (56 vs. 12%). Jones and Batteiger[99] also found that the isolation rate for chlamydiae was lower in men and women with a previous positive culture or history of a sexually transmitted disease. Similarly, Schachter et al.[100] reported that men with gonorrhea had a lower rate of recovery of chlamydiae when compared to NGU patients, despite evidence for recent exposure. These individuals also had a higher prevalence of antibody to *Chlamydia* when compared to men with NGU or no urethritis at all. These data suggested that at least in males, previous contact with *Chlamydia* resulted in the apparent development of immunity.

While there is evidence for the development of protective immunity, it should be emphasized that individuals may become reinfected. Reinfection may possibly occur from contact with a different immunotype, ineffective antibiotic treatment, persistent latent infection, or as a result of a short-lived protective immune response, as in trachoma. The latter phenomena has been reported to occur in animal models of genital infection. Johnson et al.[101] found that when marmosets, after recovering from a primary infection with *C. trachomatis*, were reinoculated, they did become reinfected but generally for a shorter period of time. However, immunity seemed to be stronger when there was less time between the primary and challenge infections, although the inoculating dose may also have been a factor. In the guinea pig, we have noted that animals were solidly immune to challenge infection shortly after recovery,[86] but when they were reinoculated considerably later, i.e., 412, and 825 days after the primary infection they also developed a second, although abbreviated, infection.[102] Similarly, mice recovered from a genital infection with MoPn were immune to challenge at 50 and 100 days, but became reinfected at 150 days after infection.[103]

When one assesses the guinea pig serum and secretory antibody response to chlamydial membrane antigens by immunoblot analysis, one finds several interesting features.[104] In serum, the IgG response to MOMP appeared about day 12 after infection and peaked at days 20 to 30, concurrent with the resolution of infection. However, the level of anti-MOMP antibodies declined between days 50 to 90 and remained relatively unchanged at low levels for as long as 825 days. In contrast, antibody to the 61K component developed at the same time as anti-MOMP but continued to increase while anti-MOMP was decreasing, finally reaching maximum levels 150 to 300 days after infection. The immune responses also recognized a variety of other membrane components and various patterns of waxing and waning of the antibodies to these components were noted during the observation period. In secretions, the IgG antibody response reflected what was occurring in serum, although IgA antibodies to MOMP and 61K appeared rather abruptly about day 20 and were already minimal by day 30. From these data, it is apparent that even though the antibody response in serum remains elevated for long periods of time when measured by enzyme-linked immunoadsorbent assay (ELISA) or fluorescent antibody assays, the response to certain chlamydial components, particularly MOMP is short-lived. Even more dramatic is the extremely short life of specific IgA in genital secretions. One could speculate that if MOMP harbors the protective antigen, as suggested by other studies,[8,9,74] then when the antibody response to MOMP wanes, the animal again becomes susceptible to reinfection. The second infection may, however, be cut short by the appearance of a strong anamnestic response, hence the lower isolation rate in patients with a history of prior chlamydial infection. While immunoblot analyses on sequential serum specimens from humans are not available, it has been observed that most sera from patients who were at one time or another isolation positive, also reacted with the MOMP and 60K membrane components of *C. trachomatis*,[105] indicating that the human and guinea pig are responding to the same outer membrane components as a result of genital infection. Moreover, it is interesting to note that antibodies in serum and genital secretions from guinea pigs infected genitally with GPIC also cross-react with the MOMP and 60K components of *C. trachomatis* (immunotypes E and L$_2$).[106]

Evidence that immunity to chlamydial infection in humans is also abbreviated has recently been reported by Jones and Batteiger.[99] When men and women who were exposed to chlamydial infection were cultured, they found that the patients were less likely to be isolation positive if they had had a previous documented infection less than 6 months before. However, if the previous documented infection was more than 6 months earlier, they were more likely to be isolation positive. Therefore, it appears that as in trachoma, immunity to genital tract infection is a relatively short-lived phenomenom, so that individuals may suffer repeated infections. As in trachoma, the reasons for the brevity of this protective period are undefined, but are of critical importance when considering the development of a vaccine.

The exact roles of the various immune components in resistance to infection have only been explored on a limited basis. Data in the GPIC:guinea pig model indicate that the humoral immune response is required for resistance to reinfection, even in the presence of functional CMI.[86] Guinea pigs deprived of their humoral immune response by cyclophosphamide treatment were infected, but then treated with tetracycline to resolve the infection. When resolution was confirmed and CMI was found to be functional, the animals were challenged. All immunosuppressed guinea pigs became reinfected while unmanipulated, but tetracylcine-treated controls, were solidly immune. These data do not rule out the possibility that a cooperative effort of both humoral immunity and CMI is required.

One nagging problem associated with chlamydial genital infections is the apparent ability of chlamydiae to persist in the presence of an active immune response, often without producing apparent clinical disease. The organism would seem to have evolved a mechanism which allows it to avoid the immune response, just as has occurred in a variety of other host-parasite relationships, such as herpes simplex, malaria, and tuberculosis, among others. "Cryptic" forms of chlamydiae have been reported to occur in vitro,[107] and it is certainly conceivable that they may also occur in vivo, and in so doing, persist untouched by the immune system. In one possible scenario, some organisms could remain intracellular in a "cryptic" form, hidden from antibody; but when the protective immune mechanisms wane, they revert to infectious forms with a resulting clinical response. An anamnestic response would preempt a longer more severe infection, causing the organism to again revert to its cryptic form. One mechanism for the maintenance of chlamydiae in a quiescent form might employ gamma interferon. Gamma interferon has been shown to have bacteriostatic activity on *C. psittaci* in vitro, and replication of the organisms in the presence of gamma interferon was demonstrated, although at a markedly reduced rate.[16,17,19] One might hypothesize that T cells elaborate gamma interferon locally, driving the organisms into a slowly replicating intracellular form, safe from the effects of antibody. At such time when few extracellular chlamydiae are available to act as an immunogenic stimulus, the T cell response would dissipate. Without the presence of gamma interferon, the organisms could again revert to normal replication until the immunogenic load is once again sufficient to activate T cells. Thus, one could conceive of the immune response as playing a role in the continuance of the infection.

REFERENCES

1. **Howard, L. V.**, Neutralization of *Chlamydia trachomatis* in cell culture, *Infect. Immun.*, 11, 698, 1975.
2. **Reeve, P. and Graham, D. M.**, A neutralization test for trachoma and inclusion blennorrhea viruses grown in HeLa cells, *J. Gen. Microbiol.*, 27, 177, 1962.
3. **Graham, D. M. and Layton, J. E.**, The induction of Chlamydia group antibody in rabbits inoculated with trachoma agents and demonstration of strain-specific neutralizing antibody in sera, in *Trachoma and Related Disorders Caused by Chlamydial Agents*, Nichols, R. L., Ed., Excepta Medica, Amsterdam 1971.
4. **Zakay-Rones, Z. and Becker, Y.**, Antibodies to trachoma elementary bodies, in *Trachoma and Related Disorders Caused by Chlamydial Agents*, Nichols, R. L., Ed., Excerpta Medica, Amsterdam 1971.
5. **Byrne, G. I. and Moulder, J. W.**, Parasite-specified phagocytosis of *Chlamydia psittaci* and *Chlamydia trachomatis* by L and HeLa cells, *Infect. Immun.*, 19, 598, 1978.
6. **Ainsworth, S., Allan, I., and Pearce, J. H.**, Differential neutralization of spontaneous and centrifuge-assisted chlamydial infectivity, *J. Gen. Microbiol.*, 114, 61, 1979.

7. **Allen, I., Spragg, S. P., and Pearce, J. H.,** Pressure and directional force components in centrifuge-assisted chlamydial infection of cell cultures, *FEMS Microbiol. Lett.*, 2, 79, 1977.

8. **Caldwell, H. D. and Perry, L. J.,** Neutralization of *Chlamydia trachomatis* infectivity with antibodies to the major outer membrane protein, *Infect. Immun.*, 38, 745, 1982.

9. **Peeling, R., Maclean, I. W., and Brunham, R. C.,** In vitro neutralization of *Chlamydia trachomatis* with monoclonal antibody to an epitope on the major outer membrane protein, *Infect. Immun.*, 46, 484, 1984.

10. **Lucero, M. E. and Kuo, C-C.,** Neutralization of *Chlamydia trachomatis* cell culture infection by serovar-specific monoclonal antibodies, *Infect. Immun.*, 50, 595, 1985.

11. **Wyrick, P. B. and Brownridge, E. A.,** Growth of *Chlamydia psittaci* in macrophages, *Infect. Immun.*, 19, 1054, 1978.

12. **Wyrick, P. B., Brownridge, E. A., and Ivins, B. E.,** Interaction of *Chlamydia psittaci* with mouse peritoneal macrophages, *Infect. Immun.*, 19, 1061, 1978.

13. **Lammert, J. K.,** Cytotoxic cells induced after *Chlamydia psittaci* infection in mice, *Infect. Immun.*, 35, 1011, 1982.

14. **Pavia, C. S. and Schachter, J.,** Failure to detect cell-mediated cytotoxicity against *Chlamydia trachomatis* infected cells, *Infect. Immun.*, 39, 1271, 1983.

15. **Qvigstad, E. and Hirschberg, H.,** Lack of cell-mediated cytotoxicity towards *Chlamydia trachomatis* infected target cells in humans, *Acta. Pathol. Microbiol. Immunol.*, 92, 153, 1984.

16. **Byrne, G. I. and Faubion, C. L.,** Lymphokine-mediated microbistatic mechanisms restrict *Chlamydia psittaci* growth in macrophages, *J. Immunol.*, 128, 469, 1982.

17. **Byrne, G. I. and Krueger, D. A.,** Lymphokine-mediated inhibition of Chlamydia replication in mouse fibroblasts is neutralized by anti-gamma interferon immunoglobulin, *Infect. Immun.*, 42, 1152, 1983.

18. **Nathan, C. F., Murray, H. W., Wiebe, M. E., and Rubin, B. Y.,** Identification of interferon-gamma as the lymphokine that activates human macrophage oxidative metabolism and antimicrobial activity, *J. Exp. Med.*, 158, 670, 1983.

19. **Rothermel, C. D., Rubin, B. Y., and Murray, H. W.,** Gamma interferon is the factor in lymphokine that activates human macrophages to inhibit intracellular *Chlamydia psittaci* replication, *J. Immunol.*, 131, 2542, 1983.

20. **Rothermel, C. D., Byrne, G. I., and Havell, E. A.,** Effect of interferon on the growth of *Chlamydia trachomatis* in mouse fibroblasts (L cells), *Infect. Immun.*, 39, 362, 1983.

21. **de la Maza, L. M., Peterson, E. M., Fennie, C. W., and Czarniecki, C. W.,** The anti-chlamydial and anti-proliferative activities of recombinant murine interferon-gamma are not dependent on tryptophane concentrations, *J. Immunol.*, 135, 4198, 1985.

22. **Byrne, G. I., Lehmann, L. K., and Landry, G. J.,** Induction of tryptophan catabolism is the mechanism for gamma interferon-mediated inhibition of intracellular *Chlamydia psittaci* replication in T24 cells, *Infect. Immun.*, 53, 347, 1986.

23. **Monnickendam, M. A. and Pearce, J. H.,** Immune responses and chlamydial infections, *Br. Med. Bull.*, 39, 187, 1983.

24. **Grayston, J. T., Wang, S-P., Yeh, L. J., and Kuo, C-C.,** Importance of reinfection in the pathogenesis of trachoma, *Rev. Infect. Dis.*, 7, 717, 1985.

25. **Taylor, H. R.,** Ocular models of chlamydial infection, *Rev. Infect. Dis.*, 7, 737, 1985.

26. **Watson, R. R., MacDonald, A. B., Murray, E. S., and Moddaber, F. Z.,** Immunity to chlamydial infections of the eye. III. Presence and duration of delayed hypersensitivity to guinea pig inclusion conjunctivitis, *J. Immunol.*, 111, 618, 1973.

27. **Malaty, R., Dawson, C. R., Wong, I., Lyon, C., and Schachter, J.,** Serum and tear antibodies to Chlamydia after reinfection with guinea pig inclusion conjunctivitis agent, *Invest. Ophthalmol. Vis. Sci.*, 21, 833, 1981.

28. **Sacks, D. L., Todd, W. J., and MacDonald, A. B.,** Cell-mediated immune responses in owl monkeys (*Aotus trivirgatus*) with trachoma to soluble antigens of *Chlamydia trachomatis*, *Clin. Exp. Immunol.*, 33, 57, 1978.

29. **Modabber, F., Bear, S. E., and Cerny, J.,** The effect of cyclophosphamide on the recovery from a local chlamydial infection — guinea pig inclusion conjunctivitis, *Immunology*, 30, 929, 1976.

30. **Mull, J. D. and Thompson, S. E., III,** The effect of antilymphocyte globulin on guinea pig resistance to ocular chlamydial infection, *Abstr. Annu. Meet. Am. Soc. Microbiol.*, May 7, 1973, Miami Beach, Fla., 200.

31. **Nichols, R. L., Oertley, R. E., Fraser, C. E. O., MacDonald, A. B., and McComb, D. E.,** Immunity to chlamydial infections of the eye. VI. Homologous neutralization of trachoma infectivity for the owl monkey conjunctivae by eye secretions from humans with trachoma, *J. Infect. Dis.*, 127, 429, 1973.

32. **Barenfanger, J. and MacDonald, A. B.,** The role of immunoglublin in the neutralization of trachoma infectivity, *J. Immunol.*, 113, 1607, 1974.

33. **Pearce, J. H., Allan, I., and Ainsworth, S.,** Interaction of chlamydiae with host cells and mucous surfaces, in *Adhesion and Microorganism Pathogenicity*, Ciba Foundation Symposium, Ed., Pittman Wells, Turnbridge Wells, England, 1981.

34. **Schachter, J. and Dawson, C. R.,** *Human Chlamydial Infections*, John Wright-PSG, Littleton, Mass., 1978.

35. **Murray, E. S. and Charbonnet, L. T.,** Experimental conjunctival infection of guinea pigs with the guinea pig inclusion conjunctivitis organism, in *Trachoma and Related Disorders Caused by Chlamydial Agents*, Nichols, R. L., Ed., Excerpta Medica, Amsterdam, 1971, 369.

36. **Murray, E. S. and Radcliffe, F. T.,** Immunologic studies in guinea pigs with guinea pig inclusion conjunctivitis (Gp-ic) Bedsonia, *Am. J. Ophthalmol.*, 63, 1263, 1967.

37. **Wang, S-P. and Grayston, J. P.,** Local and systemic antibody response to trachoma eye infection in monkeys, in *Trachoma and Related Disorders Caused by Chlamydial Agents*, Nichols, R. L., Ed., Excerpta Medica, Amsterdam, 1971.

38. **Fraser, C. E. O., McComb, D. E., Murray, E. S., and MacDonald, A. B.,** Immunity to chlamydial infections of the eye. IV. Immunity in owl monkeys to reinfection with trachoma, *Arch. Ophthalmol.*, 93, 518, 1975.

39. **MacDonald, A. B., McComb, D., and Howard, L.,** Immune response of owl monkeys to topical vaccination with irradiated *Chlamydia trachomatis*, *J. Infect. Dis.*, 149, 439, 1984.

40. **Murray, E. S., Charbonnet, L. T., and MacDonald, A. B.,** Immunity to chlamydial infections of the eye. I. The role of circulatory and secretory antibodies in resistance to reinfection with guinea pig inclusion conjunctivitis, *J. Immunol.*, 110, 1518, 1973.

41. **Watson, R. R., Mull, J. D., MacDonald, A. B., Thompson, S. E., III, and Bear, S. E.,** Immunity to chlamydial infections of the eye. II. Studies of passively transferred serum antibody in resistance to infection with guinea pig inclusion conjunctivitis, *Infect. Immun.*, 7, 597, 1973.

42. **Orenstein, N. S., Mull, J. D., and Thompson, S. E., III,** Immunity to chlamydial infections of the eye. V. Passive transfer of antitrachoma antibodies to owl monkeys, *Infect. Immun.*, 7, 600, 1973.

43. **Schachter, J., Grossman, M., Holt, J., Sweet, R., Goodner, E. and Mills, J.,** Prospective study of chlamydial infection in neonates, *Lancet*, 2(8139), 377, 1979.

44. **Nichols, R. L., Murray, E. S., and Nisson, P. E.,** Use of enteric vaccines in protection against chlamydial infections of the genital tract and the eye of guinea pigs, *J. Infect. Dis.*, 138, 742, 1978.

45. **Rank, R. G. and Barron, A. L.,** Specific effect of estradiol on the genital mucosal antibody response in chlamydial ocular and genital infections, *Infect. Immun.*, 55, 2317, 1987.

46. **Montgomery, P. C., Connelly, K. M., Cohn, J., and Skandera, C. A.,** Remote-site stimulation of secretory IgA antibodies following bronchial and gastric stimulation, *Adv. Exp. Med. Biol.*, 107, 113, 1978.

47. **McDermott, M. R. and Bienenstock, J.,** Evidence for a common mucosal immunologic system. I. Migration of B immunoblasts into intestinal, respiratory, and genital tissues, *J. Immunol.*, 122, 1892, 1979.

48. **Weisz-Carrington, P., Roux, M. E., McWilliams, M., Phillips-Quagliata, J. M., and Lamm, M. E.,** Organ and isotype distribution of plasma cells producing specific antibody after oral immunization: evidence for a generalized secretory immune system, *J. Immunol.*, 123, 1705, 1979.

49. **Taylor, H. R., Prendergast, R. A., Dawson, C. R., Schachter, J., and Silverstein, A. M.,** An animal model for cicatrizing trachoma, *Invest. Ophthalmol. Vis. Sci.*, 21, 422, 1981.

50. **Ahmad, A., Dawson, C. R., Yoneda, C., Togni, B., and Schachter, J.,** Resistance to reinfection with a chlamydial agent (guinea pig inclusion conjunctivitis), *Invest. Ophthalmol. Vis. Sci.*, 16, 549, 1977.

51. **Monnickendam, M. A., Darougar, S., Treharne, J. D., and Tilbury, A. M.,** Development of chronic conjunctivitis with scarring and pannus, resembling trachoma, in guinea pigs, *Br. J. Ophthalmol.*, 64, 284, 1980.

52. **Monnickendam, M. A., Darougar, S., and Tilbury, A. M.,** Ocular and dermal delayed hypersensitivity reactions in guinea pigs following infection with guinea pig inclusion conjunctivitis agent *(Chlamydia psittaci)*, *Clin. Exp. Immunol.*, 44, 57, 1981.

53. **Taylor, H. R., Johnson, S. L., Prendergast, R. A., Schachter, J., Dawson, C. R., and Silverstein, A. M.,** An animal model of trachoma. II. The importance of repeated reinfection, *Invest. Ophthalmol. Vis. Sci.*, 23, 507, 1982.

54. **Caldwell, H. D., Stewart, S., Taylor, H. R., and Johnson, S.,** Immunoblotting analysis of tears and sera of cynomolgus monkey with experimental chlamydial conjunctivitis, in *Chlamydial Infections*, Oriel, D., Ridgway, G., Schachter, J., Taylor-Robinson, D., and Ward, M., Eds., Cambridge University Press, London, 1986, 162.

55. **Tomasi, T. B.,** The secretory immune system, in *Basic and Clinical Immunology*, 5th Ed., Stites, D. P., Stobo, J. D., Fudenberg, H. H., and Wells, J. V., Eds., Lange Medical, Los Altos, Calif., 1984, 187.

56. **Young, E. and Taylor, H. R.,** Immune mechanisms in chlamydial eye infection: cellular immune responses in chronic and acute disease, *J. Infect. Dis.*, 150, 745, 1984.

57. **Young, E. and Taylor, H. R.,** Immune mechanisms in chlamydial eye infection. Development of T suppressor cells, *Invest. Ophthalmol. Vis. Sci.*, 27, 615, 1986.

58. **Horohov, D. W., Moore, R. N., and Rouse, B. T.,** Herpes simplex virus-specific lymphoproliferation: an analysis of the involvement of lymphocyte subsets, *Immunobiology,* 170, 460, 1985.

59. **Jones, B. R. and Collier, L. H.,** Inoculation of man with inclusion blennorrhea virus, *Ann. NY Acad. Sci.,* 98, 212, 1962.

60. **Wang, S-P. and Grayston, J. T.,** Pannus with experimental trachoma and inclusion conjunctivitis agent infection of Taiwan monkeys, *Am. J. Ophthalmol.,* 63, 1133, 1967.

61. **Monnickendam, M. A., Darougar, S., Treharne, J. D., and Tilbury, A. M.,** Guinea pig inclusion conjunctivitis as a model for the study of trachoma: clinical, microbiological, serological, and cytological studies of primary infection, *Br. J. Ophthalmol.,* 64, 279, 1980.

62. **Collier, L. H. and Blyth, W. A.,** Immunogenicity of experimental trachoma vaccines in baboons. II. Experiments with adjuvants, and tests of cross-protection, *J. Hyg.,* 64, 529, 1966.

63. **Grayston, J. T., Kim, K. S. W., Alexander, E. R., and Wang, S-P.,** Protective studies in monkeys with trivalent and monovalent trachoma vaccines, in *Trachoma and Related Disorders Caused by Chlamydial Agents,* Nichols, R. L., Ed., Excerpta Medica, Amsterdam 1971, 377.

64. **Watkins, N. G. and Caldwell, H. D.,** Delayed hypersensitivity as a pathogenic mechanism in chlamydial disease, in *Chlamydial Infections,* Oriel, J., Ridgway, G., Schachter, J., Taylor-Robinson, D., and Ward, M., Ed., Cambridge University Press, London, 1986, 408.

65. **Tipple, M. A., Beem, M. O., and Saxon, E. M.,** Clinical characteristics of the afebrile pneumonia associated with *Chlamydia tranchomatis* infection in infants less than 6 months of age, *Pediatrics,* 63, 192, 1979.

66. **Stites, D. P.,** Clinical laboratory methods for detection of antigens and antibodies, in *Basic and Clinical Immunology,* Stites, D. P., Stobo, J. D., Fudenberg, H. H., and Wells, J. V., Eds., Lange Medical, Los Altos, Calif., 1984.

67. **Harrison, H. R., Alexander, E. R., Chiang, W. T., Giddens, W. E., Boyce, J. T., Benjamin, D., and Gale, J. L.,** Experimental nasopharyngitis and pneumonia caused by *Chlamydia trachomatis* in infant baboons: histopathologic comparison with a case in a human infant, *J. Infect. Dis.,* 139, 141, 1979.

68. **Rank, R. G., Hough, A. J., Jr., Jacobs, R. F., Cohen, C., and Barron, A. L.,** Chlamydial pneumonitis induced in newborn guinea pigs, *Infect. Immun.,* 48, 153, 1985.

69. **Johnson, K. J., Chapman, W. E., and Ward, P. A.,** Immunopathology of the lung, *Am. J. Pathol.,* 95, 795, 1979.

70. **Kuo, C-C. and Chen, W. J.,** A mouse model of *Chlamydia trachomatis* pneumonitis, *J. Infect. Dis.,* 141, 198, 1980.

71. **Harrison, H. R., Lee, S. M., and Lucas, D. O.,** *Chlamydia trachomatis* pneumonitis in the C57BL/KsJ mouse: pathologic and immunologic features, *J. Lab. Clin. Med.,* 100, 953, 1982.

72. **Williams, D. M., Schachter, J., Drutz, D. J., and Sumaya, C. V.,** Pneumonia due to *Chlamydia trachomatis* in the immunocompromised (nude) mouse, *J. Infect. Dis.,* 143, 238, 1981.

73. **Williams, D. M., Schachter, J., Grubbs, B., and Sumaya, C. V.,** The role of antibody in host defense against the agent of mouse pneumonia, *J. Infect. Dis.,* 145, 200, 1982.

74. **Williams, D. M., Schachter, J., Weiner, M. H., and Grubbs, B.,** Antibody in host defense against mouse pneumonitis agent (murine *Chlamydia trachomatis*), *Infect. Immun.,* 45, 674, 1984.

75. **Williams, D. M., Schachter, J., Coalson, J. J., and Grubbs, B.,** Cellular immunity to the mouse pneumonitis agent, *J. Infect. Dis.,* 149, 630, 1984.

76. **Williams, D. M. and Schachter, J.,** Role of cell-mediated immunity in chlamydial infection: implications for ocular immunity, *Rev. Infect. Dis.,* 7, 754, 1985.

77. **Roder, J. C., Lohmann-Matthes, M. L., Domzig, W., and Wigzell, H.,** The beige mutation in the mouse. II. Selectivity of the natural killer (NK) cell defect, *J. Immunol.,* 123, 2174, 1979.

78. **Mount, D. T., Bigazzi, P. E., and Barron, A. L.,** Experimental genital infection of male guinea pigs with the agent of guinea pig inclusion conjunctivitis and transmission to females, *Infect. Immun.,* 8, 925, 1973.

79. **Rank, R. G., White, H. J., Soloff, B. L., and Barron, A. L.,** Cystitis associated with chlamydial infection of the genital tract in male guinea pigs, *Sex. Transm. Dis.,* 8, 203, 1981.

80. **Mount, D. T., Bigazzi, P. E., and Barron, A. L.,** Infection of genital tract and transmission of ocular infection to newborns by the agent of guinea pig inclusion conjunctivitis, *Infect. Immun.,* 5, 921, 1972.

81. **Barron, A. L., White, H. J., Rank, R. G., Soloff, B. L., and Moses, E. G.,** A new animal model for the study of *Chlamydia trachomatis* gential infections; infection of mice with the agent of mouse pneumonitis, *J. Infect. Dis.,* 143, 63, 1981.

82. **Ito, J. I., Harrison, H. R., Alexander, R. E., and Alexander, R. E.,** Establishment of genital tract infection in the CF-1 mouse by intravaginal inoculation of a human oculogenital isolate of *Chlamydia trachomatis, J. Infect. Dis.,* 150, 577, 1984.

83. **Johnson, A. P., Hetherington, C. M., Osborn, M. F., Thomas, B. J., and Taylor-Robinson, D.,** Experimental infection of the marmoset genital tract with *Chlamydia trachomatis, Br. J. Exp. Pathol.,* 61, 291, 1980.

84. **Darougar, S., Monnickendam, M. A., El-Sheikh, H., Treharne, J. D., Woodland, R. M., and Jones, B. R.**, Animal models for the study of chlamydial infections of the eye and genital tract, in *Nongonococcal Urethritis and Related Infections*, Hobson, D. and Holmes, K. K., Eds., American Society for Microbiology, Washington, D.C., 1977.

85. **Rank, R. G., White, H. J., and Barron, A. L.**, Humoral immunity in the resolution of genital infection in female guinea pigs infected with the agent of guinea pig inclusion conjunctivitis, *Infect. Immun.*, 26, 573, 1979.

86. **Rank, R. G. and Barron, A. L.**, Humoral immune response in acquired immunity to chlamydial genital infection of female guinea pigs, *Infect. Immun.*, 39, 463, 1983.

87. **Rank, R. G., White, H. J., Hough, A. J., Pasley, J. N., and Barron, A. L.**, Effect of estradiol on chlamydial genital infection of female guinea pigs, *Infect. Immun.*, 38, 699, 1982.

88. **Rank, R. G. and Barron, A. L.**, Prolonged genital infection by GPIC agent associated with immunosuppression following treatment with estradiol, in *Chlamydial Infections*, Mardh, P-A., Holmes, K. K., Oriel, J. D., Piot, P., and Schachter, J., Eds., Elsevier, New York, 1982.

89. **Pasley, J. N., Rank, R. G., Hough, A. J., Jr., Cohen, C., and Barron, A. L.**, Effects of various doses of estradiol on chlamydial genital infection in ovariectomized guinea pigs, *Sex. Trans. Dis.*, 12, 8, 1985.

90. **Brunham, R. C., Kuo, C-C., Cles, L., and Holmes, K. K.**, Correlation of host immune response with quantitative recovery of *Chlamydia trachomatis* from the human endocerivs, *Infect. Immun.*, 39, 1491, 1983.

91. **Barron, A. L., Rank, R. G., and Moses, E. B.**, Immune response in mice infected in the genital tract with mouse pneumonitis agent (*Chlamydia trachomatis* biovar), *Infect. Immun.*, 44, 82, 1984.

92. **Rank, R. G., Soderberg, L. S. F., and Barron, A. L.**, Chronic chlamydial genital infection in congenitally athymic nude mice, *Infect. Immun.*, 48, 847, 1985.

93. **Rank, R. G. and Soderberg, L. S. F.**, unpublished data, 1986.

94. **Tuffrey, M., Falder, P., and Taylor-Robinson, D.**, Genital tract infection and disease in nude and immunologically competent mice after inoculation of a human strain of *Chlamydia trachomatis*, *Br. J. Exp. Pathol.*, 63, 539, 1982.

95. **Tuffrey, M., Falder, P., and Taylor-Robinson, D.**, Effect on *Chlamydia trachomatis* infection of the murine genital tract of adoptive transfer of congenic immune cells or specific antibody, *Br. J. Exp. Pathol.*, 66, 427, 1985.

96. **Tuffrey, M. and Taylor-Robinson, D.**, Progesterone as a key factor in the development of a mouse model for genital-tract infection with *Chlamydia trachomatis*, *FEMS Microbiol. Lett.*, 12, 111, 1981.

97. **Rank, R. G. and Barron, A. L.**, Effect of antithymocyte serum on the course of chlamydial genital infection in female guinea pigs, *Infect. Immun.*, 41, 876, 1983.

98. **Alani, M. D., Darougar, S., Burns, D. C., Thin, R. N., and Dunn., H.**, Isolation of *Chlamydia trachomatis* from the male urethra, *Br. J. Vener. Dis.*, 53, 88, 1977.

99. **Jones, R. B. and Batteiger, B. E.**, Human immune response to *Chlamydia trachomatis* infections, in *Chlamydial Infections*, Oriel, J., Ridgway, G., Schachter, J., Taylor-Robinson, D., and Ward, M., Eds., 1986, 423.

100. **Schachter, J., Cles, L. D., Ray, R. M., and Hesse, F. E.**, Is there immunity to chlamydial infections of the human genital tract?, *Sex. Trans. Dis.*, 10, 123, 1983.

101. **Johnson, A. P., Osborn, M. F., Thomas, B. J., Hetherington, C. M., and Taylor-Robinson, D.**, Immunity to reinfection of the genital tract of marmosets with *Chlamydia trachomatis*, *Br. J. Exp. Pathol.*, 62, 606, 1981.

102. **Rank, R. G. and Barron, A. L.**, unpublished data, 1983.

103. **Ramsey, K. H. and Rank, R. G.**, unpublished data, 1986.

104. **Batteiger, B. E. and Rank, R. G.**, Analysis of the humoral immune response in chlamydial genital infection in guinea pigs, *Infect. Immun.*, 55, 1767, 1987.

105. **Batteiger, B. E.**, personal communication, 1987.

106. **Batteiger, B. E. and Rank, R. G.**, Antigenic specificity of the humoral immune response to chlamydial genital infection in guinea pigs, in *Chlamydial Infections*, Oriel, D., Ridgway, G., Schachter, J., Taylor-Robinson, D., and Ward, M., Eds., Cambridge University Press, London, 1986, 453.

107. **Moulder, J. W., Levy, N. J., and Schulman, L. P.**, Persistent infection of mouse fibroblasts (L cells) with *Chlamydia psittaci*: evidence for a cryptic chlamydial form, *Infect. Immun.*, 30, 874, 1980.

Index

INDEX

Lymphokines, 203, 204, 219, 226, see also specific
types
Lymphotoxins, 213, see also specific types
Lysine synthesis, 100
Lysosomes, 48, see also specific types
fusion of, 138, 146

M

Macchiavello's stain, 22
Macromolecular composition, 47—66, see also
specific types
lipids and, 55—58
lipopolysaccharides and, 54—55
nucleic acids and, 59—60
peptidoglycan and, 58—59, see also Peptidoglycan
proteins and, see Proteins
Macromolecular structure, see also specific types
hexagonal, 32
Macromolecules, see also specific types
synthesis of, 103
Macrophages, 199, 212
activated, 226
activation of, 219
lymphokine-mediated activation of, 219
migration inhibition of, 220
phagosomes and, 78
spleen cells enriched with, 226
Major outer membrane protein (MOMP), 9, 48, 54,
63, 65, 74
acidic isoelectric points in, 52
antibodies to, 64, 211, 218, 223, 226, 229
cross-linking of, 138
elementary bodies and, 51
IgG response to, 229
metabolism and, 102
molecular cloning and, 120—123
molecular weights of, 121
monoclonal antibody specific to, 121
regulatory role of, 89
reticulate bodies and, 51
structural gene for, 121
synthesis of, 88
Mammals, 72, 195, see also specific types
Mapping, 117, 130
chromosomal, 118
epitope, 52
Marmosets, 229
Mastitis, 180—181
McCoy cells, 10, 75, 145, 201
Mediators, see also specific types
of leukocytes, 203—205
Membrane-bound vesicles, 138
Membranes, see also specific types
cytoplasmic, 24
fluidity of, 56—58, 141
fusion of, 55
inclusion, 38—39, 41
mitochondrial, 141
outer, see Outer membranes
proteins in, 4, 9

trilaminar, 24, 29
Meningo-encephalomyelitis, 184—186
Meningopneumonitis, 36, 98, 100, 103, 104, 201,
212
cell wall thickness in, 24
elementary bodies and, 27, 28, 32
inoculation of, 41
reticulate bodies and, 30, 38
2-Mercaptoethanol, 51
Metabolism, 97—105, see also specific types
amino acid, 100—101
energy, 98—99
host-free activities in, 103—105
in vivo, 98—103
problems in identification of, 98
shut down host cell, 142
ultrastructural changes in, 125
vitamin, 101
Methacrylate embedding medium, 22
Mice, 8, 9, 174, 201
athymic, 210, 225
biovars of, 194
diabetes in, 225
L fibroblast in, 75
pneumonitis in, 100, 101, 114, 174
toxicity prevention tests in, 154
Microfilaments, 76
Micro-immunofluorescence, 9, 49, 60, 162, 211
indirect, 169
Micropannus, 158
Microscopy
electron, see Electron microscopy
immune electron, 61, 64
inverted phase contrast, 144
light, 22, 31
phase contrast, 26
Microvilli, 173
Microvillous projections, 73
Mini cell-like forms, 87
Minnows, 187
Mitochondria, 39, 85
membranes of, 141
Mobile genetic elements, 131
Molecular cloning, 119—124
Molecular weights, 6, 48, 113, 121
Molluscs, 72, 186—187, see also specific types
MOMP, see Major outer membrane protein
Monkeys, 168, 222, 224
eye infections in, 213
Monocistronic genes, 129
Monoclonal antibodies, 9, 50, 55, 61
anti-MOMP, 64
fluorescein-conjugated, 157, 163
MOMP-specific, 120, 121
serovar-specific, 63
Monocytes, 147, 199
Monodansyl cadaverine, 78
Mononuclear cells, 219, see also specific types
Mononuclear phagocytes, 137
Morphology
of cell envelopes, 32—38

Milton Keynes UK
Ingram Content Group UK Ltd.
UKHW051950071024
449327UK00026B/2251

9 780367 227074